J. F. Urban, Jr.
U.S. Department of Agriculture, ARS
Helminthic Diseases Laboratory, APX
Building 1040, BARC-East
Beltsville, Maryland 20705

MAST CELL DIFFERENTIATION
AND HETEROGENEITY

Mast Cell Differentiation and Heterogeneity

Editors

A. Dean Befus, Ph.D.
Gastrointestinal Research Group
Department of Microbiology and
Infectious Diseases
University of Calgary
Calgary, Alberta
Canada

John Bienenstock, M.D.
Host Resistance Programme
Department of Pathology
McMaster University
Hamilton, Ontario
Canada

Judah A. Denburg, M.D.
Host Resistance Programme
Department of Medicine
McMaster University
Hamilton, Ontario
Canada

Raven Press ■ New York

Raven Press, 1140 Avenue of the Americas, New York, New York 10036

Made in the United States of America

Library of Congress Cataloging-in-Publication Data

Mast cell differentiation and heterogeneity.

   Based on the Mast Cell Symposium held on Feb. 3–7,
1985 at University of British Columbia.
   Includes bibliographies and index.
   1. Mast cells—Congresses. 2. Cell differentiation.
I. Befus, D. (Dean) II. Bienenstock, John. III. Denburg,
Judah A. IV. Mast Cell Symposium (1985: University of
British Columbia). [DNLM: 1. Cell Differentiation—
congresses. 2. Mast Cells—physiology—congresses.
QS 532.5.C7 M4233 1985]
QH607.M35   1986   599'.017   86-6589
ISBN 0-88167-212-2

This book, by unanimous agreement of all who attended, is dedicated to the memory of Dr. Ellen Jarrett, who died just before the meeting.

Ellen Elizabeth Evelyn Jarrett

Ellen was born in Berlin in 1939 and died in her home in Milngavie, near Glasgow, on January 15, 1985. Her childhood was spent in Bromley, Kent, where her stepfather was a general practitioner. She came to Glasgow as a veterinary student in the autumn of 1959. When she graduated five years later, she already had made the decision to pursue an academic career. This was fully achieved with her appointments as research student, research fellow, and recent lecturer in the Department of Veterinary Parasitology in the Veterinary School of the University of Glasgow.

An overview of her 20-year research career shows how her interest in allergy and IgE evolved naturally from the subject of her PhD project, "Immunity to *Nippostrongylus brasiliensis* in the Rat." The observations by others of a remarkable potentiating effect of helminth infection on IgE responses led her into a detailed analysis of induction and regulation of IgE in rodents. Work on T-cell regulation of IgE biosynthesis overlapped with a developing interest in other T-cell factors, particularly those inducing differentiation of mast cells. Her recent work on mucosal mast cells is described in the paper that she contributed to this volume.

Ellen read widely in the biomedical and scientific literature. The depth and breadth of her knowledge shine through the 70 or so papers she has published, which include six "Letters to Nature" that document some of the highlights of her research. At the other end of the academic spectrum, she was one of the few people I have known who enjoys the challenge of preparing and writing a review, and hers were careful, critical, and provocative analyses of the literature.

Ellen's interests outside science were varied, and she pursued them with the same enthusiasm she brought to her research. Although her health was never robust, she skied well, sailed, and enjoyed swimming and walking.

She had a busy, happy family life with her husband Oswald and their two sons, Paul and Ben. Her warm and welcoming home was filled with books and paintings, plants and flowers. She was a superb cook, and very much enjoyed having dinner at home with friends, followed by an evening of stimulating conversation and debate.

Shortly before she died, Ellen read and enjoyed Leonard Wolff's biography *The Journey, Not the Arrival, Matters.* When young and idealistic, we aim to achieve all sorts of things, personal and professional. At the end of a lifetime the quality of the way things were done, and the satisfaction gained in the doing of them, are more important than the results of our endeavors, or whether there has been enough time to finish everything. Ellen Jarrett was truly a woman of quality—a loving wife and mother, a scientist of international repute, and a dear friend. Her personal courage in the 16 months of her last illness was remarkable.

Ann Ferguson

Preface

Until recently, the conceptual framework of mast cell function was based on the pivotal role of this cell type in allergic, immediate hypersensitivity reactions. The majority of studies of mast cells employed cells derived from the peritoneal cavity of rats because of their availability and ease of purification, and the widely held assumption that they were representative of other mast cell populations. However, as evidence accumulated and ultimately was widely recognized, the central paradigm in the discipline has become that mast cells appear to be heterogeneous in histological, biochemical, and functional properties, both across and within species. Unfortunately, the extent of this heterogeneity was not fully recognized initially, and mast cells were divided into only two subtypes, namely so-called "connective tissue" and "mucosal" mast cells. Rapid, essentially serendipitous advances in the culture of murine mast cells appeared to provide a mucosal mast cell source to facilitate characterization of mast cell subtypes. However, this designation was based on extrapolation of data, and at present the number and significance of mast cell subtypes is unknown. No single population can be considered to be representative of another, and thus to establish the properties and functions of mast cells in tissues such as the human intestine, lung, genital tract, or even the brain, it is essential to study cells derived from these specific sites. These types of comparative studies are only just beginning.

As some of these issues were becoming evident, it was clear that both a workshop/symposium and a book were needed, where the problems could be carefully presented and openly debated. The diversity of basic scientists and clinic researchers studying these problems included: histologists, microscopists, and pathologists; allergists; gastroenterologists, respirologists, hematologists, pharmacologists and internists, parasitologists and microbiologists, as well as cell biologists interested in the control of differentiation and myelopoiesis. Thus, a unique group of leading medical researchers representative of these disparate disciplines was selected, and a program and book outline to facilitate an open interchange of ideas and to provide guidelines for further research by clearly identifying significant, limiting gaps in current knowledge was designed.

One problem that had to be seriously debated in order to make recommendations to the medical sciences community was the often unjustified and generally inconsistent use of terminology to describe mast cell subtypes. Connotations associated with terms such as "mucosal mast cell" were misleading and, at least in some situations, incorrect. General guidelines were discussed at the meeting and informal recommendations are outlined in the text that will serve as the basis for the design of formal nomenclature proposals in the future. A subcommittee was struck to review mast cell terminology and report to the Nomenclature Committee of the International Union of Immunological Societies. It was agreed to avoid the descriptive terms "mucosal" and "connective tissue" as ascribed to mast cells. Where possible, tissue source and conditions of derivation such as culture should be given. Other special characteristics such as proteoglycan content, where known, should be stated. The manuscripts that were provided before the meeting were edited for terminological consistency. However, the editors take full responsibility for errors in the text, because to ensure an appropriately early publication date, the proofreading was almost entirely in our hands. The various contributors are thanked for their tolerance in this matter.

The chapters faithfully follow the order of presentations. The book is introduced by a historical perspective chapter by Dr. Lennert Enerback, a scientist instrumental in the initial documentation of mast cell heterogeneity. Thereafter, there are three major sections on Mast Cell Heterogeneity—I: Ontogeny and Differentiation; II: Functional Characteristics; and III: Clinical and Biological Significance. Each section is introduced by a review, is followed by a series of concise chapters on specific areas, and is closed by a chapter aimed at critical analysis of the information available, together with a series of new perspectives in the subdiscipline.

An appendix was added which summarizes the consensus reached at the meeting about nomenclature. In addition, the meeting suggested that Drs. Enerback, Miller, and Mayrhofer write some notes on histochemistry of mast cells, and this is included as a separate entity in the appendix.

With the diversity and eminence of the contributors, this volume represents a comprehensive, state-of-the-art assessment of mast cell heterogeneity, ontogeny, and function that should provide some direction for both well-established and new workers interested in mast cells and basophils. It will be valuable to basic scientists and clinical researchers, and the reader will appreciate the issues concerning mast cell function in various diseases and conceptual designs for appropriate therapeutic intervention.

The editors are optimistic that future publications in this field will recognize the contributions that the various authors have made in this volume.

The Editors

Acknowledgments

The Mast Cell Symposium and this volume were possible because of the insight and very generous contribution of Dr. Sam Orr and his associates at the Science and Technology Laboratories, Fisons Pharmaceuticals, United Kingdom. Other essential financial support was provided by: Abbott Laboratories, American Cyanamid, Astra Pharmaceuticals, Bencard Allergy Labs, Beecham Pharmaceuticals, Boehringer Ingelheim, Bristol Myers Company, Burroughs-Wellcome Company, Pharmacia, Revlon Health Care Group, Schering Corporation, and The Upjohn Company.

We would like to thank McMaster University for its continuing support in the preparation of this book, as well as in the organization of the meeting. Mr. Neil Johnston, Department of Pathology, was the organizer, and without his skills and the enthusiastic support of Margaret Best, Janice Butera, Sandy Jess, and Rona Madison, this project would never have been completed. Finally, we are grateful to the staff of the Millcroft Inn, Alton, Ontario, Canada, for creating such a warm and pleasant environment in which scholarly exchange occurred so freely.

Contents

Mast Cell Heterogeneity II: Functional Characteristics

Mast Cell Heterogeneity III: Clinical and Biological Significance

Appendix

Contributors

S. J. Ackerman
Division of Infectious Diseases
Beth Israel Hospital
Boston, Massachusetts 02215

R. M. Agius
Medicine I and Clinical Pharmacology
Southampton General Hospital
Southampton SO9 4XY
England

Hidekazu Asai
Shizuoka Laboratory Animal Center
Hamamatsu-city, Shizuoka 435
Japan

K. F. Austen
Department of Medicine
Harvard Medical School
Boston, Massachusetts 02115

Kim E. Barrett
Allergic Diseases Section
Laboratory of Clinical Investigation
National Institute of Allergy and
*   Infectious Diseases*
National Institutes of Health
Bethesda, Maryland 20205

Michael A. Beaven
Laboratory of Chemical Pharmacology
National Heart, Lung and Blood
*   Institute*
National Institutes of Health
Bethesda, Maryland 20205

A. Dean Befus
Gastrointestinal Research Group
Department of Microbiology and
*   Infectious Diseases*
University of Calgary
Calgary, Alberta T2N 4N1
Canada

John Bienenstock
Host Resistance Programme
Department of Pathology
McMaster University
Hamilton, Ontario L8N 3Z5
Canada

B. M. C. Chan
MRC Group in Allergy Research
Department of Immunology
University of Manitoba
Winnipeg, Manitoba R3E 0W3
Canada

Tania Chernov-Rogan
Howard Hughes Medical Institute
*   Laboratories*
University of California Medical Center
San Francisco, California 94143

M. K. Church
Medicine I and Clinical Pharmacology
Southampton General Hospital
Southampton SO9 4XY
England

J. Cichon
Department of Pathology
School of Medicine
St. Louis University Medical Center
St. Louis, Missouri 63104

R. Coleman
Department of Biological Structure
The Rappaport Family Institute for
*   Research in Medical Science*
Faculty of Medicine
Technion
Israel Institute of Technology
Haifa
Israel

Ayuis Corcia
Department of Membrane Research
The Weizmann Institute of Science
Rehovot 76100
Israel

S. Davidson
Department of Immunology
The Rappaport Family Institute for
  Research in Medical Science
Faculty of Medicine
Technion
Israel Institute of Technology
Haifa
Israel

Judah A. Denburg
Host Resistance Programme
Department of Medicine
McMaster University
Hamilton, Ontario L8N 3Z5
Canada

Alain L. de Weck
Institute of Clinical Immunology
University of Bern
Bern
Switzerland

Jerry Dolovich
Department of Pediatrics
McMaster University
Hamilton, Ontario L8N 3Z5
Canada

Vjekoslav Dulic
Department of Chemical Immunology
The Weizmann Institute of Science
Rehovot 76100
Israel

Ann M. Dvorak
Department of Pathology
Beth Israel Hospital
Boston, Massachusetts 02115

R. P. Eady
Fisons Plc - Pharmaceutical Division
Bakewell Road
Loughborough,
  Leicestershire L311 ORH
England

Lennart Enerback
Department of Pathology
Sahlgren's Hospital
University of Goteborg
S413 45 Goteborg
Sweden

C. C. Fox
Department of Medicine
Division of Clinical Immunology
The Johns Hopkins University School of
  Medicine
Good Samaritan Hospital
Baltimore, Maryland 21239

Marc M. Friedman
Division of Molecular Virology and
  Immunology
Georgetown University Medical Center
Rockville, Maryland 20852

A. Froese
MRC Group in Allergy Research
Department of Immunology
University of Manitoba
Winnipeg, Manitoba R3E OW3
Canada

Kiyoshi Furuichi
National Institute of Arthritis, Diabetes,
  Digestive and Kidney Diseases
Bethesda, Maryland 20205

Stephen J. Galli
Department of Pathology
Beth Israel Hospital
Boston, Massachusetts 02215

S. Gibson
Department of Pathology and
  Immunology
Moredun Institute
Gilmerton Road
Edinburgh
Scotland

H. Ginsburg
Department of Immunology
The Rappaport Family Institute for
  Research in Medical Science
Faculty of Medicine
Technion
Israel Institute of Technology
Haifa
Israel

G. J. Gleich
Department of Immunology
Mayo Medical School
Mayo Clinic and Foundation
Rochester, Minnesota 55905

Edward J. Goetzl
Howard Hughes Medical Institute
 Laboratories
Department of Medicine
University of California Medical Center
San Francisco, California 94143

Laura M. Goetzl
Howard Hughes Medical Institute
 Laboratories
University of California Medical Center
San Francisco, California 94143

R. Goodacre
Department of Medicine
McMaster University
Hamilton, Ontario L8N 3Z5
Canada

T. Goto
Third Department of Internal Medicine
Tokushima University
Kuramoto 3, Tokushima 770
Japan

B. Greenwood
Fisons Plc - Pharmaceutical Division
Bakewell Road
Loughborough,
 Leicestershire LE11 ORH
England

D. Guy-Grand
INSERM - U.132
Groupe d'Immunologie et de
 Rhumathologic Pediatriques
Hôpital Necker-Enfants Malades
75730 Paris
France

David M. Haig
Department of Veterinary Medicine
University of Glasgow
Veterinary School
Bearsden, Glasgow G61 1QH
Scotland

S. T. Harper
Fisons Plc - Pharmaceutical Division
Bakewell Road
Loughborough,
 Leicestershire LE11 ORH
England

Koichi Hirai
Institute of Clinical Immunology
University of Bern
Bern
Switzerland

Ken-ichi Hisamatsu
Department of Otorhinolaryngology
Yamanashi College of Medicine
1110 Shimogato
Tamaho-cho
Nakakoma-gun, Yamanshi-ken 409-39
Japan

S. T. Holgate
Medicine I and Clinical Pharmacology
Southampton General Hospital
Southampton SO9 4XY
England

P. H. Howarth
Medicine I and Clinical Pharmacology
Southampton General Hospital
Southampton SO9 4XY
England

J. F. Huntley
Department of Pathology and
 Immunology
Moredun Institute
Gilmerton Road
Edinburgh
Scotland

J. N. Ihle
NCI - Frederick Cancer Research
 Facility
Litton Bionetics, Inc.
Basic Research Program
Frederick, Maryland 21701

Teruko Ishizaka
Johns Hopkins University School of
 Medicine
Good Samaritan Hospital
Baltimore, Maryland 21239

Ellen E. E. Jarrett*
Department of Veterinary Medicine
University of Glasgow Veterinary
School
Bearsden, Glasgow G61 1QH
Scotland

A. Kagey-Sobotka
Department of Medicine
Division of Clinical Immunology
The Johns Hopkins University School of
Medicine
Good Samaritan Hospital
Baltimore, Maryland 21239

Michael Kaliner
Allergic Diseases Section
Laboratory of Clinical Investigation
National Institute of Allergy and
Infectious Diseases
National Institutes of Health
Bethesda, Maryland 20205

Yoshio Kanayama
Second Department of Internal
Medicine
Osaka University Medical School
Fukushima-ku-1-1-50, Osaka 553
Japan

H. R. Katz
Department of Medicine
Harvard Medical School
Boston, Massachusetts 02115

J. Keller
NCI - Frederick Cancer Research
Facility
Litton Bionetics Inc.
Basic Research Program
Frederick, Maryland 21701

A. Kinarty
Department of Immunology
The Rappaport Family Institute for
Research in Medical Science
Faculty of Medicine
Technion
Israel Institute of Technology
Haifa
Israel

S. J. King
Department of Pathology and
Immunology
Moredun Institute
Gilmerton Road
Edinburg
Scotland

Yukihiko Kitamura
Institute for Cancer Research
Osaka University Medical School
Nakanoshima 4-3-57
Kita-ku, Osaka 530
Japan

D. Lagunoff
Department of Pathology
School of Medicine
St. Louis University Medical Center
St. Louis, Missouri 63104

Johnny Y. Lee
Howard Hughes Medical Institute
Laboratories
University of California Medical Center
San Francisco, California 94143

T. Lee
Gastrointestinal Research Group
Department of Microbiology and
Infectious Diseases
University of Calgary
Calgary, Alberta T2N 4N1
Canada

K. M. Leiferman
Department of Dermatology
Mayo Medical School
Mayo Clinic and Foundation
Rochester, Minnesota 55905

L. M. Lichtenstein
Department of Medicine
Division of Clinical Immunology
The Johns Hopkins University School of
Medicine
Good Samaritan Hospital
Baltimore, Maryland 21239

J. Mann
Fisons Plc - Pharmaceutical Division
Bakewell Road
Loughborough,
Leicestershire LE11 ORH
England

*Deceased.

Diana L. Marquardt
Division of Allergy
Department of Medicine
University of California at San Diego
* School of Medicine*
University of California at San Diego
* Medical Center*
San Diego, California 92103

Graham Mayrhofer
Department of Microbiology and
* Innumology*
The University of Adelaide
Adelaide 5000
South Australia

Christine McMenamin
Department of Veterinary Medicine
University of Glasgow Veterinary
* School*
Bearsden, Glasgow G61 1QH
Scotland

K. McNeill
MRC Group in Allergy Research
Department of Immunology
University of Manitoba
Winnipeg, Manitoba R3E OW3
Canada

Dean D. Metcalfe
Allergic Diseases Section
Laboratory of Clinical Investigation
National Institute of Allergy and
* Infectious Diseases*
National Institutes of Health
Bethesda, Maryland 20205

H. R. P. Miller
Department of Pathology and
* Immunology*
Moredun Institute
Gilmerton Road
Edinburgh
Scotland

R. M. Naclerio
Department of Medicine
Division of Clinical Immunology
The Johns Hopkins University School of
* Medicine*
Good Samaritan Hospital
Baltimore, Maryland 21239

Toru Nakano
Institute for Cancer Research
Osaka University School of Medicine
Nakanoshima 4-3-57
Kita-ku, Osaka 530
Japan

G. F. J. Newlands
Department of Pathology and
* Immunology*
Moredun Institute
Gilmerton Road
Edinburgh
Scotland

Makio Ogawa
Department of Medicine
Medical University of South Carolina
Charleston, South Carolina 29403

T. S. C. Orr
Fisons Plc - Pharmaceutical Division
Bakewell Road
Loughborough,
* Leicestershire LE11 ORH*
England

Hirokuni Otsuka
Host Resistance Programme
Department of Medicine
McMaster University
Hamilton, Ontario L8N 3Z5
Canada

Frederick L. Pearce
Department of Chemistry
University College London
London WC1H OAJ
England

Israel Pecht
Department of Chemical Immunology
The Weizmann Institute of Science
Rehovot 76100
Israel

D. Peizner
Department of Pathology
School of Medicine
St. Louis University Medical School
St. Louis, Missouri 63104

Pamela N. Pharr
Department of Medicine
Medical University of South Carolina
Charleston, South Carolina 29403

Jackie Pierce
Laboratory of Cellular and Molecular
* Biology*
National Institutes of Health
Bethesda, Maryland 20205

Rosanne Pitts
The Clinical Immunology Research
* Unit*
Princess Margaret Hospital
Subiaco 6008
Western Australia

D. Proud
Department of Medicine
Division of Clinical Immunology
The Johns Hopkins University School of
* Medicine*
Good Samaritan Hospital
Baltimore, Maryland 21239

M. Rao
MRC Group in Allergy Research
Department of Immunology
University of Manitoba
Winnipeg, Manitoba R3E OW3
Canada

Ulf Rapp
National Cancer Institute
Frederick, Maryland 21701

Alan Rein
NCI - Frederick Cancer Research
* Facility*
Litton Bionetics Inc.
Basic Research Program
Frederick, Maryland 21707

Frederick Renold
Howard Hughes Medical Institute
* Laboratories*
University of California Medical Center
San Francisco, California 94143

A. Reshef
Department of Immunology
The Rappaport Family Institute for
* Research in Medical Science*
Faculty of Medicine
Technion
Israel Institute of Technology
Haifa
Israel

A. Rickard
Department of Pathology
School of Medicine
St. Louis University Medical School
St. Louis, Missouri 63104

Benjamin Rivnay
Department of Membrane Research
The Weizmann Institute of Science
Rehovot 76100
Israel

C. Robinson
Medicine I and Clinical Pharmacology
Southampton General Hospital
Southampton SO9 4XY
England

P. A. Roth
MRC Group in Allergy Research
Department of Immunology
University of Manitoba
Winnipeg, Manitoba R3E OW3
Canada

Thomas Schaffner
Institute of Pathology
University of Bern
Bern, Switzerland

R. P. Schleimer
Department of Medicine
Division of Clinical Immunology
The Johns Hopkins University School of
* Medicine*
Good Samaritan Hospital
Baltimore, Maryland 21239

John W. Schrader
The Walter and Elisa Hall Institute of
* Medical Research Post Office*
Royal Melbourne Hospital
Victoria 3050
Australia

D. C. Seldin
Department of Medicine
Harvard Medical School
Boston, Massachusetts 02115

Fergus Shanahan
Center for Ulcer Research and
 Education
Department of Medicine
University of California at Los Angeles
 School of Medicine and Medical and
 Research Services
Wadsworth Veterans Administration
 Hospital Center
Los Angeles, California 90073

Andrew H. Soll
Center for Ulcer Research and
 Education
Department of Medicine
University of California at Los Angeles
 School of Medicine and Medical and
 Research Services
Wadsworth Veterans Administration
 Hospital Center
Los Angeles, California 90073

Takashi Sonoda
Institute for Cancer Research
Medical School
Osaka University
Nakanoshima
Kita-ku, Osaka 530
Japan

Beda M. Stadler
Institute of Clinical Immunology
University of Bern
Bern
Switzerland

R. L. Stevens
Department of Medicine
Harvard Medical School
Boston, Massachusetts 02115

Yasuo Tanno
Host Resistance Programme
Department of Medicine
McMaster University
Hamilton, Ontario L8N 3Z5
Canada

Lovick Thomas
Center for Ulcer Research and
 Education
Department of Medicine
University of California at Los Angeles
 School of Medicine and Medical and
 Research Services
Wadsworth Veterans Administration
 Hospital Center
Los Angeles, California 90073

Mary Toomey
Center for Ulcer Research and
 Education
Department of Medicine
University of California at Los Angeles
 School of Medicine and Medical and
 Research Services
Wadsworth Veterans Administration
 Hospital Center
Los Angeles, California 90073

P. Vassalli
Department of Pathology
University of Geneva
1211 Geneva 4
Switzerland

Stephen I. Wasserman
Division of Allergy
Department of Medicine
University of California at San Diego
 School of Medicine
University of California at San Diego
 Medical Center
San Francisco, California 92103

E. Wells
Fisons Plc - Pharmaceutical Division
Bakewell Road
Loughborough,
 Leicestershire LE11 ORH
England

Elizabeth WoldeMussie
Laboratory of Chemical Pharmacology
National Heart, Lung and Blood
 Institute
Bethesda, Maryland 20205

R. G. Woodbury
Department of Biochemistry
University of Washington
Seattle, Washington 98195

Teiichi Yamamura
Department of Dermatology
Osaka University School of Medicine
Fukushima-ku, Osaka 553
Japan

M. Zarka
Department of Pathology
School of Medicine
St. Louis University Medical Center
St. Louis, Missouri 63104

Mast Cell Differentiation and Heterogeneity,
edited by A. D. Befus et al.
Raven Press, New York © 1986.

Mast Cell Heterogeneity: The Evolution of the Concept of a Specific Mucosal Mast Cell

Lennart Enerback

Department of Pathology, Sahlgren's Hospital, University of Goteborg, S413 45 Goteborg, Sweden

I should like to thank the organizers of this symposium, John Bienenstock, Dean Befus and Judah Denburg, for inviting me to this conference and for entrusting me with the privilege of giving this introductory lecture. The best way I can see of addressing the problem of mast cell heterogeneity, and of providing some historical perspectives on this problem, is to give a brief account of the evolution of our knowledge of the mucosal mast cell. The identification of this specific mast cell phenotype is a result of the work of several laboratories including my own. The study of the mucosal mast cell has hopefully helped to focus interest on mast cell heterogeneity, but also to identify some problems that lie ahead in future studies of this problem.

In order to recognize possible heterogeneities in the mast cell system we must obviously first consider how to define the mast cell. Ehrlich, in his remarkable thesis of 1878 (1), made it clear that the specific feature of the mast cell was its cytoplasmic metachromatic granules distinguishing it from other

Parts of the investigations reported in this review were carried out at the author's laboratory with grants from the Swedish Medical Research Council (Project 2235) and from Walter, Ellen and Lennart Hesselman's Foundation.

Acknowledgement: The author wishes to thank Gun Augustsson, Antia Olofsson and Marie Svensson for their devoted and skillful technical assistance and Annika Dahlqvist for expert secretarial aid.

Abbreviations: MMC, mucosal mast cell; CTMC, connective tissue mast cell; BB, blood basophil; GAG, glycosaminoglycan; RMCP I and RMCP II, rat mast cell protease I and II; 5-HT, 5-hydroxy-tryptamine; ECC, enterochromaffin cells.

cells of similar morphology, at that time collectively referred to as plasma cells. He also pointed out that the definition of the mast cell was a chemical, as opposed to a morphological one, and due to a compound awaiting identification. We still adhere to this basic definition of the mast cell but have come to realize that there is a diversity of metachromatic granulated cells and that additional criteria are therefore necessary. We lack a good definition of the mast cell, obviously due to the absence of precise knowledge. However, I believe that a majority of people would agree that mast cells are characterized by their content of numerous large and electron-dense cytoplasmic granules, made up of a proteoglycan matrix to which small molecules such as histamine and 5-hydroxytryptamine are bound. An additional important property of mast cells is that they possess IgE receptors and that they respond very rapidly by secretion to challenges with antigen, anti-IgE or a number of chemical secretagogues.

This definition will include both the blood basophil (BB) and the tissue mast cell in the mast cell system. A common derivation of these two cell types from bone marrow precursors has been clearly established by the work of Kitamura and associates (2). The BB should therefore best be considered as a mast cell whose final differentiation takes place in the blood rather than in the tissues. It is usually distinguished from the tissue mast cell by morphological criteria (see 3). When BB are found in the tissue as a result of migration from the blood stream, the validity of such distinguishing criteria is of course critical, especially when it comes to differentiation between BB and MMC. Functional interrelations between BB and tissue mast cells are suggested by the well known inverse relationship between tissue mast cells and blood basophils in many species, first pointed out by Maximow (4). Thus, the rat lacks, or contains extremely few, BB but has numerous tissue mast cells. Finally, I would like to point out that before we begin to examine the possibilities of defining different mast cell subsets, we must appreciate that the criteria we use, whether structural, biochemical or functional, may be the subject of interspecies variations. This problem will be further discussed below.

History of the MMC. Alexander Maximow was, as far as I can tell, the first to observe mucosal mast cells of specific character. In a paper describing the various cell types of the loose connective tissue (4) he briefly noted that the small intestine of the rat contained many mast cells of strange appearance and differing from mast cells of other sites. They were smaller, their granulation less regular and the granules badly preserved in tissues fixed in ethanol and stained with thionin. The cells could not be found at all in tissue sections stained with methylene blue. A period of disagreement about the existence of mucosal mast cells followed, and they were considered by some to be plasma cells (5). The conflicting early literature on gastrointestinal mast cells has been reviewed by Michels (6) who used the term "atypical mast cell" for a group of cells with reddish granules previously described by many authors in a variety of species and apparently including eosinophils, globule leukocytes and mast cells.

My interest in MMC dates back to the mid-sixties when I was studying amine storing cells in the rat intestinal mucosa. At that time the opinion prevailed that mast cells were few or absent in the rat intestine. Intestinal histamine was therefore considered to be a non-mast cell histamine pool and this was further supported by the finding that compound 48/80, while depleting other tissues, did not reduce the histamine content of the intestine (7). It was found that the gastrointestinal mucosa of the rat contained a very large number of mast cells which differed from the classical connective tissue mast cells (CTMC) of other sites in terms of morphological appearance, reactivity to compound 48/80 and monoamine storage capacity. The cells were found to be unusually susceptible to fixation and differed in dye-binding properties from rat CTMC. Techniques for their demonstration in tissue sections were therefore worked out (8-11).

Distribution and Structure. After adequate fixation and staining, numerous mucosal mast cells are found in the lamina propria of the entire rat gastrointestinal tract. The gastro-intestinal mucosa is one of the tissues of the rat which is richest in mast cells. The mast cell density is higher than that of dorsal skin and about the same as in the ears. The cell numbers per unit area are somewhat higher in the ileum and jejunum than they are in the duodenum and colon (10). In normal rats, MMC are never found in the epithelium. They can be distinguished from CTMC by morphological criteria. MMC are smaller, of more variable shape and as a rule contain fewer granules of more variable size and shape (Figs. 1-4). The variable form of the MMC, which often show cytoplasmic processes and projections, can perhaps be taken as an expression of motility. That MMC have, in fact, a migratory capacity is shown during the "self cure" reaction of nematode infections (see below), when numerous MMC are found in the epithelium, between the individual epithelial cells. In the electron microscope the cells are found to be tightly interspersed between mesenchymal and epithelial cells, and, unlike CTMC, to have a smooth surface, lacking microvilli (12).

MMC are often found near nerve fibers of the lamina propria. In a recent study, we observed nerve terminals seemingly in direct contact with the plasma membrane of MMC in the rat ileum (13). Some of these terminals contained many clear and a few dense core vesicles (diameter 80-100 nm), while others contained mainly larger dense core vesicles (diameter around 150 nm). No attempts have been made so far to specifically identify adrenergic nerve terminals. The boutons with many clear and some larger dense core vesicles may therefore represent adrenergic and/or cholinergic nerve terminals, while the ones with predominantly larger dense core vesicles may represent peptidergic terminals. These findings strongly suggest that MMC may be directly

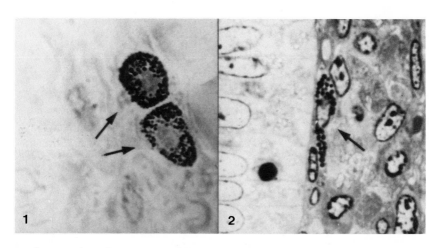

Figures 1 & 2. Mast cells (arrows) in the tongue (Fig. 1) and lamina propria of the duodenum (Fig. 2) of the rat. Tissues fixed in glutaraldehyde, embedded in Epon and stained with toluidine blue at pH 1 for 120 min. Note differences in size, shape and granulation density between CTMC (Fig. 1) and MMC (Fig. 2).

innervated under certain conditions. More work is obviously needed before this finding can be fully evaluated. It is interesting, however, that several investigators have previously observed a close association between CTMC and nerves (for references see 13). Furthermore, it has been reported that both acetylcholine (14) and substance P (see 15) release histamine from isolated peritoneal mast cells of the rat *in vitro*.

Granule Matrix: Proteoglycan and Proteins

Fixation of tissue mast cells. MMC were identified as a separate entity after a systematic study of the fixation and staining of rat mast cells (8,9). When a number of different standard fixatives were tested it was found that the mast cells of the rat intestinal mucosa and those of the skin differed greatly in behaviour. Dermal mast cells showed only minor variations and retained a strong stainability with toluidine blue after fixation in all the tested solutions. MMC, on the other hand, could be stained only after fixation in Carnoy's solution (containing methanol, chloroform and acetic acid) and some fixatives containing lead salts. Further trials showed that low concentrations of formaldehyde preserved the stainability of MMC granules, especially in combination with low concentrations of

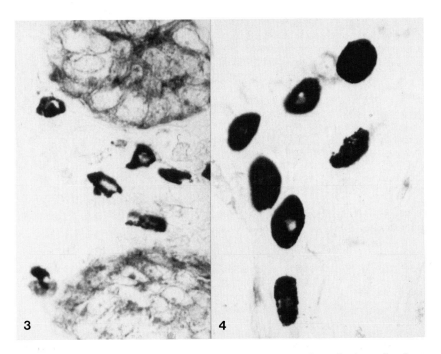

Figures 3 & 4.   Mast cells in lamina propria of the duodenum (Fig. 3) and skin (Fig. 4) of the rat. Tissues fixed in IFAA, embedded in paraffin and stained with toluidine blue at pH 0.5 for 30 min.

acetic acid. A combination of 0.6% formaldehyde and 0.5% acetic acid was selected as an optimal fixative for both types of mast cell. This solution is approximately isotonic and its pH is 2.9. It provides a very good preservation of structure and dye binding of mast cells when studied by light microscopy.   The tissue preservation is adequate in small tissue samples if the fixation time is kept short (12 to 24 h) and followed by immersion of the tissues in 70% ethanol for 12 h.   The addition of methanol to the fixative was reported to improve the fixation of epithelium (16). We have now used the isotonic, acetic acid formaldehyde (IFAA) fixative successfully on a variety of tissues of a number of species to demonstrate mast cells with toluidine blue staining.
     It was originally suggested that the failure of higher concentrations of formaldehyde to preserve the stainability of MMC granules was due to the extraction of a soluble glycosamino- glycan by the fixative rather than to a blocking of dye binding groups of the GAG by the aldehyde (8).   It was subsequently

found, however, that MMC could be stained with toluidine blue after normal aldehyde fixation provided that very long staining times were used (own observation), and also by alcian blue in the critical electrolyte concentration staining with magnesium chloride (17). We therefore re-evaluated the fixation and staining properties of MMC with a view to obtaining an improved method for their demonstration (18). We analyzed the critical staining time of duodenal MMC and found that they could be stained after aldehyde fixation both with toluidine blue and alcian blue if the staining times were prolonged to at least 3 days. Trypsination of fixed sections reduced the staining times significantly. These results suggested that the aldehyde blocking was caused by a diffusion barrier of protein nature, absent in the granules of CTMC which instead show an increased affinity to cationic dyes after aldehyde fixation (19). MMC and CTMC thus appear to differ in terms of the spatial arrangement of GAG and protein in their granules.

We also analyzed the solubility in situ of the granules of the two types of mast cell by fixation of tissues in solutions of low ionic strength containing 0.6% formaldehyde and of different pH. We found that the dye binding of both MMC and CTMC was not well preserved by fixation at pH 6 or higher. In sections of skin CTMC numbers were not reduced after fixation at high pH but the cells were badly preserved and surrounded by metachromatic halos. MMC, on the other hand, gradually disappeared from the sections. Plots of arbitrary staining units (graded on coded sections) versus pH had a shape suggesting dissociation curves with apparent pK values in the range of 4–6 for both types of cell.

The effect of the ionic strength of the fixative was tested in a similar way by the addition of NaCl to solutions containing 0.6 to 4% formaldehyde. The mast cells reacted in a similar way. Above a "critical electrolyte concentration" of 0.7M NaCl, stainable MMC were absent and CTMC weakly stained and surrounded by halos.

These results thus suggest that the structural integrity and cationic dye binding of the granules of both type of mast cell is strongly dependent on ionic linkages between GAG and protein. This is in agreement with previous findings of Lagunoff and coworkers showing that both heparin (20) and proteinase (21) can be extracted from intact mast cell granules by solutions of high ionic strength. The findings further suggest that the matrix of MMC granules is more soluble in solutions of high salt concentrations and pH than that of CTMC granules, but the similar apparent pK and critical electrolyte concentration would suggest that similar groups are involved in the binding. Finally, it may be suggested, on the basis of the low pH required for an adequate fixation, that sulfates rather than carboxyls are the contribution of the GAG to this hypothetical ionic binding.

Based on these findings a long toluidine blue staining technique was devised for the demonstration of mast cells in tissue section (18). It involves staining at pH 0.5 for 5 to 7 days and visualizes MMC as well as CTMC. It has the additional advantage that eosinophils can be stained in the same section. The counting of mast cells in the same tissue section after a short (30 min) and a subsequent long (6 days) staining with toluidine blue can be used to assess the degree of aldehyde blocking of the dye binding of the mast cells in a tissue (to be published).

TABLE I. Staining Properties of Rat MMC and CTMC

Method	MMC	CTMC
Thiazine dyes pH 0.5 and 4	Violet metachromasia	
Aldehyde fixation	Blocking	No blocking
Alcian blue/MgCl$_2$	Blue <0.7M MgCl$_2$	Blue <1.0M MgCl$_2$
Berberine	Non-fluorescent	Fluorescent

Staining properties. In appropriately fixed tissue, MMC and CTMC also differ in staining properties (Table I). In general, these differences suggest that MMC granules contain a GAG with a lower degree of sulfation than that of CTMC which is known to be heparin. Thus, rat MMC granules stain blue with copper phthalocyanin dyes such as astra blue or alcian blue, in a staining sequence with safranin, while rat CTMC granules stain red (9). A preferential affinity to alcian blue in this staining sequence may be associated with low N-sulphate, since selective removal of sulfamido sulfate suppresses the safranin O staining of CTMC but not of cartilage (22). Furthermore, the critical electrolyte concentration (23) of the staining with alcian blue is substantially lower for MMC than for CTMC (17). Rat MMC granules, unlike those of CTMC, do not exhibit a fluorescent binding with the dye berberine (18). This dye forms a very strongly fluorescent complex with heparin in CTMC granules (24). The fluorescence intensity is proportional to heparin content and the dye has therefore been used to measure heparin in mast cells by microscope fluorometry (24) or flow cytometry (25). The finding that MMC did not exhibit the characteristic reaction with berberine prompted a study of its binding mechanism. The results, to be reported in detail elsewhere, have shown that the fluorescent berberine binding has a certain degree of specificity for heparin and heparin-like heparan sulfates. Complexes are formed with the various GAG, characterized by a lower absorption than that of the free dye with a small shift in the main absorption peak from 420 to 435 nm. Heparin-bound berberine has a much more intense fluorescence than the free dye, corresponding to a 25-fold increase in its relative fluorescence quantum yield, without any change in the shape of the emission spectrum. Among the major classes of GAG, heparin forms the complex of by far the strongest fluorescence (Table II).

TABLE II. Relative fluorescence quantum yields (Q_{rel}) of berberine–GAG complexes

GAG	Q_{rel}
Heparin	1.00
N-acetylheparin	0.20
Chondroitin 6-sulfate	0.20
Chondroitin 4-sulfate	0.10
Dermatan sulfate	0.10
Keratan sulfate	0.04
Hyaluronic acid	0.04
Berberine	0.04

The properties of the complexes with heparan sulfates are of special interest since the members of this group of GAG are closely related to heparin structurally. We have studied the berberine complexes formed with 5 different heparan sulfate fractions which have previously been classified in terms of sulfate content and uronic acid (26). The fluorescence quantum yields were lower than that of heparin but increased with increasing sulfate (and iduronic acid) content (Table III).

TABLE III. Relative fluorescence quantum yields (Q_{rel}) of berberine complexes with heparan sulfate fractions

Fraction		Total sulfate/ Hexosamine	N-sulfate/ Hexosamine	Q_{rel}
Heparan Sulfate Fractions	1	0.44	0.21	0.22
	2	0.56	0.26	0.10
	3	1.00	0.40	0.20
	4	1.15	0.47	0.40
	5	1.23	0.62	0.50
Heparin		2.40	–	1.00
Berberine				0.04

Detailed information about the binding of berberine in situ is required before its use as a histochemical reagent can be fully evaluated. However, among the rat mast cells only CTMC of various sites show the strongly fluorescent binding, while MMC are negative. Peritoneal CTMC are negative during the early period of granule regeneration after compound 48/80 treatment (to be published), suggesting that immature granules do not react. A staining sequence of berberine followed by toluidine blue can be used to distinguish CTMC from MMC in the same specimen (Wingren and Enerback, to be published).

Glycosaminoglycans of MMC. We have recently attempted to identify the GAG of rat MMC (27). For this purpose we took advantage of the fact that MMC proliferate in response to infections with the nematode N. brasiliensis, resulting in a dramatic increase in mast cell numbers and the histamine content of the mucosa (see below). Infected and control rats were injected with sodium ^{35}S-sulfate, four times daily over a four day period, 10–13 days after infection, when the MMC of infected rats proliferate and increase greatly in number. After a further 10 days the rats were killed and tissues taken for analysis. We found a parallel increase in mast cell numbers, histamine content and ^{35}S-labelled GAG (Table IV).

TABLE IV. Mucosal mast cells, histamine content and ^{35}S-labelled glycosaminoglycans in the jejunum of nematode-infected rats

Animals[a]	Mast Cells[a,c]	Histamine[b] (μg/g tissue)	(^{35}S) Glycosaminoglycans[b] (cpm/g tissue x 10^{-3})
Infected	651 ± 10	20.4 ± 2.0	599 ± 89
Controls	138 ± 16	4.5 ± 1.5	177 ± 12
Infected controls	4.7	4.5	5.1

a. Four rats infected with N. brasiliensis and 4 uninfected controls
b Mean ± SE
c Number of cells/unit mucosal length

Autoradiography demonstrated a selective labelling of the mast cells. Analysis of the labelled polysaccharide from infected animals showed that almost 60% of this material consisted of oversulfated galactosaminoglycan, while heparin-related polysaccharides accounted for only 13% (Table V).

The galactosaminoglycan contained 4-monosulfated and 4,6-disulfated N-acetylgalactosamine residues in a molar ratio of approximately 5:1, both being linked to D–glucuronic acid residues. It is concluded that the major GAG produced by rat MMC is an oversulfated galactosaminoglycan rather than heparin. Our analytical data are in agreement with the presence of oversulfated chondroitin sulphate of the type sometimes referred to as chondroitin sulphate E (60).

TABLE V. Composition of rat intestinal ^{35}S–labelled glycosaminoglycan preparations

Animals[a]	Mast Cells[a,c]	Histamine[b] (μg/g tissue)	(^{35}S) Glycosaminoglycans[b] (cpm/g tissue x 10^{-3})
Infected	13 (n = 3)	57 (n = 4)	30
Controls	28 (n = 2)	53 (n = 2)	19

a. Same experiment as in Table 4. Percent of total radioactivity; averages from 2-4 preparations, as indicated in parenthesis.

Granular proteins.   Mast cell granules contain several proteins which have been partially characterized (21). In rat mast cells there are two distinct serine proteases (28,29) differing in solubility, structure and antigenicity. One of these, referred to as rat mast cell protease II (RMCP II) is found in the intestinal mucosa and localized in MMC (30); the other enzyme, which is the well–known mast cell chymase (31), is referred to as rat mast cell protease I and mainly located in mast cells of CTMC type. Since RMCP II can be both localized microscopically and quantified by immunological methods (32) it may serve as a useful marker for MMC and their secretory activity.

Biogenic Amines: Storage and Secretion

Histochemical findings indicate that rat MMC, like CTMC, contain histamine (17,33). No additional storage site for histamine was found in the intestinal mucosa (34) in contrast to the gastric mucosa of the rat which also contains a histamine storing enterochromaffin–like cell (33). Parallel changes in MMC numbers and histamine content during the response to N. brasiliensis infections (35,36) and after repeated injections of polymyxin B (37) further indicate that MMC are the main repository for histamine in rat intestinal mucosa. However, the histamine content of the individual MMC appears to be much lower than that of CTMC. From tissue histamine content and mast cell densities the relative histamine content of a rat MMC was calculated to be only about one-tenth of that of a CTMC (38). This value agrees reasonably well with the analytical data of Befus and coworkers (39) who measured the histamine content of isolated lymphoid cells containing about 12% MMC, and calculated that the histamine content was 1.3 pg/MMC. In contrast, about 15 pg/CTMC was found in peritoneal cell suspension. Rat MMC, like CTMC, normally contain 5-hydroxytryptamine (5-HT) and take up exogenous dopamine. Both amines are stored by a reserpine sensitive mechanism (11,40). Again, the concentration of 5-HT was found to be lower in the individual MMC than in CTMC, but increased during infection with N. brasiliensis (40). Entero-chromaffin cells (ECC) are the major storage site for intestinal 5-HT. Using cytofluorometric measurements it was calculated that a MMC contains about 20% of the 5-HT content of an ECC in normal rats. Since MMC are about three times as frequent per unit volume of tissue, it may be estimated that up to 1/4 of the gut mucosal 5-HT can be stored in MMC.

A notable property of rat MMC is their resistance to the histamine-releasing action of polyamines, such as compound 48/80 and polymyxin B, both in vivo (10,37) and in vitro (39). Repeated injections of compound 48/80 or polymyxin B for 5 days resulted in a 50% to 100% increase in MMC numbers and a parallel increase in mucosal histamine content. The MMC numbers then returned slowly to the control level following an exponential course. From this we calculated the half-life of the newly formed rat MMC to be about 40 days (41). CTMC, on the other hand, have a very long lifespan and do not show a significant turnover in terms of cell death and renewal.

MMC isolated from the small intestine of rats infected with N. brasiliensis responded by histamine secretion after challenge from worm-specific antigen, anti-IgE, concanavalin A and calcium ionophores, but were refractory to compound 48/80 and bee venom peptide 401 (42). These authors also showed that disodium cromoglycate and theophylline did not inhibit antigen-induced histamine release of MMC, although effective against CTMC. These findings show that MMC differ from CTMC not only in response to secretagogues but also to modulators of secretion.

We have attempted to trace the amine secretion from MMC in vivo. An intraluminal secretion of 5-HT has been previously demonstrated both in vitro and in vivo and in several species, including the rat (43). We therefore performed a series of intestinal perfusion experiments and measured histamine and 5-HT in the perfusates (Ahlman and co-workers, to be published). Rats in nembutal anesthesia were ventilated on air using a rodent respirator and kept at a controlled body temperature. The proximal or distal half of the small intestine was isolated with intact vascular supply and plastic tubings inserted at each end. The lumen was perfused with saline at a rate of 0.1 ml/min and the perfusates collected every 10 min. The amines were identi-fied and measured with sensitive HPLC methods. Rats injected with nematode larvae and age-matched controls injected with saline were perfused 14 days later, when MMC numbers and mucosal histamine content was increased 5-fold compared to controls. At this stage worms are found in the proximal segment of the intestine where many MMC have migrated into the epithelium. In the distal segment, MMC numbers and histamine content are elevated to about the same extent, but MMC are largely confined to the lamina propria and worms are absent.

After an initial washout period of 30 min, a steady state level of histamine secretion was obtained both in controls and in infected rats, but the histamine concentration was about 10 times higher in the latter (Fig. 5). There was no difference in histamine secretion rate between the distal and proximal segments of the gut. In another experiment we measured histamine and 5-HT in jejunal perfusates of normal and infected rats. The concentrations of the two amines were well correlated. These results thus suggest that histamine may be released from MMC into the gut lumen in vivo. Histamine release appears to be related to the total quantity of MMC, but does not seem to depend on the presence of intraepithelial MMC. The strong correlation between histamine and 5-HT concentration in the perfusates is of considerable interest and will be further investigated. The explanation may be that both amines are released from MMC or that amine secretion from MMC and ECC is functionally inter-related. There are no indications that the amine content of the perfusates is the result of a migration of MMC into the lumen, although this possibility cannot be wholly excluded at present. It should be noted in this context that the release of 5-HT into the gut lumen is increased by vagal stimulation (43).

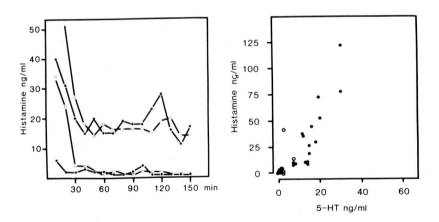

Figure 5. Intestinal perfusion of normal and N. brasiliensis-infected rats. (a) Intraluminal histamine release from proximal (filled symbols) and distal (open symbols) segments of the gut. Symbols represent means of three experiments in infected (circles) and normal (squares) animals. (b) Correlation between the concentration of histamine and 5-HT in the perfusates.

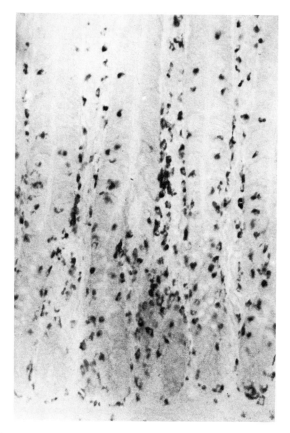

Figure 6. Marked increase of MMC in the jejunal mucosa of a rat infected 14 days previously with N. brasiliensis. The tissue was fixed in IFAA, embedded in paraffin and stained with toluidine blue at pH 0.5 for 30 min.

MMC Proliferation in Immune Responses

Intestinal helminth infections are accompanied by a massive production of IgE antibodies and a proliferation of MMC (44). The MMC response has been studied in detail in rats infected with the nematode N. brasiliensis by Miller and associates (reviewed in 45). After skin penetration of subcutaneous injection, infective larvae migrate to the intestine via the lungs, trachea, and esophagus, molt to adult worms and start to lay eggs, which are passed in the feces. The worms predominantly localize in the proximal part of the jejunum (46). At 8 to 10 days after the infection, MMC are virtually absent from the lamina propria (47). This is perhaps an effect of a non-specific mast cell degranula-

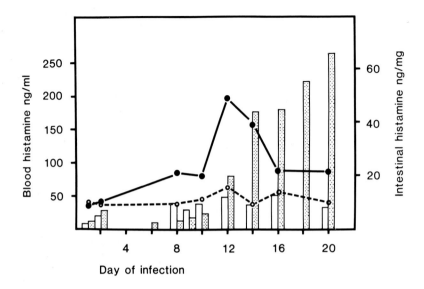

Figure 7. Concentration of histamine in whole blood (circles) and intestine (bars) during the course of an infection with N. brasiliensis. Symbols represent means of three infected (filled symbols) and control animals (open symbols, broken line).

tor produced by the worms (see 44). The mast cell degranulation is followed by a proliferation of MMC reaching a maximum level of about five times the normal number on day 12 to 14 after the infection (Fig. 6), when the worms are expelled from the gut, a mechanism called "self cure". During this stage MMC also migrate into the epithelium (47). Cells referred to as globule leukocytes are probably MMC which have migrated into the epithelium and partly discharged their granules (48). Rat MMC proliferation is a property of the whole intestine, but the intraepithelial migration only occurs at the site of the worms (46). The proliferative MMC response is immunologically mediated and dependent on T-lymphocytes (49,50,51).

The proliferative rat MMC response is accompanied by a parallel increase in the content of histamine and 5-HT of the gut, as discussed above. A 9-fold increase in mucosal mast cell protease content was obtained at the peak of the mast cell response (32). Concomitant immunoperoxidase studies showed that only a proportion of the MMC of the lamina propria and none of the intraepithelial MMC contained the enzyme. A release of protease into the circulation has also been reported (52).

MMC, Basophils and Blood Histamine

Circulating blood basophils (53) as well as elevated blood histamine levels (54,55) have been reported as occurring during the N. brasiliensis infection in rats. Blood basophils are normally absent or extremely rare in rat blood, but whole blood histamine, as well as free plasma histamine levels, have been reported as being quite high (56). The reported values of 50 ng/ml blood and 17 ng/ml plasma indicate that a significant proportion of the blood histamine in normal rats is contained in the cellular fraction.

We have studied histamine levels in the gut, whole blood and plasma in relation to MMC of the gut and blood cell morphology during the course of the infection in an attempt to clarify the possible relationship between rat MMC and circulating blood basophils (57). The results (Fig. 7) showed an increase of whole blood histamine from a control level of about 40 ng/ml to 200 ng/ml on day 12 after the infection, coinciding with the early phase of the proliferative MMC response. There was also an increase in free plasma histamine from the normal level of 15-20 ng/ml to a maximum of 80 ng/ml on day 12. Both rapidly returned towards the control level during day 16-20 while gut histamine content remained high.

The total number of blood neutrophils increased during the early phase of the infection and there was a pronounced eosinophilia from day 10 to 16. No cells containing granules showing metachromasia or cationic dye binding at low pH levels were found in a screening which comprised 10,000 cells per specimen. During day 10 to 16 after the infection we found a number of coarsely granulated and vacuolated cells. The granules stained dark blue with the Giemsa stain, but did not stain metachromatically with toluidine blue or with alcian blue at a low pH. We interpret this change as being equivalent to the so-called toxic granulation occurring in human neutrophils during certain infections. It may have been mistaken for a specific basophilic granulation by previous investigators. Apparently, mature granulated basophils are absent from or extremely rare in rat blood. The cellular repository for the blood histamine in normal as well as nematode infected rats remains unclear. We have suggested, as a working hypothesis, that the histamine is contained in a circulating progenitor cell for MMC which may have the ability to synthesize the amine and store it in a loosely bound state, but which has not yet acquired the ability to assemble the specific cytoplasmic granules. The absence of such granules would make the histamine highly susceptible to leakage into the plasma which would in turn explain the unexpectedly high plasma histamine levels that are obtained in the rat. An alternative explanation of the high blood histamine levels during the nematode infection could be that these are the result of a release from MMC. The time course of the changes in blood histamine, as well as the absence of evidence of histamine uptake by leukocytes, tend to make this alternative the less likely of the two.

Derivation of Mast Cells

The origin of mast cells has been the subject of controversy for a long time. However, recent studies by Kitamura and associates on mutant mice genotypes have shown that there are two hierarchical classes of mast cell precursor in the mouse (2). The primary progenitor is a circulating, bone marrow-derived precursor originating from the multipotential hemopoietic stem cell (CFU-S). This primary progenitor gives rise to a localized, tissue-bound mast cell precursor whose differentiation may be locally regulated. Kitamura's work concerns mouse CTMC, since MMC were not specifically studied. However, the results neatly explain the existence of specific mast cell phenotypes such as MMC in special tissue sites.

Another important recent development is the demonstration, by several groups, of the growth of mast cells from hemopoietic tissues of mice in vitro (reviewed in 58). These mast cells proliferate in the presence of interleukin 3 (59). The cultured mouse cells resemble the MMC rather than the CTMC phenotype of the rat, and at least one such cell line produces chondroitin sulfate E instead of heparin (60). Moreover, Jarrett and associates have cultured mast cells from normal rat bone marrow, stimulated by factors released from lymphocytes by N. brasiliensis antigen, or by concanavalin A. These mast cells were identified as MMC on the basis of their histochemical and biochemical properties, including the content of a chondroitin sulfate-like proteoglycan and a specific serine endoprotease, RMCP II (58).

Mucosal Mast Cells of the Rat and Man

It would thus appear that the identification of MMC and CTMC as different phenotypes rests on the fact that the lamina propria of the gastrointestinal mucosa and connective tissue sites, such as those in the skin and tongue, contain practically pure populations of one or the other cell type in the adult rat. Major points of difference between MMC and CTMC of the rat are summarized in Table 6.

The properties of the MMC listed in the table may perhaps be used as markers for their identification in other tissues and mucous membranes of the rat. The most useful of the biochemical MMC markers appears to be its specific proteoglycan and proteinase which can also be visualized by staining methods in tissue sections. The blocking of dye binding by aldehyde may also be potentially useful. It must be realized, however, that it is not known to what extent the mast cell characteristics

TABLE VI. <u>Points of difference between two mast cell</u>
<u>phenotypes in the rat</u>

Mucosal Mast Cells	Connective Tissue Mast Cells
Small cells of variable shape with few variable-size granules	Larger cells of uniform shape with many uniform-size granules
Migratory capacity	Non-migratory
Soluble granular proteoglycan matrix containing chondroitin sulfate E; Anionic sites blocked by aldehyde fixation	Less soluble proteoglycan matrix containing macromolecular heparin; anionic sites not blocked by aldehydes
Contain Rat Mast Cell Protease II	Contain Rat Mast Cell Protease I
Low, age-dependent histamine content (<2 pg/cell)	High, age-dependent histamine content (<35 pg/cell)
Low 5-HT content, increased in nematode response	Age-dependent 5-HT content (<1.5 pg/cell)
Non-secretory, proliferative response to polyamines	Secretory response to polyamines
Thymus-dependent proliferation in immune responses (nematode infections)	No proliferative immune response
Short lifespan, half-life <40 days	Long lifespan, half-life >6 months

listed in the table are the subject of interspecies variations. Properties of MMC of the rat, such as its content of chondroitin sulfate E, or its dye binding properties, can therefore not be used to identify MMC of other species in <u>vivo</u> or in <u>vitro</u>.

Mast cells are numerous in the mucous membranes of a number of species and have a morphological appearance similar to that of rat MMC. The staining properties of human intestinal MMC were found to be influenced by the method of fixation, much like those of the rat (61,62). We have compared the degree of aldehyde blocking of dye binding of human mast cells in different tissues. In tissues fixed in 4% neutral formaldehyde, mast cells were

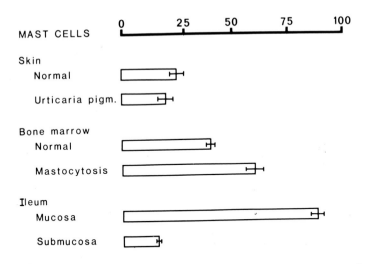

Figure 8. Formaldehyde-blocking of the dye-binding of human mast cells. Tissues were fixed in 4% neutral buffered formaldehyde and embedded in paraffin. The sections were stained with 0.5% toluidine blue at pH 0.5 for 30 min or 6 days and the number of cells counted. The degree of blocking was defined as the fraction of the cells counted in sections stained for 6 days but not for 30 min.

counted after staining with toluidine blue at pH 0.5 for 30 min and after a subsequent staining for 6 days. The degree of blocking was defined as the proportion of the cells that did not stain at 30 min but after 6 days. The results (Fig. 8) show that the mast cells were subject to a variable degree of blocking. Mast cells in the lamina propria of the mucosa, unlike those of the submucosa or skin, showed an almost complete blocking of dye binding. The cases of urticaria pigmentosa showed a 6-fold increase in mast cell numbers per unit area compared to controls but the degree of blocking of the mast cells was similar (Olafsson et al., to be published). In systemic mastocytosis, the bone marrow contains an increased number of mast cells whose dye-binding is to a large extent blocked by aldehyde (unpublished observation). This finding is of interest in relation to the recent demonstration of mast cell/basophil precursors in the blood of patients with mast cell proliferative disease (71).

MMC in Human Allergy

The problem of how contact is established between allergen and mast cells is of fundamental importance to the understanding of

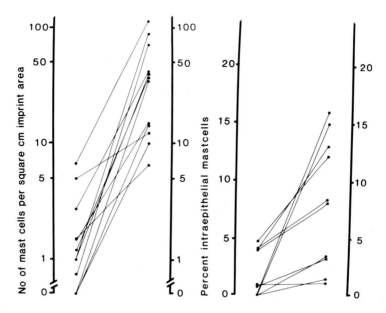

Figure 9. Nasal mucosal mast cells in birch pollen allergic individuals before and during the pollen season. Imprints and tissues were fixed in IFAA and stained with toluidine blue at pH 0.5 for 30 min and mast cells counted for the unit area of imprint or section. (a) Mast cells in mucosal imprints. (b) Fraction of intraepithelial mucosal mast cells in tissue sections.

allergic reactions of mucous membranes. That mast cell mediated reactions may take place at the mucosal surface is suggested by the presence in bronchial lavages of such cells, capable of releasing histamine by contact with allergen in vitro (63). Basophil cells have also been found both in nasal secretions and smears and in epithelial scrapings from patients with allergic rhinitis (64,65). Based on structural criteria, basophil cells of smears were identified as BB (66,67) while those found in epithelium and stroma had the appearance of tissue mast cells (67,68,69).

We have studied patients with birch pollen allergy before and during the pollen season (Pipkorn and Enerback, to be published). Imprints were taken from the nasal mucosa with soft polyester strips (70). The majority of the cells of the imprints were granulocytes and epithelial cells; their numbers did not change with the onset of the pollen season. Mast cells were extremely rare in the imprints taken before the season but numerous during the season (Fig. 9). Most of these cells were found in association with epithelial cells and had the morphological appearance of mast cells rather than BB judging by the size and density of granules and their nuclear shape. The number of BB as

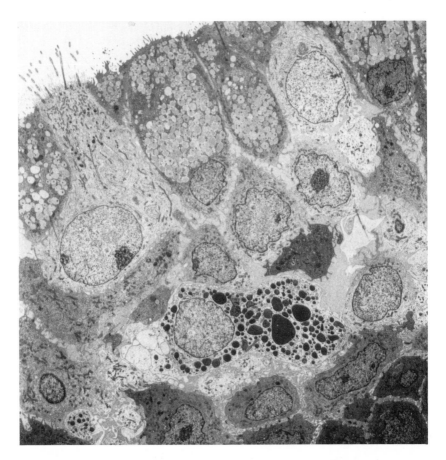

Figure 10.   Electron micrograph of the nasal mucosa in a
patient with birch pollen allergy during the pollen season. Note
intraepithelial mast cell with granules of greatly varying size
and shape.

well as whole blood or plasma histamine levels did not change
significantly in relation to the allergic season. In biopsies
taken before the pollen season, the nasal mucosa contained many
mast cells located in the connective tissue stroma and only
occasionally in the epithelium. These MMC were of variable shape
and contained relatively few granules, thus resembling MMC of
rats rather than CTMC. With the onset of the pollen season there
was a redistribution of MMC into the epithelium (Fig. 9). On the
other hand, neither total mast cell number nor histamine content
of the biopsies changed significantly. We interpret this as a
migration of MMC into the epithelium, much like the reaction of

the rat MMC during nematode infections. Ultrastructural observations tend to support this assumption and have demonstrated a certain resemblance between these intraepithelial, presumably migrating MMC, and those of the rat (Fig. 10).

CONCLUSIONS

In this review I have attempted to describe the evolution of our knowledge of the MMC to the present day. Accumulated evidence from studies in the rat shows that the MMC represents a phenotype different from the classical CTMC. The demonstration of a specific rat mucosal mast cell is of obvious interest in view of the role of mucosal barriers in the IgE-mediated immunological reactions. It also emphasizes the necessity of recognizing that the mast cell system may in fact constitute a heterogeneous group of cells with significant differences in structural and chemical organization and in physiological properties.

REFERENCES

1. **Ehrlich, P.** 1878. Beitrage zur Kenntnis der Anilinfarbungen und ihrer Verwendung in der mikroskopischen Technik. Arch. mikr. Anat. (Bonn) 13, 263: In: Collected papers of Paul Ehrlich. Edited by F. Himmelweit, London/New York, Pergamon Press, 1956.
2. **Kitamura, Y., T. Sonoda, and M. Yokoyama.** 1983. Differentiation of tissue mast cells. In: Haemopoietic Stem Cells. Edited by S.-A. Killmann, E.P. Cronkite, and C.N. Muller-Berat, Copenhagen, Munksgaard, p.350.
3. **Dvorak, H.F., and A. Dvorak.** 1975. Basophilic leucocytes: Structure, function and role in disease. Clin. Haematol. 4:651.
4. **Maximow, A.** 1906. Uber die Zellformen des lockeren Bindegewebes. Arch. f. mikr. Anat. (Bonn) 67:680.
5. **Ballantyne, E.N.** 1929. Differentiation of plasma cells from mast cells in the intestinal mucosa of the white rat. Can. Med. Assoc. J. 21:195.
6. **Michels, N.A.** 1938. The mast cells. In: Downey's Handbook of Haematology 1:232. Republished in: Mast Cells and Basophils. 1963. Edited by J. Padawer. Ann. N.Y. Acad. Sci. 103, Appendix.
7. **Mota, I., W.I. Beraldo, A.G. Ferri, and L.C.U. Junueira.** 1956. Action of 48/80 on the mast cell population and histamine content of the wall of the gastrointestinal tract of the rat. In: Ciba Symposium on Histamine. London, J & A Churchill, p.47.
8. **Enerback, L.** 1966. Mast cells in rat gastrointestinal mucosa. I. Effect of fixation. Acta Pathol. Microbiol. Scand. 66:289.
9. **Enerback, L.** 1966. Mast cells in rat gastrointestinal

mucosa. ii. Dye binding and metachromatic properties. Acta Pathol. Microbiol. Scand. 66:303.

10. **Enerback, L.** 1966. Mast cells in rat gastrointestinal mucosa. III. Reactivity towards compound 48/80. Acta Pathol. Microbiol. Scand. 66:313.

11. **Enerback, L.** 1966. Mast cells in rat gastrointestinal mucosa. IV. Monoamine storing capacity. Acta Pathol. Microbiol. Scand. 67:365.

12. **Enerback, L., and P.M. Lundin.** 1974. Ultrastructure of mucosal mast cell in normal and compound 48/80 treated rats. Cell Tissue Res. 150:95.

13. **Newson, B., A. Dahlstrom, L. Enerback, and H. Ahlman.** 1983. Suggestive evidence for a direct innervation of mucosal mast cells. An electron microscopic study. Neuroscience 10:565.

14. **Blandina, P., F. Fantozzi, P.F. Mannaioni, and E. Masini.** 1980. Characteristics of histamine release evoked by acetylcholine in isolated rat mast cells. J. Physiol. 301:281.

15. **Foreman, J., and C. Jordan.** 1983. Histamine release and vascular changes induced by neuropeptides. Agents Actions 13:105.

16. **Mayrhofer, G.** 1980. Fixation and staining of granules in mucosal mast cells and intraepithelial lymphocytes in the rat jejunum, with special reference to the relationship between the acid glycosaminoglycans in the two cell types. Histochem. J. 12:513.

17. **Miller, H.R.P., and R. Walshaw.** 1972. Immune reactions in mucous membranes. IV. Histochemistry of intestinal mast cells during helminth expulsion in the rat. Am. J. Path. 69:195.

18. **Wingren, U., and L. Enerback.** 1983. Mucosal mast cells of the rat intestine: a re-evaluation of fixation and staining properties, with special reference to protein blocking and solubility of the granular glycosaminoglycan. Histochem. J. 15:571.

19. **Spicer, S.S.** 1963. Histochemical properties of mucopolysaccharide and basic protein in mast cells. In: Mast Cells and Basophils. Edited by J. Padawer. Ann. N.Y. Acad. Sci. 103:322.

20. **Lagunoff, D., M.M. Phillips, O.A. Iseri, and E.P. Benditt.** 1964. Isolation and preliminary characterization of rat mast cell granules. Lab. Invest. 13:1331.

21. **Lagunoff, D., and P. Pritzl.** 1975. Characterization of rat mast cell granule proteins. Arch. Biochem. Biophys. 173:554.

22. **Combs, J.W., D. Lagunoff, and E.P. Benditt.** 1965. Differentiation and proliferation of embryonic mast cells of the rat. J. Cell Biol. 25:577.

23. **Scott, J.E., and J. Dorling.** 1965. Differential staining of acid glycosaminoglycans (mucopolysaccharides) by alcian blue in salt solutions. Histochemie 5:221.

24. **Enerback, L.** 1974. Berberine sulphate binding to mast cell polyanions: A cytoflorometric method for quantitation of heparin. Histochemistry 42:301.

25. **Enerback, L., G. Berlin, I. Svensson, and I. Rundquist.** 1976. Quantitation of mast cell heparin by flow cytofluorometry. J. Histochem. Cytochem. 24:1231.

26. **Fransson, L.A., I. Sjoberg, and B. Havsmark.** 1980. Structural studies on heparan sulphates. Characterization of oligosaccharides; obtained by periodate oxidation and alkaline elimination. Eur. J. Biochem. 106:59.

27. **Enerback, L., S.O. Kolset, M. Kusche, A. Hjerpe, and U. Lincahl.** 1985. Glycosaminoglycans in rat mucosal mast cells. Biochem. J. 227:661.

28. **Lagunoff, D.** 1981. Neutral proteases of mast cells. In: Biochemistry of Acute Allergic Reactions, Vol. 14. Edited by E.L. Becker, S. Simon, and K.F. Austen. Korc Foundation Ser. New York: Liss, p.350.

29. **Woodbury, R.G., and H. Neurath.** 1980. Structure, specificity and localization of the serine proteases of connective tissue. FEBS Lett. 114:189.

30. **Woodbury, R.G., G.M. Gruzenski, and D. Lagunoff.** 1978. Immunofluorescent localization of a serine protease in rat small intestine. Proc. Natl. Acad. Sci. USA 75:2785.

31. **Benditt, E.P., and M. Arase.** 1959. An enzyme in mast cells with properties like chymotrypsin. J. Exp. Med. 110:451.

32. **Woodbury, R.G., and H.R.P. Miller.** 1982. Quantitative analysis of mucosal mast cell protease in the intestines of Nippostongylus infected rats. Immunology 46:487.

33. **Hakansson, R., and C. Owman.** 1967. Concomitant histochemical demonstration of histamine and catecholamines in entero-chromaffin-like cells of gastric mucosa. Life Sci. 6:759.

34. **Enerback, L., and U. Wingren.** 1980. Histamine content of peritoneal and tissue mast cells of growing rats. Histochemistry 66:113.

35. **Befus, A.D., N. Johnston, and J. Bienenstock.** 1979. Nippostrongylus brasiliensis: Mast cells and histamine levels in tissues of infected and normal rats. Exp. Parasitol. 48:1.

36. **Wingren, U., A. Wastesson, and L. Enerback.** 1983. Storage and turnover of histamine, 5-hydroxytryptamine and heparin in rat peritoneal mast cells in vivo. Int. Archs. Allergy Appl. Immunol. 70:193.

37. **Enerback, L., G.-B. Lowhagen, O. Lowhagen, and U. Wingren.** 1981. The effect of polymyxin B and some mast-cell constituents on mucosal mast cells in the duodenum of the rat. Cell Tissue Res. 214:239.

38. **Enerback, L.** 1981. The gut mucosal mast cell. Monogr. Allergy 17:222.

39. **Befus, A.D., F.L. Pearce, J. Gauldie, P. Horsewood, and J. Bienenstock.** 1982. Mucosal mast cells. I. Isolation and functional characteristics of rat intestinal mast cells. J. Immunol. 128:2475.

40. Wingren, U., L. Enerback, H. Ahlman, S. Allenmark, and A. Dahlstrom. 1983. Amines of the mucosal mast cell of the gut in normal and nematode infected rats. Histochemistry 77:145.

41. Enerback, L., and G.-B. Lowhagen. 1979. Long term increase of mucosal mast cells in the rat induced by administration of compound 48/80. Cell Tissue Res. 198:209.

42. Pearce, F.L., A.D. Befus, J. Gauldie, and J. Bienenstock. 1982. Mucosal mast cells. II. Effects of anti-allergic compounds on histamine secretion by isolated intestinal mast cells. J. Immunol. 128:2481.

43. Ahlman, H., K. Gronstad, O. Nilsson, and A. Dahlstrom. 1984. Biochemical and morphological studies on the secretion of 5-HT into the gut lumen of the rat. Biogenic Amines 1:63.

44. Jarrett, E.E.E., and H.R.P. Miller. 1982. Production and activities of IgE in helminth infection. Prog. Allergy 31:178.

45. Miller, H.R.P. 1980. The structure, origin and function of mucosal mast cells. A brief review. Biologie Cell 39:229.

46. MacDonald, T.T., M. Murray, and A. Ferguson. 1980. Nippostrongylus brasiliensis: Mast cell kinetics at small intestinal sites in infected rats. Exp. Parasitol. 49:9.

47. Miller, H.R.P., and W.F.H. Jarrett. 1971. Immune reactions in mucous membranes. I. Intestinal mast cell response during helminth expulsion in the rat. Immunology 20:277.

48. Murray, M., H.R.P. Miller, and W.F.H. Jarrett. 1968. The globule leukocyte and its derivation from the subepithelial mast cell. Lab. Invest. 19:222.

49. Mayrhofer, G. 1979. The nature of the thymus dependency of mucosal mast cells. II. The effect of thymectomy and of depleting recirculating lymphocytes on the response to Nippostrongylus brasiliensis. Cell. Immunol. 47:312.

50. Befus, A.D., and J. Bienenstock. 1979. Immunologically-mediated intestinal mastocytosis in Nippostrongylus brasiliensis infected rats. Immunology 38:95.

51. Nawa, Y., and H.R.P. Miller. 1979. Adoptive transfer of the intestinal mast cell response in rats infected with Nippostrongylus brasiliensis. Cell. Immunol. 42:225.

52. Miller, H.R.P., R.G. Woodbury, J.F. Huntley, and G. Newlands. 1983. Systemic release of mucosal mast cell-protease in primed rats challenged with Nippostrongylus brasiliensis. Immunology 49:471.

53. Ogilvie, B.M., P.M. Hesketh, and M.E. Rose. 1978. Nippostrongylus brasiliensis: Peripheral blood leucocyte response of rats, with special reference to basophils. Exp. Parasitol. 46:20.

54. Roth, R.L., and D.A. Levy. 1980. Nippostrongylus brasiliensis: Peripheral leukocyte responses and correlation of basophils with blood histamine concentration during infection in rats. Exp. Parasitol. 50:331.

55. Heatley, R.V., A.D. Befus, and J. Bienenstock. 1982.

Nippostrongylus brasiliensis infection in the rat: Effects of surgical removal of Peyer's patches, mesenteric lymph nodes and spleen. Int. Arch. Allergy Appl. Immunol. 68:397.

56. Almeida, A.P., W. Flye, D. Deveraux, Z. Horakova, and M.A. Beaven. 1980. Distribution of histamine and histaminase (diamine oxidase) in blood of various species. Comp. Biochem. Physiol. 67C:187.

57. Enerback, L., G. Lindenger, T. van Loo, and G. Granerus. 1985. Cellular repository for blood histamine in normal and nematode infected rats. Agents Actions 16:314.

58. Jarrett, E.E.E., and D.M. Haig. 1984. Mucosal mast cells in vivo and in vitro. Immunol. Today 5:115.

59. Ihle, J.N., J. Keller, S. Oroszlan, L.E. Henderson, T.D. Copeland, F. Fitch, M.B. Prystowsky, E. Goldwasser, J.W. Schrader, E. Palaszynski, M. Dy, and B. Lebel. 1983. Biologic properties of homogeneous interleukin 3. I. Demonstration of WEHI-3 growth factor activity, mast cell growth factor activity, P cell-stimulating factor activity, colony-stimulating factor activity, and histamine-producing cell-stimulating factor activity. J. Immunol. 131:282.

60. Razin, E., R.L. Stevens, F. Akiyama, K. Schmid, and K.F. Austen. 1982. Culture from mouse bone marrow of a subclass of mast cells possessing a distinct chondroitin sulfate proteoglycan with glycosaminoglycans rich in N-acetylgalac-tosamine-4,6-disulfate. J. Biol. Chem. 257:7229.

61. Strobel, S., H.R.P. Miller, and A. Ferguson. 1981. Human intestinal mucosal mast cells: evaluation of fixation and staining. J. Clin. Path. 34:851.

62. Ruitenberg, E.J., L. Gustowska, A. Elgersma, and H.M. Ruitenberg. 1982. Effects of fixation on the light micro-scopical visualization of mast cells in the mucosa and connective tissue of human duodenum. Int. Arch. Allergy Appl. Immunol. 67:233.

63. Patterson, R., Y. Oh, S. Tomita, I.M. Suszko, and J.J. Pruzansky. 1974. Respiratory mast cells and basophiloid cells. Clin. Exp. Immunol. 16:223.

64. Bryan, W.T.K., and M.P. Bryan. 1959. Significance of mast cells in nasal secretion. Trans. Am. Acad. Ophthalmol. Otolaryngol. 63:613.

65. Mygind, N., and J. Thomsen. 1973. Cytology of the nasal mucosa. A comparative study between replica-method and a smear-method. Arch. klin. exp. Ohr.-, Nas.- u. Kehlk. Heilk. 204:123.

66. Okuda, M., S. Kawabori, and H. Ohtsuka. 1978. Electron microscopy study of basophilic cells in allergic nasal secretions. Arch. Otorhinolaryngol. (NY) 221:215.

67. Hastie, R., J.H. Heroy III, and D.A. Levy. 1979. Basophil leukocytes and mast cells in human nasal secretions and scrapings studied by light microscopy. Lab. Invest. 40:554.

68. Kawabori, S., T. Unno, M. Okuda, and H. Ohtsuka. 1981. Electronmicroscopic studies of basophilic cells in allergic

nasal secretions and mucous membranes. Rhinology 19(1):115.

69. Okuda, M., H. Ohtsuka, and S. Kawabori. 1983. Basophil leukocytes and mast cells in the nose. Eur. J. Resp. Dis. 64, suppl. 128:7.

70. Pipkorn, U., and L. Enerback. 1984. A method for the preparation of imprints from the nasal mucosa. J. Immunol. Meth. 73:133.

71. Denburg, J.A., J. Olafsson, G. Roupe, S. Ahlstedt, S. Telizyn, and J. Bienenstock. 1985. Basophil/mast cell precursors in mast cell proliferative disorders. Clin. Invest. Med. (in press).

Mast Cell Differentiation and Heterogeneity,
edited by A. D. Befus et al.
Raven Press, New York © 1986.

The Ontogeny and Differentiation of Mast Cells: A Review

John W. Schrader

The Walter and Eliza Hall Institute of Medical Research, Royal Melbourne Hospital, Victoria 3050, Australia

In the last 5 years direct experimental approaches have unequivocally shown that mast cells are derived from the pluripotential hemopoietic stem cell. The generation from pluripotential stem cells of hemopoietic progenitor cells committed to differentiate along various lineages and their multiplication and maturation are normally regulated through interactions between hemopoietic cells and cells in the stroma of the bone marrow (1,2). Hemopoietic stem and progenitor cells are also influenced by series of soluble growth factors, namely granulocyte–macrophage colony stimulating factor (CSF), granulocyte–CSF, macrophage–CSF (or CSF–1) and P–cell stimulating factor or IL–3 (1). Indeed, the role of these soluble hemopoietic growth factors seems to lie not in steady–state hemopoiesis but in enhancing the production and function of hemopoietic stem cells in response to pathological stimuli, such as antigenic activation of the immune system or the presence of endotoxin.

Analysis of the generation of mast cells from the pluripotential hemopoietic stem cell is complicated by the fact that mast cells exhibit a diversity of histological, biochemical and functional properties (4,5). Most of the experimental evidence on the generation of mast cells derives from studies in rodents in which there are at least two classes of mast cells, those at serosal surfaces and in connective tissues, and those at mucosal sites (4,5). There is still considerable uncertainty about whether

The work was supported by the National Health and Medical Research Council, Canberra, Australia; the Bushell Trust, the Windemere Hospital Foundation and the Wenkert Foundation.

Acknowledgements: I thank my colleagues, Drs. Ian Clark–Lewis, Richard Crapper, Keven Leslie, Sabariah Schrader, and Grace Wong for their collaboration, Miss Joanne Ringham and Denise Galatis for their exellent technical assistance, and Professor G.V.J. Nossal for his continued support and encouragement.

such distinctions between subtypes of mast cells, based principally on studies in the rat, apply to mast cells in other species.

At a practical level, experiments in mice have established the hemopoietic origin of mast cells and the existence of committed progenitor cells and regulatory factors for mast cells. One line of investigation has focused on the in vitro generation of murine mast cells showing that such cells remain absolutely dependent on the presence of a T-cell–derived lymphokine and are thus operationally "thymus–dependent". Another line of investigation utilizing methods of cell transfer in vivo and conventional histochemical techniques has not revealed obvious involvement of T-cells or a T-cell–derived lymphokine in mast cell ontogeny in the mouse. Although it has been assumed that these two experimental systems deal with two different types of murine mast cells, referred to in many publications as "thymus–dependent or atypical" and "thymus–independent or typical" mast cells, this has not been established unequivocally, nor has the basis for any such dichotomy been fully clarified.

The generation in vitro of pure populations of lymphokine-dependent mast cells. A major breakthrough in the analysis of mast cells came 5 years ago with the development of reproducible techniques for the generation of homogeneous populations of murine mast cells in tissue cultures (6–10). Provided that a specific T-cell lymphokine was present, cultures of cells from the bone marrow or lymphoid organs of mice reproducibly gave rise to homogeneous populations of non-adherent cells, all of which contained metachromatic granules, histamine or serotonin and had cell-surface receptors for IgE. These cells were diploid and grew relatively slowly for prolonged but not unlimited periods. Mindful of the difficulties in identifying cells in vitro (see 1), we initially preferred the operational title of "persisting cell" (P cell) (6,7) in reference to their characteristic persistence in serial cultures long after other cell-types had disappeared; the T-cell lymphokine which was required for the growth and survival of these cells in vitro was termed "persisting cell–stimulating factor" (PSF) (6,12). Work from a number of laboratories has now established some similarities between cytological, cell-surface and biochemical features of PSF-dependent, in vitro–derived mast cell lines and those of mast cells found in vivo at certain mucosal surfaces. For example, PSF-dependent, in vitro–derived murine mast cells and in vivo-derived rat intestinal mast cells lack heparin but contain unique proteoglycan (13,16) and the former can produce large quantities of leukotriene C4 (17). In the rat, mast cells generated from bone-marrow in vitro in response to medium conditioned by activated T-cells contain the protease characteristic of in vivo intestinal mast cells (18).

Biological data indicate that in vitro–derived, PSF–dependent murine mast cells could be counterparts of intestinal as well as other mast cells in vivo. Correlations exist at various sites in vivo between the number of mast cells and cells which when cultured in vitro with PSF would form a clone of P-cells (i.e., P-cell precursors). In the gut mucosa, which is well-recognized as a site for the thymus-dependent accumulation of mast cells, the frequency of P-cell precursors amongst intramucosal lymphocytes is about 1 in 300 (19), or about 10-fold higher than in the bone marrow, or 100-fold higher than in the spleen (20). Moreover, in W^I/W^I mutant mice, which have a genetically determined inability to mount a mastocytosis in the gut wall, the frequency of P-cell precursors is at least 1,000-fold lower than in normal litter mates (21). Likewise, in the vicinity of T-cells activated by an immune response, e.g. in a lymph node draining an immunization site, there are increases in both histologically detectable mast cells and P-cell precursors (20). As expected from data on the thymus-dependence of mastocytosis in vivo, immunization of athymic mice did not lead to an increase in P-cell precursors (21).

In vitro-derived P-cells retain their mast cell-like properties when transferred in vivo (22). One week after P-cells are injected into the dermis of mast-cell deficient W^I/W^I mice, alcian-blue positive mast cells of the same genotype as the injected cells are detectable at the injection site. Furthermore, whereas in normal W^I/W^I mice, the mast cells derived from the injected P-cells disappear by the second week after injection, in mice bearing a tumour secreting PSF into the serum, mast cells derived from the injected P-cells are still present at 2 weeks.

This result raises the important possibility that through the secretion of PSF, activated T cells regulate not only the generation of mast cells at sites of inflammation, but also their continued survival. Dependence for survival on the presence of PSF released by locally-activated T cells may thus provide a useful operational definition for a subset of what can be termed lymphokine-dependent mast cells in vivo. Some evidence in support of this notion comes from presentations at this symposium indicating that the administration of glucocorticoids results in a rapid reduction in a number of mast cells in the gut of parasitized rats or in human nasal mucosa, and from previous information that steroids block the release of lymphokines by T lymphocytes (66) but have relatively little direct effect on the variability of PSF-dependent mast cells (23).

P-cells can be sensitized with specific IgE antibodies in vitro and injected locally into the dermis of mast-cell deficient W/W^V mice (22), which are unable to mount an immediate-type hypersensitivity response. Intravenous challenge with specific antigen results in a local cutaneous anaphylactic reaction at the site of injection of the sensitized P-cells, demonstrating that the in vitro-derived P-cells behave like mast cells when transferred in vivo. Experiments of this general design, i.e. of reconstitution of mast-cell deficient W/W^V mice with pure populations of in vitro–derived mast cells, may be useful in analyzing the function of a subtype of mast cell in immunity to parasites.

Despite the structural and functional evidence that P–cells may be the in vitro counterparts of mast cells that are associated with immune responses in vivo, it has yet to be excluded that some or all PSF–dependent mast cells may have the capacity to differentiate into mast cells that contain heparin or no longer require PSF for their survival. In this respect the data presented at this symposium by Kitamura and colleagues are of great interest. It will be important to determine by the transfer of single cells whether heparin–containing, lymphokine–independent mast cells arise from PSF–dependent mast cell progenitors, from PSF–dependent mast cells themselves, or from a distinct, undifferentiated progenitor cell.

P–cell Stimulating Factor (PSF). A great deal is now known about the structure and physiological and pathological function of the murine T–cell lymphokine responsible for the generation and differentiation of mast cells. The factor is produced by murine T–cells following immunological or mitogenic activation (41,44,45) and is a glycoprotein with an apparent molecular weight on gel filtration of 30–45,000 daltons (6,24,25). PSF is secreted in vivo in lymph nodes of mice undergoing immune reactions and in vivo, as in vitro, the production of PSF depends on the presence of T cells (21). Usually the effects of PSF are confined to the vicinity of the activated T–cell; immunization in one footpad for example, results in the release of PSF and an increase in the number of thymus–dependent mast cells in the ipsilateral but not contralateral popliteal lymph node (20). With intense and widespread T–cell activation, such as occurs in murine graft–versus–host disease, or in parasitic infections (26), PSF can be detected in the serum. PSF is not bound to serum proteins and is cleared rapidly from the serum (t1/2 = 40 min) (21). It is not degraded by serum and is probably broken down in the kidney in the proximal tubules, since urinary levels are far lower than those expected in the serum (21).

Murine PSF has been purified to homogeneity (27) and the N–terminal amino acid sequence shown to resemble closely that determined earlier by Ihle et al. (28) for IL–3, a lymphokine assayed by the induction of the synthesis of the enzyme 20–α–hydroxysteroid dehydrogenase in spleen cells from congenitally athymic mice. The only difference between the reported sequences of IL–3 and PSF is that PSF has an additional six amino acids at the N–terminus. The amino acid sequences predicted from two cDNA clones isolated using the assays based on the stimulation of growth of lines of mast cells (29,30) correspond with the N–terminal amino acid sequences of both purified PSF and IL–3. Cleavage of the signal peptide from the initial translation product probably produces the N–terminal amino acid sequence determined for PSF: studies using antibodies specific for these six N–terminal amino acid peptides suggest that they are present on at least the bulk of PSF molecules produced by T–cells

(Ziltner, Clark-Lewis, Kent and Schrader, unpublished data). Further proteolytic cleavage resulting in the loss of the six N-terminal amino acids probably accounts for the sequence determined by Ihle et al., and additional studies will be necessary to assess its physiological significance.

Experiments in which purified PSF has been added to isolated single cells have conclusively shown that PSF acts directly on the progenitors of murine megakaryocytes, neutrophils and macrophages as well as on those of mast cells (31). There is evidence that the addition of purified PSF to cultures of murine bone marrow cells stimulates directly or indirectly the division of pluripotential hemopoietic stem cells (CFU-S), the progenitors of erythroid cells and eosinophils, and also the survival in vitro of cells able to repopulate irradiated mice with T lymphocytes, thymocytes and B lymphocytes (although there is no evidence that PSF has any direct effects on cells committed to the lymphoid lineage). The action of PSF on cells of multiple hemopoietic lineages has meant that different groups have studied the same molecule using different assays, e.g. involving the growth in agar of colonies of various cells including erythro-cytes ("burst-promoting activity") (32), megakaryocytes (megakaryocyte-colony stimulating activity) (32), and colonies containing a mixture of hemopoietic cells (mixed or multi-colony stimulating factor) (32), or the in vitro growth of lines of hemopoietic progenitor cells (WEHI-3 factor) (33). Its stimulation of the growth and differentiation of mast cells and their progenitors has also led to the names "mast-cell growth factor" (10,24,30) and "histamine-producing cell stimulating factor" (34) for PSF (IL-3).

One unresolved question is whether PSF (IL-3) also stimulates the generation of basophils. Earlier, Denburg and colleagues (35) demonstrated in the guinea pig that supernatants of cultures from antigen-stimulated T-cells stimulated the in vitro produc-tion of basophils from guinea pig bone-marrow. The paucity of basophils in the mouse constitutes a major practical obstacle to experiments with purified murine PSF (IL-3), although Dvorak et al. have reported that basophils are produced when bone marrow is cultured with medium conditioned by Concanavalin A-stimulated spleen cells (36). There is one report (37) of a permanent factor-dependent murine cell-line that resembles a basophil ultrastructurally and bears high affinity receptors for IgE, but also has natural killer (NK) activity. However, the factor required for the growth of this line appears to be IL-2 and it is likely that this line is best characterized as an NK-cell and is not representative of a basophil (S. Galli, personal communica-tion). Overall, given that PSF (IL-3) stimulates the generation of all other cells of hemopoietic origin that have been studied, it seems highly likely that PSF (IL-3) also stimulates the production of basophils.

The origin of PSF (IL-3)-dependent mast cells from the pluri-potential hemopoietic stem cell. Early experiments using chimeric mice indicated that the cells in the spleen and thymus that give rise to P-cells in vitro are ultimately derived from the bone marrow (6,7). Further in vitro experiments have shown that single cells that are capable of generating neutrophils, macrophages and erythrocytes also give rise to PSF-dependent mast cells (11,38). It seems clear that the link between the thymus and the proliferation of mast cells, once argued to reflect a progenitor-progeny relationship between T-cells and mast cells, can now be adequately accounted for by the secretion from activated T-cells of the lymphokine (PSF or IL-3) that drives the generation of mast cells from myeloid precursors.

Committed progenitor cells of PSF (IL-3)-dependent mast cells. Limiting dilution techniques have been used to obtain minimal estimates of the frequency of cells in various tissues that responded to PSF (IL-3) in vitro by generating clones of PSF (IL-3)-dependent mast cells (Table 1). In normal bone marrow, the precursors of PSF (IL-3)-dependent mast cells are negative for antigens such as Thy-1 or Ia (Crapper and Schrader, unpublished data). Mast cell progenitors, however, resemble pluripotential hemopoietic stem cells and the progenitors of neutrophils and macrophages in that they show a transient expression of the Thy-1 antigen during stimulation with PSF (IL-3) (42). The Thy-1 antigen is lost during maturation, since well-differentiated PSF-dependent macrophages are Thy-1 negative.

The best evidence for the existence of a committed, restricted progenitor cell for PSF (IL-3)-dependent mast cells comes from experiments on cells from the murine gut mucosa (19). Whereas the frequency of progenitors of PSF (IL-3)-dependent mast cells in this cell population was 10-fold higher than in the bone marrow, pluripotential hemopoietic stem cells or committed progenitors of other hemopoietic lineages were not present at detectable levels. The relative absence of progenitors of other cells of hemopoietic lineage in the gut mucosa may explain the paradox that, although PSF (IL-3) is capable of stimulating the generation of cells of multiple hemopoietic lineages, when released in the gut it stimulates the generation of mast cells alone.

It remains to be determined whether the committed progenitor cells that occur in murine mucosal and lymphoid tissues are generated in the bone-marrow and home via the blood to these tissues, or alternatively are generated from pluripotential hemopoietic stem cells at these sites. The mast cell-deficient mice with mutations of the W and S1 loci may be useful in analyzing this question. For example, whereas W^f/W^f mice have T-cells that can produce PSF (11) and have a normal frequency of progenitors of PSF-dependent mast cells in the bone-marrow, committed progenitors of mast cells are virtually undetectable in the gut mucosa (20), suggesting that these mice have a defect in either the homing of the committed progenitor or (less likely) in its generation in the mucosa.

The committed progenitors of PSF (IL-3)-dependent mast cells in the gut mucosa are negative for Thy-1 and Lyt 2 antigens and lack cytoplasmic granules, thus differing from both T lymphocytes and the granulated intramucosal lymphocytes found at this site. The lineage relationships of these granulated lymphocytes illustrate well the problems that arise in attempts to assign lineage using histochemical techniques. Although the granulated intramucosal lymphocytes are negative for the Thy-1 antigen and contain granules which stain with astra blue and metachromatically with toluidine blue, they exhibit the Lyt 2 antigen characteristic of T-cells (19). Thus, in contrast to earlier speculation that such cells are progenitors of mast cells, granulated intramucosal lymphocytes are clearly distinct from progenitors of P-cells, and by inference, distinct from mast cells. It seems likely that these granulated intramucosal lymphocytes belong to the T-cell or natural killer (NK) lineages.

Influence of cells other than T-cells on the generation of PSF (IL-3)-dependent mast cells. Congenitally athymic animals show no deficiency in steady-state myelopoiesis or erythropoiesis; in athymic rats there are likewise some intestinal mast cells in response to parasites (43) and normal numbers of precursors of PSF (IL-3)-dependent mast cells are found in the gut and spleen of athymic mice (19,20). The critical defect in T-cell deficient animals is their inability to amplify numbers of mast cells and their precursors in response to immunological stimuli (20,21). Thus, in a strict sense the term "thymus-dependent" mast cell is a misnomer, as the generation of these cells is not absolutely thymus-dependent and it is only the expansion of their numbers in response to immunological stimuli that depends on the presence of T-cells.

The basal-level generation of PSF (IL-3)-dependent mast cells and their precursors in the absence of a thymus seems to depend on cell-cell interaction of hemopoietic stem cells and stromal cells. Mast cells and precursors of PSF (IL-3)-dependent mast cells are generated in long-term bone-marrow cultures in which PSF (IL-3) is undetectable (11). Moreover, PSF (IL-3)-dependent cell lines that had been deliberately seeded into such long-term bone-marrow cultures can be recovered after intervals of up to 6 months (44), suggesting that the stromal cells in these cultures in some way have substituted for PSF (IL-3) in the maintenance of PSF (IL-3)-dependent cells. Ongoing work on the mechanism(s) responsible for maintaining normal hemopoiesis in the bone-marrow may also shed light on steady-state production of mast-cell progenitors in the absence of a thymic influence.

Another question is whether cells other than T lymphocytes produce factors similar to PSF (IL-3). There are reports that murine keratinocytes (45) and astroglia (46) release factors which stimulate in vitro growth of permanent lines of PSF (IL-3)-dependent mast cells. It will be important to confirm that these factors also act on normal mast cell progenitors or freshly derived PSF (IL-3)-dependent mast cells.

The possibility of non-T lymphocyte sources of PSF (IL-3) is also raised by the fact that the myelomonocytic leukemia WEHI-3B secretes PSF (IL-3) constitutively, and indeed has been used as a convenient source of the molecule both for biochemical and in vivo studies (25,27,28,32,33). However, based on direct experimental evidence (3,47) we have argued that the secretion of PSF (IL-3) by myeloid leukemias such as WEHI-3B is pathological and reflects the aberrant, oncogenic activation of the PSF (IL-3) gene in a hemopoietic cell, i.e. a neutrophil-macrophage progenitor in the case of WEHI-3B, that is normally the target for PSF (IL-3). The autogenous production of PSF (IL-3) by a PSF (IL-3)-responsive cell results in autostimulation and uncontrolled growth (47).

Do mast cells grow in response to factors other than PSF (IL-3)? At present there is no good evidence that freshly derived cultures of murine PSF (IL-3)-dependent mast cells can be supported by any factors other than PSF (IL-3). However, one permanent factor-dependent mast-cell line has been reported to grow in response to interleukin-2 (48).

However, some caution is necessary in extrapolating conclusions based upon permanent, immortal cell lines to normal mast cells. Immortal lines of mast cells or of other hemopoietic cells probably result from a genetic abnormality; thus, these lines occur only infrequently, and their emergence may involve the action of murine leukemia viruses (see 1). In contrast, the P-cells that were originally described are generated reproducibly at high frequency from normal bone marrow or lymphoid tissue, are diploid, and do not grow for indefinite periods, their limited capacity for self-renewal probably faithfully reflecting that of certain murine mast cells in vivo. Although permanent lines of factor-dependent mast cells offer a convenient source of material for experiments on the structure and function of mast cells, it is important that results of these experiments be confirmed using freshly derived populations of PSF (IL-3)-dependent mast cells.

Regulation of the differentiation of PSF (IL-3)-dependent mast cells. Two T-cell lymphokines, gamma-interferon and PSF (IL-3) influence the differentiation in vitro of murine mast cells, in particular having antagonistic effects on the levels of expression of antigens coded by the major histocompatibility complex such as H-2, Ia, TL and Qa (23,49-51). Because both gamma-interferon and PSF (IL-3) are commonly secreted by the same clone of activated T-cells, the level of major histocompatibility complex antigens expressed by IL-3-dependent mast cells at sites of inflammation will depend on the balance of the levels of PSF (IL-3), gamma-interferon and probably of other factors.

The functional implications of changes in expression of major histocompatibility complex antigens on mast cells and in particular the induction of Ia antigens by gamma-interferon, remain to be explored but are potentially of great significance in the qualitative and quantitative regulation of immune responses at mucosal surfaces. It will also be important to determine whether gamma-interferon and PSF (IL-3) regulate other aspects of differentiation of IL-3-dependent mast cells, e.g. sensitivity to degranulation, arachidonic acid metabolism, etc.

Factors released by T lymphocytes and other cell types are likely to have important effects on the differentiation of PSF (IL-3)-dependent mast cells, and thus will contribute to the overall heterogeneity of mast cells at different sites under different conditions.

Lymphokine-Independent Mast Cells. A completely different experimental approach that involves cell-transfer in vivo, has been used by Kitamura and colleagues to show that a subtype of murine mast cell is derived from the bone-marrow via blood-born precursors generated from the pluripotential hemopoietic stem cell (40,41,51). A key feature of these experiments that distinguishes them from the in vitro experiments with PSF (IL-3)-dependent mast cells, is that the recipient mice (usually W/W^v) were killed at least 5 weeks after cell-transfer, and there was no evidence that T-cells were activated or were necessary. Thus, the mast cells in these experiments appeared to be generated and capable of survival in the absence of PSF (IL-3).

Nevertheless, PSF (IL-3) may in some circumstances be able to influence at least the early stages of development of the mast cells detected in this system. Thus, single colonies containing erythroid and myeloid cells grown in vitro from bone-marrow cells by stimulation with medium conditioned by mitogen-activated spleen cells contain cells that give rise to mast cells when injected directly into the skin of W/W^v mice (52). The factor in spleen-cell conditioned medium that stimulates the development of colonies of erythroid and myeloid cells is known to be identical with PSF (IL-3). Further studies are required to determine whether the cells that give rise to skin mast cells in the above system were pluripotential hemopoietic stem cells or were committed progenitors of mast cells generated in vitro under the influence of PSF (IL-3).

Evidence for committed progenitors of lymphokine (PSF, IL-3)-independent mast cells. In most tissues the frequency of cells that give rise to mast cells when injected directly into the skin of W/W^v mice (40,41) about equals that of pluripotential hemopoietic stem cells (Table II), although there are tissues such as mesenteric lymph node, where a clear dissociation between the frequencies of these cells points to the existence of a distinct committed progenitor of a PSF (IL-3)-independent type of mast cell (41). Comparison of results from different laboratories (Table I) suggests that in the mesenteric lymph node there is also a dissociation between the frequency of precursors of cells that gave rise to PSF (IL-3)-dependent mast cells in vitro and cells that gave rise to mast cells when injected into the skin of W/W^v mice. Confirmation of this result by parallel assays on the same cell population would be necessary to conclude that the progenitors measured in these two assays are distinct. Interestingly, the frequency in mesenteric lymph node of progenitors measured in the in vivo cell transfer assay was increased by immunization (41), again raising the possibility that progenitors measured in this assay are generated or proliferate in response to PSF (IL-3).

TABLE I. Comparison of the frequencies[a] in various tissues of precursors of "thymus-dependent" mast cells (P cells) and "connective-tissue" mast cells

Tissue	Frequency per 10^6 cells			
	"Thymus-dependent" mast cells	"Connective-tissue" mast cells		Pluripotential hemopoietic stem cells (CFUs)
Bone marrow	200,500[20]	800,900[39]	100[40]	200[41]
Blood (mononuclear cells)	11[20]		20[40]	30[41]
Spleen	30[20]	34,70[39]	19[40]	29[41]
Lymph node (popliteal)	0.5[20]	<1.5[39]		
Lymph node (mesenteric)	18[20]	6[39]	0.4[41]	0.05[41]
Gut mucosal lymphocytes	1800[19,20]	700,3000[39]		<1[19]

a. The frequencies of precursors of thymus-dependent mast cells were determined in vitro, and the frequency of precursors of connective-tissue mast cells in vivo. Duplicate figures refer to results with different mouse strains.

Kitamura has recently shown that the cells in the peritoneal cavity that generate colonies of mast cells when injected into the skin of W/W^v mice include not only undifferentiated cells, but also well-differentiated, peritoneal mast cells (53). Thus, under appropriate circumstances, peritoneal mast cells retain a considerable capacity for division and self-renewal. Other evidence for the existence of a distinct committed progenitor for PSF (IL-3)-independent mast cells comes from experiments which suggest that new mast cells which appear in a skin graft are generated from undifferentiated progenitors derived from the bone-marrow which become fixed in the grafted skin (54).

Factors regulating the development of PSF (IL-3)-independent skin mast cells. The regulation of the differentiation of PSF (IL-3)-independent, skin mast cells from committed progenitors or from pluripotential hemopoietic stem cells is not well understood, although there is evidence for the influence of local regulatory factors. In W/W^v mice, normal, unstimulated skin appears to provide adequate stimulation for the differentiation of such mast cells from bone-marrow progenitors (40), although the skin of normal mice does not (55). One possibility is that mature mast cells, which are absent in W/W^v skin, normally suppress the division and differentiation of progenitors. The positive stimulus for the differentiation of mast cells in the skin may be related to that which is exerted by normal stromal cells upon hemopoietic cells; for example, the skin of mice of the Sl/Sl^d genotype, which have a defect in the hemopoietic stromal microenvironment, fails to be repopulated by mast cells when grafted to normal mice (68). In addition, control of mast cell development appears to be exerted at the level of homing of blood-borne progenitors, since entry of these to the skin appears to be inhibited by pre-existing progenitors fixed in the skin (54).

In vitro studies of the generation of PSF (IL-3)-independent, peritoneal mast cells. In vitro systems have not yet proved useful in analyzing the differentiation of PSF (IL-3)-independent mast cells such as those in the peritoneal cavity or skin. In early studies (56) in which mouse thymocytes were cultured with feeder layers of embryonic cells, it is likely that the cultures contained PSF (IL-3), since use of cells from non-inbred mice or mixtures of cells from inbred strains would probably lead to allogeneic activation of T lymphocytes. In later experiments in which lymph-node cells from mice immunized with horse serum were cultured in the presence of horse serum (57), it seems likely that the mast cells generated were dependent on PSF (IL-3). Interestingly, the ultrastructural development of these mast cells (58) closely resembled those of mast cells during embryogenesis (59). Some of the mast cells generated when rat thymocytes were cultured alone or with feeder layers of embryonic cells stained with safranin, like peritoneal mast cells (60). The adherent cells that were a feature of these cultures may have been important in promoting the growth of this type of mast cell. However, PSF (IL-3)-dependent mast cells may also have been present, for reasons mentioned above, or because the heterologous serum proteins may have led to the generation of PSF (IL-3).

Recently, we have observed that the addition of PSF (IL-3) to cultures of non–adherent peritoneal cells, enriched in safranin–positive peritoneal mast cells, permits their survival for periods of at least 8 weeks. Further experiments using single, isolated cells will be necessary to demonstrate that PSF (IL–3) is acting directly on these cells; nevertheless, such experiments raise the possibility that heparin–containing mast cells have receptors for and are affected by PSF (IL–3).

Do PSF (IL–3)–dependent mast cells give rise to PSF (IL–3)–independent mast cells? At present there is no direct evidence that PSF (IL–3)–dependent mast cells are able to differentiate into cells that no longer require PSF (IL–3) for their survival, or have other characteristics of such mast cells. Galli et al. (15) have shown that sodium butyrate increases the granulation and histamine content of PSF (IL–3)–dependent mast cells in vitro, although heparin was not produced. One technical problem is that immature murine mast cells can resemble PSF (IL–3)–dependent mast cells in some respects, for example in the granules staining with alcian blue rather than safranin and possibly in containing proteoglycans other than heparin.

Mastocytosis and neoplastic diseases of the mast cell. Analysis of the biochemical and biological properties of the mast cells in the various forms of human or experimental mastocytosis may prove instructive in the analysis of mast cell heterogeneity and the factors influencing mast cell differentiation. In some cases cells in mastocytomas (61,65) or disseminated mastocytosis (62) have been reported to contain heparin although the fact that chondroitin sulphate is also produced in mastocytosis (63), and mastocytomas in the mouse have been reported to generate leukotriene C (67), indicate that neoplastic mast cells can resemble lymphokine–dependent mast cells in some respects. It may be significant that a prominent infiltration of mononuclear cells, including lymphocytes, often occurs in mastocytosis (64) and it will be interesting to determine whether these cells include T cells that are releasing molecules resembling PSF (IL–3).

The notion that the proliferation of abnormal mast cells in mastocytosis may be dependent on PSF (IL–3) is important as it could offer a new approach to therapy. There is also the possibility that progression of disease could involve aberrant activation of a PSF (IL–3) gene with consequent autostimulation by autogenously produced PSF (IL–3) as has been demonstrated in the mouse (3,47).

REFERENCES

1. **Schrader, J.W.** 1983. Bone marrow differentiation in vivo. CRC Crit. Rev. Immunol. 4:197.
2. **Dexter, T.M., T.D. Allen, and L.G. Lajtha.** 1977. Conditions controlling the proliferation of haemopoietic stem cells in vitro. J. Cell Physiol. 91:335.
3. **Schrader, J.W., I. Clark–Lewis, R.M. Crapper, and G.H.W. Wong.** 1983. P cell stimulating factor: characterization,

action on multiple lineages of bone–marrow derived cells and role in oncogenesis. Immunol. Rev. 76:79.

4. Askenase, P. 1980. Immunopathology of parasitic diseases: involvement of basophils and mast cells. Springer Symp. Immunopathol. 2:417.

5. Metcalfe, D.D., M. Kaliner, and M.A. Donlon. 1981. The mast cell. CRC Crit. Rev. Immunol. 3:23.

6. Schrader, J.W., and G.J.V. Nossal. 1980. Strategies for the analysis of accessory cell function: the in vitro cloning and characterization of the P cell. Immunol. Rev. 53:61.

7. Schrader, J.W. 1981. The in vitro production and cloning of the P cell, a bone marrow–derived null cell that expresses H–2 and Ia-antigens, has mast cell-like granules, and is regulated by a factor released by activated T cells. J. Immunol. 126:452.

8. Razin, E., C. Cordon–Cardo, and R.A. Good. 1981. Growth of a pure population of mast cells in vitro with conditioned medium derived from concanavalin–A stimulated splenocytes. Proc. Natl. Acad. Sci. USA 78:2559.

9. Tertian, G., Y.–P. Yung, D. Guy–Grand, and M.A.S. Moore. 1981. Long-term in vitro culture of murine–mast cells. I. Description of a growth factor dependent culture technique. J. Immunol. 127:788.

10. Nabel, G., S.J. Galli, A.M. Dvorak, H.F. Dvorak, and H. Cantor. 1981. Inducer T lymphocytes synthesize a factor that stimulates proliferation of cloned mast cells. Nature (London) 291:333.

11. Schrader, J.W., S.J. Lewis, I. Clark–Lewis, and J.G. Culvenor. 1981. The persisting (P) cell: histamine content, regulation by a T–cell derived factor, origin from a bone-marrow precursor and relationship to mast cells. Proc. Natl. Acad. Sci. USA 78:323.

12. Clark–Lewis, I., and J.W. Schrader. 1981. P cell–stimulating factor: biochemical characterization of a new T cell-derived factor. J. Immunol. 127:1941.

13. Razin, E., R.L. Stevens, F. Akiyama, K. Schmid, and K.F. Austen. 1982. Culture from mouse bone marrow of a subclass of mast cells possessing a distinct chondroitin sulphate proteoglycan with glycosaminoglycans rich in N–acetyl-galactosamine–4,6–disulphate. J. Biol. Chem. 257:7229.

14. Sredni, B., M.M. Friedman, C.E. Blank, and D.D. Metcalfe. 1983. Ultrastructural, biochemical, and functional characteristics of histamine–containing cells cloned from mouse bone marrow: tentative identification as mucosal mast cells. J. Immunol. 131:915.

15. Galli, S.J., A.M. Dvorak, J.A. Marcum, T. Ishizaka, G. Nabel, H. Der Simonian, K. Pyne, J.M. Goldin, R.D. Rosenberg, H. Cantor, and H.F. Dvorak. 1982. Mast cell clones: a model for the analysis of cellular maturation. J. Cell Biol. 95:435.

16. Bland, C.E., K.L. Rosenthal, D.H. Pluznik, G. Dennert et al.

1984. Glycosaminoglycan profiles in cloned granulated lymphocytes with natural killer function and in cultured mast cells; their potential use as biochemical markers. J. Immunol. 132:1937.

17. **Razin, E., J.M. Mencia-Huerta, R.A. Lewis, E.J. Corey, and K.F. Austen.** 1982. Generation of leukotriene-C4 from a subclass of mast cells differentiated in vitro from mouse bone-marrow. Proc. Natl. Acad. Sci. USA 79:4665.

18. **Haig, D.M., T.A. McKee, E.E.E. Jarrett, R. Woodbury, and H.R.P. Miller.** 1982. Generation of mucosal mast cells is stimulated in vitro by factors derived from T cells of helminth-infected rats. Nature 300:188.

19. **Schrader, J.W., R. Scollay, and F. Battye.** 1983. Intra-mucosal lymphocytes of the gut: Lyt-2 and Thy-2 phenotype of the granulated cells and evidence for the presence of both T cells and mast cell precursors. J. Immunol. 130:558.

20. **Crapper, R.M., and J.W. Schrader.** 1983. Frequency of mast cell precursors in normal tissues determined by an in vitro assay: antigen induces parallel increases in the frequency of P cell precursors and mast cells. J. Immunol. 131:923.

21. **Crapper, R.M., I. Clark-Lewis, and J.W. Schrader.** 1984. The in vivo functions and properties of persisting cell-stimulating factor. Immunology 53:33.

22. **Crapper, R.M., W.R. Thomas, and J.W. Schrader.** 1984. In vivo transfer of persisting (P) cells; further evidence for their identity with T-dependent mast cells. J. Immunol. 133:2174.

23. **Wong, G.H.W., I. Clark-Lewis, J.A. Hamilton, and J.W. Schrader.** 1984. P-cell stimulating factor and glucocorticoids oppose the action of gamma-interferon in inducing Ia-antigens on T-dependent mast cells (P cells). J. Immunol. 133:2043.

24. **Yung, Y.-P., R. Egar, G. Tertian, and M.A.S. Moore.** 1981. Long-term in vitro culture of murine mast cells. II. Purification of a mast cell growth factor and its dissociation from TCGF. J. Immunol. 127:794.

25. **Hasthorpe, S.** 1980. A hemopoietic cell line dependent on a factor in pokeweed mitogen-stimulated spleen-cell conditioned medium. J. Cell. Physiol. 105:379.

26. **Filho, M.A., M. Dy, B. Lebel, G. Luffau, and J. Hamburger.** 1983. In vitro and in vivo histamine-producing cell-stimulating factor (or IL-3) production during Nippostrongylus brasiliensis infection: coincidence with self-cure phenomenon. Eur. J. Immunol. 13:841.

27. **Clark-Lewis, I., S.B.H. Kent, and J.W. Schrader.** 1984. Purification to apparent homogeneity of a factor stimulating the growth of multiple lineages of haemopoietic cells. J. Biol. Chem. 259:7488.

28. **Ihle, J.N., J. Keller, S. Oroszlan, L.E. Henderson, T.D. Copeland, F. Fitch, M.B. Prystowsky, E. Goldwasser, J.W. Schrader, E. Palaszynski, M. Dy, and B. Lebel.** 1983. Biological properties of homogeneous interleukin 3. I.

Demonstration of WEHI-3 growth factor activity, mast cell growth factor activity, P cell–stimulating factor activity, colony–stimulating factor activity, and histamine–producing cell–stimulating factor activity. J. Immunol. 131:282.

29. **Fung, M.C., A.J. Hapel, S. Ymer, D.R. Cohen, R.M. Johnson, H.D. Campbell, and I.G. Young.** 1984. Molecular cloning of cDNA for murine interleukin–3. Nature 307:233.

30. **Yokota, T., F. Lee, D. Rennick, C. Hall, et al.** 1984. Isolation and characterization of a mouse cDNA clone that expresses mast cell growth factor activity in monkey cells. Proc. Natl. Acad. Sci. USA 81:1070.

31. **Clark–Lewis, I., R.M. Crapper, K. Leslie, and J.W. Schrader.** 1985. P–cell stimulating factor: Molecular and biological properties. In: Cellular and Molecular Biology of Lymphokines. Edited by Sorg, C. and A. Schimpl. New York, Academic Press (in press).

32. **Iscove, N.N., C.A. Roitsch, N. Williams, and L.J. Guilbert.** 1982. Molecules stimulating early red cell, granulocyte, macrophage and megakaryocyte precursors in culture: similarity in size, hydrophobicity and charge. J. Cell Physiol. 1:65.

33. **Bazill, G.W., M. Haynes, J. Garland, and T.M. Dexter.** 1983. Characterization and partial purification of a haemopoietic growth factor in WEHI–3 conditioned medium. Biochem. J. 210:747.

34. **Dy, M., B. Lebel, P. Kamoun, and J. Hamburger.** 1981. Histamine production during the anti–allograft response. Demonstration of a new lymphokine enhancing histamine synthesis. J. Exp. Med. 153:293.

35. **Denburg, J.A., M. Davison, and J. Bienenstock.** 1980. Basophil production: stimulation by factors derived from guinea pig splenic T–lymphocytes. J. Clin. Invest. 65:390.

36. **Dvorak, A.M., G. Nabel, K. Pyne, H. Kantor, H.F. Dvorak, and S.J. Galli.** 1982. Ultrastructural identification of the mouse basophil. Blood 59:1279.

37. **Galli, S.J., A.M. Dvorak, T. Ishizaka, G. Nobel, H. Der Simonian, H. Cantor, and H.F. Dvorak.** 1982. A cloned cell with NK function resembles basophils by ultrastructure and expresses IgE receptors. Nature (London) 298:288.

38. **Nakahata, T., S.S. Spicer, J.R. Canter, and M. Ogawa.** 1982. Clonal assay of mouse mast cell colonies in methylcellulose culture. Blood 60:352.

39. **Guy–Grand, D., M. Dy, L. Luffau, and P. Vassalli.** 1984. Gut mucosal mast cells: origin, traffic and differentiation. J. Exp. Med. 160:12.

40. **Sonoda, T., T. Ohno, and Y. Kitamura.** 1982. Concentration of mast cell progenitors in bone–marrow, spleen and blood of mice determined by limiting dilution analysis. J. Cell. Physiol. 112:136.

41. **Hayashi, C. T. Sonoda, and Y. Kitamura.** 1983. Bone marrow origin of mast cell precursors in mesenteric lymph nodes of

mice. Exp. Hematol. 11:772.

42. **Schrader, J.W., F. Battye, and R. Scollay.** 1982. Expression of Thy-1 antigen is not limited to T cells in cultures of mouse hemopoietic cells. Proc. Natl. Acad. Sci. USA 79:4161.

43. **Mayrhofer, G., and R. Fisher.** 1979. Mast cells in severely T cell depleted rats and the response to infestation with Nippostrongylus brasiliensis. Immunology 37:145.

44. **Schrader, J.W., S. Schrader, I. Clark-Lewis, and R.M. Crapper.** 1983. In vitro approaches to lymphopoiesis, hemopoiesis and oncogenesis. In: Longterm bone-marrow Culture. Edited by Wright, D.G., and J.S. Greenberger. Alan R. Liss, Inc., p. 243.

45. **Luger, T.A., U. Wirth, and A. Kock.** 1985. Epidermal cells synthesize a cytokine with interleukin-3-like properties. J. Immunol. 134:915.

46. **Free, K., B. Stadler, and A. Fontana.** 1984. Astrocyte-derived interleukin-3-like factors. Lymphokine Res. 3:243 (Abst.).

47. **Schrader, J.W., and R.M. Crapper.** 1983. Autogenous production of a hemopoietic growth factor "P cell stimulating factor" as a mechanism for transformation of bone-marrow derived cells. Proc. Natl. Acad. Sci. USA 80:6892.

48. **Hapel, A.J., H.S. Warren, and D.A. Hume.** 1984. Different colony stimulating factors are detected by the "Interleukin-3" dependent cell lines FDC-P1 and 32Dcl-23. Blood 64:786.

49. **Wong, G.H.W., I. Clark-Lewis, J. McKimm-Breschkin, and J.W. Schrader.** 1982. Interferon-gamma-like molecule induces Ia antigens on cultured mast cell progenitors. Proc. Natl. Acad. Sci. USA 79:6989.

50. **Koch, N., G.H. Wong, and J.W. Schrader.** 1984. Both Ia antigens and associated invariant chains are induced simultaneously in lines of T-dependent mast cells by recombinant interferon-gamma. J. Immunol. 132:1361.

51. **Kitamura, Y., Y. Yokoyama, H. Matsuda, and T. Ohno.** 1981. Spleen colony-forming cell as a common precursor for tissue mast cells and granulocytes. Nature (London) 291:159.

52. **Sonoda, T., Y. Kitamura, Y. Haku, H. Hara, and K.J. Mori.** 1983. Mast cell precursors in various hemopoietic colonies of mice produced in vivo and in vitro. Brit. J. Hematol. 53:611.

53. **Sonoda, T., Y. Kanayama, H. Hara, C. Hayashi, M. Tadokoro, T. Yonezawa, and Y. Kitamura.** 1984. Proliferation of peritoneal mast cells in the skin of W/W^v mice that genetically lack mast cells. J. Exp. Med. 160:138.

54. **Matsuda, H., Y. Kitamura, T. Sonada, and T. Imori.** 1981. Precursors of mast cells fixed in the skin of mice. J. Cell. Physiol. 108:409.

55. **Hatanaka, K., Y. Kitamura, and Y. Nishimure.** 1979. Local development of bone marrow-derived precursors in the skin of

mice. Blood 53:142.

56. **Ginsburg, H., and L. Sachs.** 1963. Formation of pure suspension of mast cells in tissue culture by differentiation of lymphoid cells from mouse thymus. J. Natl. Cancer Inst. 31:1.

57. **Ginsburg, H., I. Nir, I. Hammel, R. Evan, B.A. Weissman, and Y. Naot.** 1978. Differentiation and activity of mast cells following immunization in cultures of lymph nodes. Immunology 35:485.

58. **Coombs, J.W., D. Lagunoff, and E.P. Benditt.** 1985. Differentiation and proliferation of embryonic mast cells of the rat. J. Cell Biol. 25:577.

59. **Coombs, J.W. 1971.** An electron microscopic study of mouse mast cells arising in vivo and in vitro. J. Cell Biol. 48:676.

60. **Ishizaka, T., H. Okudaira, L. Mauser, and K. Ishizaka.** 1976. Development of rat mast cells in vitro. I. Differentiation of mast cells from thymus cells. J. Immunol. 116:747.

61. **Oliver, J., F. Bloom, and C. Mangieri. 1947.** On the origin of heparin. An examination of the heparin content and specific cytoplastic particles of neoplastic mast cells. J. Exp. Med. 86:107.

62. **Drennan, J.M.** 1951. The mast cells in urticaria pigmentosa. J. Path. Bact. 63:513.

63. **Astoe-Hanse, G., and J. Clausen.** 1964. Mastocytosis (urticaria pigmentosa) with urinary excretion of hyaluronic acid and chondroitin sulphate. Changes induced by polymyxin B. Amer. J. Med. 36:144.

64. **Rhoner, H.G., G. Klingmuller, R. Burkhardt, O.-E. Rodermund, and L.S. Giesler.** 1979. Mastocytosis. In: The Mast Cell. Edited by Pepys, S. and A.M. Edwards. Tunbridge Wells, Pitman Medical, p.110.

65. **Green, J.P. and M. Day.** 1960. Heparin, 5-hydroxy-tryptamine and histamine in neoplastic mast cells. Biochem. Pharmacol. 3:190.

66. **Kelso, A., and A. Munck.** 1984. Glucocorticoid inhibition of lymphokine secreted by alloreactive T lymphocyte clones. J. Immunol. 133:784.

67. **Murphy, R.C., S. Hammarstrom, and B. Samuelson.** 1979. Leukotriene C: a slow reacting substance from murine mastocytoma cells. Proc. Natl. Acad. Sci. USA 76:4275.

68. **Matsuda, H., and Y. Kitamura.** 1981. Migration of stromal cells supporting mast-cell differentiation into open wound produced into the skin of mice. Exp. Hematol. 9:38.

Mast Cell Differentiation and Heterogeneity,
edited by A. D. Befus et al.
Raven Press, New York © 1986.

The Effects of Transforming Retroviruses on Mast Cell Growth and IL-3 Dependence

*J. N. Ihle, *Jonathan Keller, *Alan Rein, **Jackie Pierce, and †Ulf Rapp

*NCI-Frederick Cancer Research Facility, Litton Bionetics Inc., Basic Research Program, Frederick, Maryland 21701; **Laboratory of Cellular and Molecular Biology, National Institutes of Health, Bethesda, Maryland 20205; and †National Cancer Institute, Frederick, Maryland 21701

During the past several years, considerable information has been obtained concerning the factors that regulate normal mast cell differentiation and growth. Initially it was demonstrated that there existed T cell–derived factors which could support the proliferation of murine mast cells in vitro which were termed mast cell growth factors (1–3). Subsequently, it became clear that mast cell growth factors were similar to factors such as P–cell stimulating factor (4) and histamine–producing cell-stimulating factor (5). The demonstration that a single T cell factor could regulate a number of aspects of hematopoietic stem cell differentiation including mast cell differentiation and growth came from studies of biochemically homogenous murine interleukin 3 (IL–3). These studies demonstrated that IL–3 allowed the differentiation in vitro of functional mast cells (6) and was as active in assays for factors related to mast cell function as in assays specific for IL–3 (7). Studies with an antiserum against IL–3 demonstrated that among the T cell–derived lymphokines in the mouse, IL–3 was uniquely able to induce the proliferation of mature mast cells (6,8).

IL–3 was initially detected and subsequently purified to homogeneity based on its ability to induce the expression of the enzyme 20–α–hydroxysteroid dehydrogenase (20 α–SDH) in cultures containing hematopoietic/lymphoid cells (9,10). This assay was developed to study the factors which regulate early stages of T cell differentiation based on the expression of high levels of 20 α–SDH by murine T cells (11). Biochemically homogenous IL–3 is a glycoprotein with an apparent molecular weight of 28,000 daltons by SDS–PAGE. The amino terminal sequence of the protein has been

This research was sponsored by the National Cancer Institute under contract No. N01–CO–75380 with Litton Bionetics, Inc.

established (7). In addition, cDNA clones have been obtained for
IL-3 (12,13). The cDNA clones encode a protein with an identical
predicted amino terminal sequence and a similar spectrum of
biological activities (14). From the sequence of the cDNA
clones, the mature protein consists of 134 amino acids with a
predicted size of 15,000 daltons. Four potential sites for amino
glycosylation exist and by protein sequence analysis the first
site at residue 10 of the mature protein is substituted. The
carbohydrate content of IL-3 (12 μg of glucosamine/100 μg
protein) suggests that all of the potential sites may be
modified. A single cellular gene for IL-3 has been localized on
chromosome 11 in mice (J.N. Ihle and Kosack, C., in preparation).

In addition to its effects on 20 α-SDH expression and mast
cell proliferation, purified IL-3 has been shown to mediate a
variety of biologicl phenomena in vitro. IL-3 is equivalent to a
factor required for the growth of a series of cell lines derived
from long-term murine bone marrow cultures (15). In addition,
IL-3 has colony-stimulating factor activity (CSF) and accounts
for approximately 5% of the CSF activity in conditioned media
from activated murine T cells (7). IL-3 also uniquely induces
the expression of Thy-1 antigen in cultures containing murine
lymphoid precursors (16). In cultures of murine bone marrow
cells, IL-3 also increases the number of cells responding to
erythropoietin and thus has the activity of erythroid burst-
promoting factor (17). The spectrum of biological activities
observed with IL-3 is related to its ability to induce the
differentiation of early hematopoietic/lymphoid stem cells.

The ability of purified IL-3 to support the differentiation of
a variety of hematopoietic cell types in vitro has allowed
experiments to address the effects of transforming retroviruses
on hematopoietic cells. Such studies are of interest for
comparing the effects of transforming genes on hematopoietic
cells with those on fibroblasts to better understand the
mechanisms of transformation. In addition, recent studies have
shown that virus-associated oncogenes are activated in primary
hematopoietic tumors. In vitro studies to look at the effects of
these oncogenes on hematopoietic cells should provide basic
information to understand the significance of their activation in
vivo.

MATERIALS AND METHODS, RESULTS AND DISCUSSION

In in vitro cultures of murine fetal liver cells, IL-3 induces
the differentiation of a variety of hematopoietic cell types.
Initially, the cultures contain large numbers of cells differen-
tiating to mature granulocytes and macrophages. With differentiation these cells lose the ability to proliferate in vitro and

after 2 weeks in culture most are lost. In addition, IL-3 induces the differentiation of mature mast cells. Under normal conditions the cultures become progressively homogenous populations of mast cells due to the unique ability of differentiated mast cells to continue to proliferate in vitro. After approximately 4-6 weeks homogenous populations are obtained which can persist for periods of up to 4-5 months, although the ability to proliferate decreases with time in culture.

TABLE I. Establishment[a] of long-term cell lines by murine transforming viruses

Virus	Oncogene	Frequency of Establishing Lines	IL-3
−	−	0/5	−
Moloney Leukemia Virus	−	0/5	−
Harvey Sarcoma Virus	v-ras^h	5/5	Dependent
Moloney Sarcoma Virus	v-mos	2/2	Dependent
3611 Transforming Virus	v-raf	2/2	Dependent
Abelson Leukemia Virus	v-abl	5/5	Independent

a. Fetal liver cells were obtained from 18 day old BALB/c embryos and were cultured in RPMI 1640 containing 10% FCS and 20 units/ml IL-3. For infection, approximately 10^4 - 10^5 PFU were used in 1 ml containing 2 x 10^6 cells. To optimize the infection, polybrene (25 μg/ml) was added to the cultures. The cultures were either split or refed with fresh media at approximately 4 day intervals. Continuous cell lines were characterized by having cell doubling times of 14-20 h and in all cases were maintained in culture for at least 1 year. The requirement of IL-3 was assessed by the ability to grow in the absence of IL-3 and by ^{3}H-thymidine incorporation assays.

To initially examine the effects of transforming retroviruses, cultures of fetal liver cells were infected with various viruses and the cultures were watched for gross alterations in growth properties. As shown in Table I, with any of the transforming viruses examined there was a consistent and striking effect that was evident after 8–12 weeks in culture. In particular, unlike uninfected cultures or cultures infected with non–transforming helper viruses in which growth began to slow and the cultures died out, the transforming virus infected cultures gave rise to continuously proliferating cell lines. Once established, the lines were examined for their dependence on IL–3 for growth. As also shown in Table I, the lines obtained with Harvey sarcoma virus (HaSV), Moloney sarcoma virus (MSV) or the 3611 murine sarcoma virus (3611–MSV) continued to require IL–3 for growth in vitro. In contrast, the lines obtained with Abelson MuLV (Ab–MuLV) were factor–independent for growth. These results demonstrate that all transforming viruses could "immortalize" hematopoietic cells for growth in vitro but that only Ab–MuLV could also abrogate the requirement for IL–3 for growth.

The properties of a number of the cell lines were examined to determine the possible lineages of cells that were transformed. A summary of the characteristics is given in Table II. All the cell lines had properties expected of mature mast cells and by morphology of stained cytocentrifuge preparations, lines were composed of homogenous populations of mast cells. All the cell lines examined expressed 20 α–SDH at levels which are comparable to those seen with cultures of normal mast cells. Similarly, the levels of histamine and IgE receptors were comparable to normal mast cells. All of the cell lines examined expressed receptors for IL–3 irrespective of whether or not they continued to require IL–3 for growth. Cells transformed by HaSV expressed high levels of the v–ras p21 transforming gene product, whereas the lines obtained with other viruses did not, demonstrating that the cellular homologue of this transforming gene is not normally expressed in mast cells. Similarly, the v–abl p120 transforming gene product was only detected in Ab–MuLV transformed cells.

The observation that all the transforming viruses gave immortalized mast cell lines is hypothesized to be due to the absence of an effect on hematopoietic stem cell differentiation. In particular, in these cultures the differentiation of granulo–cytes and macrophages was not visibly affected by the presence of the transforming viruses. In these cases differentiation is normally associated with the loss of the ability to proliferate. Among the cells induced to differentiate in vitro by IL–3, only mast cells retain the ability to proliferate as fully differenti–ated cells. Thus, we propose that this is the basis for the apparent specificity of these transforming viruses for mast cells. Whether there exist murine transforming viruses that could block differentiation in vitro and transform other cell types is not currently known. In our experiments, we have not observed B cell transformation, which might have been anticipated based on previous results with HaSV (16) or Ab–MuLV (19). This, in part, is due to the absence of mercaptoethanol in the cultures which, for reasons that are not known, appears essential for the growth of transformed B cells.

TABLE II. Phenotypic characteristics of retrovirus transformed hematopoietic cell lines[a]

Property	Transforming Virus			
	Ab–MuLV	HaSV	MSV	3611–MSV
Morphology	Mast	Mast	Mast	Mast
Toluidine blue staining	+	+	+	+
20 α–SDH (pmoles/hr/10^8)	3000	200	400	700
IgE receptors (no/cell)	4.6×10^4	1.5×10^5	3.3×10^5	2.4×10^4
Histamine (ng/10^6)	2000	30	800	400
v-$\underline{ras}^h$ p21 expression	–	+	–	–
v-$\underline{abl}$ p120 expression	+	–	–	–

a. A number of the cell lines derived from fetal liver cells infected with the indicated viruses were examined for the indicated properties. Average values are presented.

The ability of viruses such as HaSV, MSV, and the 3611–MSV to immortalize mast cells without abrogating the requirement for IL–3 for growth was unexpected. The results suggest that the cellular functions that control the ability to progress through the cell cycle and to maintain a proliferative state are independent of those events which are regulated by growth factors such as IL–3. The concept that transformation and immortalization may be separate events in fibroblasts has been suggested by the effects of different oncogenes on primary fibroblasts (20). Whether the effects seen on mast cells represent comparable phenomena and are possibly due to similar effects on the regulation of cell proliferation is not known.

The unique ability of Ab–MuLV to immortalize mast cells and to abrogate their requirement for IL–3 for growth was also unexpected. One possible mechanism for the elimination of a requirement for IL–3 for growth is the production of a growth factor by the cells themselves. This type of autocrine mechanism has been previously suggested to be important in the spontaneous emergence of factor–independent murine mast cell lines (21). To explore this possibility we examined representative cell lines for various properties consistent with such a model. As shown in Table III, none of the cell lines produced mitogenic factors that were capable of inducing proliferation of various IL–3–dependent cell lines. To determine whether low levels of IL–3 might be produced, we examined the effects of an antiserum against IL–3 on proliferation. As shown, purified IgG at concentrations capable of completely inhibiting IL–3–dependent cell lines had no effect on the proliferation of any of the cell lines. The ability of the WEHI–3 cell line to constitutively produce IL–3 (22) has been shown to be associated with a rearrangement of one of the cellular alleles for IL–3 (D. Gilbert and J.N. Ihle, in preparation). We therefore examined the lines for comparable rearrangements, as indicated in Table III. None were evident by the Southern blot analysis of EcoR1 restricted DNA. Taken together, the results suggest that the abrogation of IL–3–dependence by Ab–MuLV is not due to an autocrine type mechanism and suggest that v–abl may directly affect the pathway by which IL–3 induces proliferation.

TABLE III. Lack of evidence of an autocrine mechanism in Ab–MuLV
abrogation of IL–3 dependence

Assay[a]	ABFTL–1	ABFTL–2	ABFTL–3
Production of mitogenic activity	–	–	–
Inhibition by anti–IL–3 IgG	–	–	–
Rearrangement of IL–3 gene	–	–	–

a. The indicated cell lines were examined for the production of mitogenic activity detectable on a series of IL–3–dependent cells as previously described (15). The ability of an anti–IL–3 IgG to inhibit proliferation was done by published procedures (8). Rearrangement of the IL–3 clone was assessed by Southern blot analysis of EcoR1 restricted DNAs using a cDNA clone obtained from the WEHI–3 cell line (D. Gilbert and J.N. Ihle, unpublished data).

HaSV has been shown to induce hematopoietic tumors in vivo which have been classified as erythroleukemias (23). We were therefore interested in whether the in vitro results were related to the effects seen in vivo. To explore this aspect, Balb/c mice were inoculated with HaSV as newborns and at 2–3 weeks when the animals were morbid, the enlarged spleens were taken and the cells cultured. In the absence of any growth factors there was no growth, whereas in IL–3 there was considerable growth.

Examination of the cultures indicated the presence of a variety of lineages of cells comparable to that seen with fetal liver cultures. As with cultures from normal animals, the majority of the cells differentiated to mature granulocytes and macrophages and died out. In contrast to control cultures, however, cultures of cells from HaSV infected mice gave continuously growing lines of mast cells which were phenotypically identical to those obtained by in vitro transformation. These results are most consistent with the hypothesis that the splenomegaly is due to the ability of HaSV to increase the proliferation rate of hematopoietic cells without affecting their factor dependence for growth or their ability to differentiate, consistent with previous studies (24). In vivo, the absence of high levels of IL-3 does not allow the accumulation of potentially immortalized mast cells.

The predominant tumors induced by Ab-MuLV are pre-B cell tumors. However, in the initial description (25) of the pathology associated with Ab-MuLV it was noted that under conditions in which the incidence of B lineage tumors was reduced, mastocytomas were induced. These tumors had the properties of mature mast cells and could be passaged without the requirement for factors. We therefore examined these tumor cells to determine whether they were comparable to the in vitro transformed cell lines. In all aspects examined we could detect no differences. In particular, morphologically and phenotypically, the lines were identical and in particular the in vivo-derived lines had IL-3 receptors but did not require IL-3 for growth. Moreover, there was no evidence that the cells produced their own growth factor. Therefore, in vivo as well as in vitro Ab-MuLV can immortalize mast cells and abrogate their requirement for IL-3 for growth.

In summary, the availability of IL-3 has allowed culture conditions which permit studies of the effects of transforming viruses on hematopoietic cells. Among the various lineages that are potentially at risk for transformation, the major effects seen are on the ability of fully differentiated mast cells to proliferate and to give rise to continuous long-term lines. In most cases this does not concomitantly involve the loss of the requirement for normal growth factors such as IL-3. The ability to abrogate factor dependence appears to be a unique property of Ab-MuLV. These observations will be of importance in understanding the general effects of oncogenes on the regulation of cell growth and should provide a basis to begin to specifically evaluate transformation within hematopoietic lineages.

REFERENCES

1. **Tertian, G., Y.-P. Yung, D. Guy-Grand, and M.A.S. Moore.** 1981. Long-term in vitro culture of murine mast cells. I. Description of a growth factor-dependent culture technique. J. Immunol. 127:788.

2. **Nagao, K., K. Yokoro,** and **S.A. Aaronson.** 1981. Continuous lines of basophil/mast cells derived from normal bone marrow. Science 212:333.
3. **Nabel, G., S.J. Galli, A.M. Dvorak, H.F. Dvorak,** and **H. Cantor.** 1981. Inducer T lymphocytes synthesize a factor that stimulates proliferation of cloned mast cells. Nature 291:332.
4. **Schrader, J.W., S.J. Lewis, I. Clark–Lewis,** and **J.G. Culvenor.** 1981. The persisting (P) cell: Histamine content, regulation by a T cell–derived factor, origin from a bone marrow precursor, and relationship to mast cells. Proc. Natl. Acad. Sci. USA 78:323.
5. **Dy, M., B. Lebel, P. Kamoun,** and **J. Hamburger.** 1981. Histamine production during the anti–allograft response. J. Exp. Med. 153:293.
6. **Razin, E., J.N. Ihle, D. Seldin, J.-M. Mencia–Huerta, H.R. Katz, P.A. Leblanc, A. Hein, J.P. Caulfield, K.F., Austen,** and **R.L. Stevens.** 1984. Interleukin 3: A differentiation and growth factor for the mouse mast cell that contains chondroitin sulfate E proteoglycan. J. Immunol. 132:1479.
7. **Ihle, J.N., J. Keller, S. Oroszlan, L. Henderson, T. Copeland, F. Fitch, M.B. Prystowsky, E. Goldwasser, J.W. Schrader, E. Palaszynski, M. Dy,** and **B. Lebel.** 1983. Biological properties of homogenous interleukin 3: I. Demonstration of WEHI-3 growth factor activity, mast cell growth factor activity, P–cell stimulating factor activity, colony stimulating factor activity and histamine producing cell stimulating factor activity. J. Immunol. 131:282.
8. **Bowlin, T.A., A. Scott,** and **J.N. Ihle.** 1984. Biologic properties of homogenous interleukin 3. II. Comparison of 20 α–SDH inducing activity, colony–stimulating activity and WEHI-3 growth factor activity using an antiserum against IL-3. J. Immunol. 133:2001.
9. **Ihle, J.N., L. Pepersack,** and **L. Rebar.** 1981. Regulation of T cell differentiation: In vitro induction of 20 α–hydroxy-steroid dehydrogenase in splenic lymphocytes is mediated by a unique lymphokine. J. Immunol. 126:2184.
10. **Ihle, J.N., J. Keller, L. Henderson, F. Klein,** and **E.W. Palaszynski.** 1982. Procedures for the purification of interleukin 3 to homogeneity. J. Immunol. 129:2431.
11. **Weinstein, Y.** 1977. 20 α–hydroxysteroid dehydrogenase: A T lymphocyte associated enzyme. J. Immunol. 119:1223.
12. **Fung, M.C., A.J. Hapel, S. Ymer, D.R. Cohen, R.M. Johnson, H.D. Campbell,** and **I.G. Young.** 1984. Molecular cloning of cDNA for murine interleukin-3. Nature 307:233.
13. **Yokota, T., F. Lee, D. Rennick, C. Hall, N. Arai, T. Mosmann, G. Nabel, H. Cantor,** and **K. Arai.** 1984. Isolation and characterization of a mouse cDNA clone that expresses mast-cell growth factor activity in monkey cells. Proc. Natl. Acad. Sci. 81:1070.
14. **Rennick, D.M., F.D. Lee, T. Yokoto, K.-I. Arai, H. Cantor,**

and G.J. Nabel. 1985. A cloned MCGF cDNA encodes a multilineage hematopoietic growth factor: Multiple activities of interleukin 3. J. Immunol. 134:910.

15. **Ihle, J.N., J. Keller, J.S. Greenberger, L. Henderson, R.A. Yetter, and H.C. Morse, III.** 1982. Phenotype characteristics of cell lines requiring interleukin-3 for growth. J. Immunol. 129:1377.

16. **Ihle, J.N., J. Keller, and E.W. Palaszynski.** 1983. Interleukin 3 regulation of a lineage of lymphocytes expressing 20 α-SDH. In: Interleukins, Lymphokines, and Cytokines. Edited by J.J. Oppenheim and S. Cohen. New York, Academic Press, p.113.

17. **Goldwasser, E., J.N. Ihle, M.B. Prystowsky, I. Rich, and G. Van Zant.** 1983. The effect of the interleukin-3 on hemopoietic precursor cells. In: Normal and Neoplastic Hematopoiesis, UCLA Symposia on Molecular and Cellular Biology, New Series, Vol. 9. Edited by D.W. Gold and P.A. Marks. New York, N.Y., Alan R. Liss, Inc., p.301.

18. **Pierce, J.H., and S.A. Aaronson.** 1982. BALB- and Harvey-murine sarcoma virus transformation of a novel lymphoid progenitor cell. J. Exp. Med. 156:873.

19. **Rosenberg, N., D. Baltimore, and C.D. Scher.** 1975. In vitro transformation of lymphoid cells by Abelson murine leukemia virus. Proc. Natl. Acad. Sci. USA 72:1932.

20. **Land, H., L.F. Parada, and R.A. Weinberg.** 1983. Tumorigenic conversion of primary embryo fibroblasts requires at least two cooperating oncogenes. Nature 304:596.

21. **Schrader, J.W., and R.M. Crapper.** 1983. Autogenous production of a hemopoietic growth factor, persisting cell-stimulating factor, as a mechanism for transformation of bone marrow-derived cells. Proc. Natl. Acad. Sci. USA 80:6892.

22. **Lee, J.C., A.J. Hapel, and J.N. Ihle.** 1982. Constitutive production of a unique lymphokine (IL-3) by the WEHI-3 cell line. J. Immunol. 128:2393.

23. **Ellis, R.W., D.R. Lowy, and E.M. Scolnick.** 1982. The viral and cellular p21 (ras) gene family. Adv. Viral Oncol. 1:107.

24. **Hankins, W.D., and E.M. Scolnick.** 1981. Harvey and Kirsten sarcoma viruses promote the growth and differentiation of erythroid precursor cells in vitro. Cell 26:91.

25. **Risser, R., M. Potter, and W.P. Rowe.** 1978. Abelson virus-induced lymphomagenesis in mice. J. Exp. Med. 148:714.

Mast Cell Differentiation and Heterogeneity,
edited by A. D. Befus et al.
Raven Press, New York © 1986.

Mast Cell Development in the Rat

David M. Haig, Christine McMenamin, and Ellen E. E. Jarrett

*Department of Veterinary Medicine, University of Glasgow Veterinary School,
Bearsden, Glasgow G61 1QH, Scotland*

The experimental system used in the following studies is infection of the rat with the nematode, Nippostrongylus brasiliensis, whose life cycle is rather like that of a hookworm and has been delineated in a previous review (1). This infection is associated with immunological events which reveal influences that helminth parasites may have on the host's immune system and provides a prototype for study.

There are three features of nematode infection in rats which we would like to stress. First, IgE responses were originally shown by Orr and Blair to be potentiated during N. brasiliensis infection, leading to the finding that the increases in IgE production consist of polyclonal mixture of IgE, mostly to non-parasite antigens (1–3). This IgE response was demonstrated to be a T cell–dependent phenomenon (4); subsequently, various T and B cell–derived factors involved in the control of IgE synthesis were discovered to be elaborated by cultured N. brasiliensis-infected rat cells (5). The T cell–dependent peripheral blood and tissue eosinophilia of the helminth infected animal is another prominent event (6,7). Lymphocytes of Trichinella spiralis–infected mice are also known to produce factors which stimulate eosinophil growth in mouse bone marrow cultures (8,9).

Finally, during the T cell–dependent hyperplasia of mast cells in the mucous membranes of helminth infected animals, T-cell-derived lymphokines are produced that affect a variety of hemopoietic cell types (10–13,22).

Our work was supported by grants from the Medical Research Council, The Wellcome Trust and the Cancer Research Campaign. Grants towards this work were also kindly made by Fisons Pharmaceuticals Division and the Smith–Kline–French Foundation. We are grateful to Mrs. E. Gault for preparing the manuscript.

RESULTS AND DISCUSSION

Cultures of Rat Mesenteric Lymph Nodes and Bone Marrow. These culture systems have been previously described (14–16). Briefly, conditioned medium (CM) prepared from immune or normal rat mesenteric lymph node (MLN) cells stimulated with worm antigen or concanavalin A (Con A) contain soluble factors capable of stimulating mast cell growth (17). To assay for mast cell growth activity, CM is added to bone marrow cultures consisting of 2.5 or 5 x 10^5 cells/ml in Iscove's medium supplemented with 20% horse serum and cultured for up to 4 wk in a humidified atmosphere at 37°C flushed with 5% CO_2 in air. Typical responses of stimulated cultures are shown in Figure 1 and are representative of previous results (14,15). In stimulated cultures, only cells with sparse granulation can be seen by day 2, progressing to definable cells by 1 wk in vitro – criteria used include metachromatic granular dye-binding, ultrastructural appearance and uptake of IgE (14,18). Eosinophils, monocyte/macrophages and neutrophils (but not lymphoid or erythroid cells) are also stimulated by CM in these cultures, disappearing approximately by day 10 (19); at 2–3 weeks, cultures are often composed of relatively pure populations of granulated mast cells which can be maintained for at least 8 wk in the presence of CM.

Appearance and Properties of Cultured Rat Mast Cells. Properties of cultured rat mast cells are summarized in Table I. The cultured cells are small (6–10 μm) in comparison to rat peritoneal mast cells (15–19 μm) and are more sparsely granulated than the latter, with varying granule size. Ultrastructural features of these cells (15) appear to be those which help distinguish rat intestinal from rat peritoneal mast cells in vivo (20,21). The granules of both cultured in vivo intestinal mast cells stain strongly and exclusively blue with copper phthalocyanine dyes such as astra blue at pH 0.3 in the astra blue/safranin staining sequence (after fixation in acid/ethanol (22); rat peritoneal mast cells, on the other hand, which have the organelle structure of immature cells, may show a mixed blue/red granulation, reflecting mixed granule contents of weakly and strongly sulfated glycosaminoglyclans, respectively. Maturation of rat peritoneal MC involves an ultrastructural reorganization of the granule matrices, associated with a shift from astra (or alcian) blue to safranin staining, recognized as a pink reaction (23). No equivalent granule reorganization occurs during the progression of rat intestinal MC in vivo (21) or in vitro; these cells which have the ultrastructural appearance of mature cells continue to demonstrate astra (or alcian) blue-staining granules.

Rat cultured mast cells are well granulated by 2 weeks with apparent necessity for either stromal cells or sodium butyrate, which have been found necessary for granule maturation of cultured murine mast cells (24–27). Addition of sodium butyrate

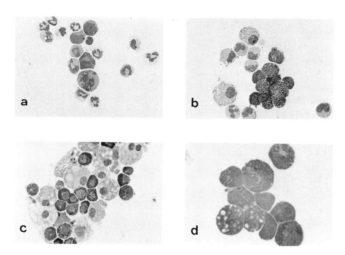

Figure 1.   Time course of appearance of MC in rat bone   marrow
cultures.   Cytocentrifuge preparations   of   bone marrow   culture
stimulated with conditioned medium.   Samples harvested after   (a)
2 days; (b) 4 days; (c) 7 days; (d) 4 weeks. Stain: Leishman.
Magnification X160 (a,b,c) and 320 (d).

(1–2 mM)   induced no   change   in the   storage of   histamine (1–2
pg/cell) or rat mast cell   protease II (RMCP II) (20–60   pg/cell)
in cultured rat   mast cells and   had no effect   on the nature   of
their proteoglycans   as assessed by staining   reaction.   Microspectrophotometric analysis   has shown that   rat intestinal   mast
cells in vivo (28)   and   cultured rat   mast cells   (16)   contain
insignificant amounts of heparin–proteoglycan, containing instead
a glycosaminoglycan with a lower   degree of sulfation.   This
feature again   distinguishes rat intestinal from peritoneal   MC,
since the latter produce   heparin (20,21,28).   Although   cultured
rat mast   cells   bear a   resemblance   to intestinal   rather   than
peritoneal MC, this issue is far from resolved, especially   since
rat   peritoneal   MC   have   been   shown   to   exhibit   a   range   of
phenotypes and functions as they mature (23,29,30).
    Further evidence of a difference   between two MC types in   the
rat has been provided   by Woodbury et al. who have   demonstrated
variant serine proteases in two types of rat mast cell (31):   the
first, mast cell   protease I (RMCP   I), originally isolated   from
skeletal muscle, was found   to be identical to that   subsequently
isolated from   peritoneal   mast   cells;   the   second,   mast   cell
protease II   (RMCP II),   differing in   solubility, structure   and
antigenicity from the first, was   localized to mast cells in   the
small intestine. Rat RMCP   II is contained in large amounts   and

secreted by rat cultured mast cells (14,15); secretion of RMCP II occurs without degranulation by day 2 in vitro.   This finding supports the proposal (32) that intestinal MC have secretory and degranulatory functions. The precise function of RMCP II is not yet known.

The solubility of RMCP II (in 0.15M KCl) has also formed the basis for its detection in the serum of N. brasiliensis- and Trichinella spiralis-infected rats, in which it is produced

TABLE I. Properties of Cultured Rat Mast Cells

PHYSICAL	Non-adherent, non-phagocytic mononuclear cells of 6-10 um diameter.
GRANULES	Variable size and density in cytoplasm; metachromatic staining with toluidine blue at pH 0.5. Astra blue positive at pH 0.3.
PROTEOGLYCAN	Oversulfated chondroitin sulfate (a percentage of the disaccharides are in the form of chondroitin sulfate di-B).
HISTAMINE	1-2 pg per cell; 30% can be released immunologically (Haig DM, Wells T et al., in preparation).
RMCP-II*	20-60 pg/cell. High rate of secretion.
SURFACE PHENOTYPE	Thy 1^+; $W3/13^-$; a small component is Ia^+ (McMenamin C et al., in preparation).
GROWTH DEPENDENCY	Absolutely dependent on factor(s) produced by activated T $OX19^+$ $W3/25^+$ $OX8^-$ cells. Probably IL-3.

* RCMP-II = rat mast cell protease II

during early infection (32).   RMCP II is also present in large amounts in serum and liver extracts of Fasciola hepatica-infected rats (33). At certain stages of infection these rats have large numbers of MC resembling those found in the intestine, in the liver parenchyma and lamina propria of bile ducts (33), together with a prolonged elevation of total serum IgE (34).

T Cell control of Intestinal and Cultured Mast Cell Growth. Low but near normal numbers of intestinal MC are present in uninfected athymic rodents (35); however, it is apparent from experiments in nude rats in vitro that intestinal MC precursors demonstrated T cell-dependence for their proliferation in immune reactions during helminth infection (15). For example, CM from nu/nu rat MLN does not contain growth-stimulating activity for MC growth from normal or athymic bone marrow. On the other hand, CM from nu/+ or control (LIS X BN) F_1 immune MLN stimulates mast cell growth from both nu/+ and nu/nu rat bone marrow. In conjunction with the information that the nu/nu genotype is associated with congenital aplasia of the thymus and a gross deficiency of mature T-cells (36,37), these experiments reveal that MLN T-cells are necessary for production of MC growth factor and that T-cell-depleted bone marrow is capable of responding to T-cell-dependent MLN-derived growth factor. Cultured rat mast cells from normal or athymic rat bone marrow are morphologically and histologically identical and produce comparable amounts of RMCP II resembling the in vivo intestinal rat MC (15). Neither IgE or IgE-binding factor produced by MLN of N. brasiliensis-infected rats and present in CM influences MC growth since either can be removed by immunoabsorption without loss of stimulating activity (15).

T-cells from MLN of infected rats produce a more potent CM after Con A activation than do T-cells from normal rats, suggesting that infection activates T-cells and enlarges the population of cells potentially able to produce MC growth factor. This may be because such cells express more receptors for the Con A induced T-cell growth factor (38) and hence undergo a more rapid in vitro expansion than the resting T-cells of normal animals.

T-cells producing rat MC growth factor have the phenotype $OX-19^+$, $W3/25^+$, $OX-8^-$ (39), which is equivalent to mouse $Ly1^+$ $Ly2^-$ or to human $T1^+$ $T4^+$ $T5^+$ cells (40). Such T-cells may be producing equivalents of the murine lymphokine interleukin 3 (IL-3) (41) which has numerous effects on hemopoietic cells. The murine IL-3 gene has recently been cloned (44,45) abrogating the necessity and difficulties in cloning IL-3-producing T-cells (42,43), and paving the way for the production of large quantities of pure murine and perhaps other species, IL-3 or equivalent.

REFERENCES

1. **Jarrett, E.E.E., and H.R.P. Miller.** 1982. Production and activities of IgE in helminth infection. Prog. Allergy 31:178.
2. **Orr, T.S.C., and A.M.J.N. Blair.** 1969. Potentiated reagin response to egg-albumin and conalbumin in N. brasiliensis infected rats. Life Sci. 8:1073.
3. **Jarrett, E.E.E., and H. Bazin.** 1974. Elevation of total serum IgE in rats following helminth parasite infection. Nature 251:613.

4. Jarrett, E.E.E. and A. Ferguson. 1974. Effect of T cell depletion on the potentiation–reagin response. Nature 250:420.

5. Ishizaka, K., J. Yodoi, M. Suemara, and M. Hirashima. 1983. Isotype–specific regulation of IgE response by IgE–binding factors. Immunol. Today 4:192.

6. Wilmott, S. 1980. The eosinophil in tropical disease. Trans. Roy. Soc. Trop. Med. Hyg. 74:1.

7. Basten, A., and P.B. Beeson. 1970. Mechanism of eosinophilia. II. Role of the lymphocyte. J. Exp. Med. 131:1288.

8. Ruscetti, F.W., R.H. Cypess, and P.A. Chervenick. 1976. Specific release of neutrophilic and eosinophilic stimulating factors from sensitized lymphocytes. Blood 47:757.

9. Bartlemetz, S.H., N.H. Dodge, and D.A. Bass. 1981. Antigen–mediated release of eosinophil growth stimulating factor from Trichinella spiralis sensitized spleen cells: a comparison of T. spiralis stage–specific antigen preparations. Immunology 45:605.

10. Murray, M. 1972. Immediate hypersensitivity effector mechanisms. II. In vivo reactions. In: Immunity to Animal Parasites. Edited by E. Soulsby. New York, Academic Press, p.155.

11. Ruitenberg, E.J., and A. Elgersma. 1976. Absence of intestinal mast cell responses in congenitally athymic mice. Nature 264:258.

12. Mayrhofer, G. 1979. The nature of the thymus dependency of mucosal mast cells. II. The effect of thymectomy and of depleting recirculating lymphocytes in the response to N. brasiliensis. Cell. Immunol. 47:312.

13. Nawa, Y., and H.R.P. Miller. 1979. Adoptive transfer of the intestinal mast cell response in rats infected with N. brasiliensis. Cell. Immunol. 42:225.

14. Haig, D.M., T.A. McKee, E.E.E. Jarrett, R.G. Woodbury, and H.R.P. Miller. 1982. Generation of mucosal mast cells is stimulated in vitro by factors derived from T cells of helminth–infected rats. Nature 300:188.

15. Haig, E.M., C. McMenamin, C. Gunneberg, R. Woodbury, and E.E.E. Jarrett. 1983. Stimulation of mucosal mast cell growth in normal and nude rat bone marrow cultures. Proc. Natl. Acad. Sci. USA 80:4499.

16. Haig, D.M., E.E.E. Jarrett, and J. Tas. 1984. In vitro studies on mast cell proliferation in N. brasiliensis infection. Immunology 51:643.

17. Iscove, N.N., and F. Melchers. 1978. Complete replacement of serum by albumin, transferrin and soybean lipid in culture. J. Exp. Med. 147:923.

18. Galli, S.J., and H.F. Dvorak. 1979. Basophils and mast cells: structure, function and role in hypersensitivity. In: Cellular, Molecular and Clinical Aspects of Allergic

Disorders. Edited by A. Gupta, and R.A. Good. New York and London, Plenum Press, p.1.

19. **Haig, D.M.** 1982. The generation of mucosal mast cells in rat bone marrow cultures. Ph.D. Thesis, University of Glasgow.
20. **Enerback, L.** 1981. The gut mucosal mast cell. Monogr. Allergy 17:222.
21. **Miller, H.R.P.** 1980. The structure, origin and function of mucosal mast cells. Biol. Cell. 39:229.
22. **Jarrett, E.E.E., and D.M. Haig.** 1984. Mucosal mast cells in vivo and in vitro. Immunol. Today 5:115.
23. **Combs, J.W., D. Lagunoff, and E.P. Benditt.** 1965. Differentiation and proliferation of embryonic mast cells in the rat. J. Cell. Biol. 25:577.
24. **Davidson, S., A. Mansour, R. Gallily, M. Smolarski, M. Rofolovitch, and H. Ginsburg.** 1983. Mast cell differentiation depends on T cells and granule synthesis on fibroblasts. Immunology 48:439.
25. **Courtoy, R., N. Schaaf-Lafontaine, G. Goffinet, and J. Bonvier.** 1984. In vitro generated mast cells: role of cell-to-cell contact with a Thy-1+ large granular adherent cell. Proc. Soc. 16th Int. Leuk. Cult. Conf. Abstracts. Immunobiology 167:28.
26. **Galli, S.J., A.M. Dvorak, J.A. Marcum, T. Ishizaka, G. Nabel, H. Der Simonian, K. Pyne, J.M. Goldin, R.D. Rosenberg, H. Cantor, and H.F. Dvorak.** 1982. Mast cell clones: a model for the analysis of cellular maturation. J. Cell. Biol. 95:435.
27. **DuBuske, L., K.F. Austen, J. Czop, and R.L. Stevens.** 1984. Granule-associated serum neutral proteases of the mouse bone marrow-derived mast cell that discharges fibronectin: their increase after sodium butyrate treatment of the cells. J. Immunol. 133:1535.
28. **Tas, J., and R.G. Berndsen.** 1977. Does heparin occur in mucosal mast cells of the rat small intestine? J. Histochem. Cytochem. 25:1058.
29. **Pretlow, T.G., and I.M. Cassidy.** 1970. Separation of mast cells in successive stages of differentiation using programmed gradient sedimentation. Am. J. Path. 61:323.
30. **Beaven, M.A., D.L. Aiken, E. WoldeMussie, and A.H. Soll.** 1983. Changes in histamine synthetic activity, histamine content and responsiveness to compound 48/80 with maturation of rat peritoneal mast cells. J. Pharmacol. Exp. Therap., 224:620.
31. **Woodbury, R.G., and H. Neurath.** 1981. Mast cell proteases. In: Metabolic Interconversion of Enzymes 1980. Edited by H. Holzer. Berlin, Springer Verlag, p. 145.
32. **Woodbury, R.G., H.R.P. Miller, J.F. Huntley, G.F.J. Newlands, A.C. Palliser, and D. Wakelin.** 1984. Mucosal mast cells are functionally active during spontaneous expulsion of intestinal nematode infections in rats. Nature 313:450.
33. **Pfister, K., C. McMenamin, E. Hall, and E.E.E. Jarrett.**

1985. Mast cells and RMCP II in the blood, liver and small intestines of <u>Fasciola</u> <u>hepatica</u> infected rats. (In preparation).

34. **Pfister, K., K. Turner, A. Currie, E. Hall, and E.E.E. Jarrett.** 1983. IgE production in rat fascioliasis. <u>Parasite Immunol</u>. 5:587.

35. **Mayrhofer, G., and H. Bazin.** 1981. Nature of thymus dependency of mucosal mast cells. III. Mucosal mast cells in nude mice and nude rats, in B rats and in a child with the Di George Syndrome. <u>Int. Arch. Allergy Appl. Immunol</u>. 64:320.

36. **Festing, M.F.W., D. May, T.A. Connors, D. Lovell, and S. Sparrow.** 1978. An athymic nude mutation in the rat. <u>Nature</u> 274:365.

37. **Brooks, C.G., P.J. Webb, R.A. Robins, G. Robinson, R.W. Baldwin, and M.F.W. Festing.** 1980. Studies on the immunobiology of rnu/rnu "nude" rats with congenital aplasia of the thymus. <u>Eur. J. Immunol</u>. 10:58.

38. **Larson, E.L., and A. Coutinho.** 1979. The role of mitogenic lectins in T-cell triggering. <u>Nature</u> 280:239.

39. **McMenamin, C., E.E.E. Jarrett, and A. Sanderson.** 1985. Surface phenotype of T cells producing the growth of mucosal mast cells in normal rat bone-marrow culture. (In preparation).

40. **Mason, D.W., R.P. Arthur, M.J. Dallman, J.R. Green, G.P. Spickett, and M.L. Thomas.** 1983. Functions of rat T lymphocyte subsets isolated by means of monoclonal antibodies. <u>Immunol. Rev</u>. 73:57.

41. **Ihle, J.N.** 1984. Biochemical and biological properties of interleukin 3: a lymphokine mediating the differentiation of a lineage of cells which includes prothymocytes and mast cell-like cells. In: <u>Contemporary Topics in Molecular Immunology</u>. Edited by F.P. Inman. New York, Plenum Press, in press.

42. **Pawelec, G., E.M. Schneider, and P. Wernet.** 1983. Human T cell clones with multiple and changing functions: indications of unexpected flexibility in immune response networks? <u>Immunol. Today</u> 4:295.

43. **Freeman, G.J., C. Clayberger, R. Dekruyff, D.S. Rosenblum, and H. Connors.** 1983. Sequential expression of new gene programs in inducer T cell clones. <u>Proc. Natl. Acad. Sci. USA</u> 80:4094.

44. **Fung, M.C., A.J. Halpel, S. Ymer, D.R. Cohen, R.M. Johnston, H.D. Campbell, and I.G. Young.** 1984. Molecular cloning of cDNA for murine interleukin 3. <u>Nature</u> 307:233.

45. **Yokota, T.F., F. Lee, D. Rennick, C. Hall, H. Arai, G. Masmann, G. Nabel, H. Cantor, and K. Arai.** 1984. Isolation and characterisation of a mouse cDNA that expresses mast-cell growth-factor activity in monkey cells. <u>Proc. Natl. Acad. Sci. USA</u> 81:1070.

Mast Cell Differentiation and Heterogeneity,
edited by A. D. Befus et al.
Raven Press, New York © 1986.

A Stochastic Model for Mast Cell/Basophil Differentiation from Pluripotent Hemopoietic Stem Cells

Pamela N. Pharr and Makio Ogawa

Department of Medicine, Medical University of South Carolina and Veterans Administration Medical Center, Charleston, South Carolina 29403

Both basophils and mast cells have IgE receptors and contain histamine, and therefore, share functional characteristics. Basophils and mast cells can be distinguished by ultrastructural differences (1). The question whether mast cells and basophils come from the same or separate lineages remains controversial and will be amply dealt with by other contributors to this monograph. The aim of our review is to summarize the studies in culture of the proliferation and differentiation of these cells from hemopoietic stem cells.

Basophils are known to be derived from hemopoietic stem cells. For example, the Philadelphia (Ph^1) chromosome has been identified in the basophils of a patient with chronic myelogenous leukemia (2). Recently, we have shown that basophils are present in human mixed hemopoietic colonies derived from single cells (3). Convincing evidence that mast cells are of hemopoietic origin came from a series of studies by Kitamura and colleagues. For example, they were able to transplant mast cells into mice of genotype W/W^v which are deficient in mast cells, using cells from individual spleen colonies (4). In our laboratory, Nakahata et al. (5) developed a methylcellulose culture assay for mouse mast cell progenitors and found that the replating of mast cell colonies containing blast cells gave a variety of types of hemopoietic colonies. Further evidence for a hemopoietic origin came from our observation that mixed hemopoietic colonies containing mast cells can be derived from single blast cells obtained by micromanipulation techniques (6).

Several culture methods have been developed for mast cells and basophils. Ginsburg and Sachs (7) first cultured mast cells by seeding mouse thymus cells on mouse embryonic fibroblast monolayers. Subsequently, mouse mast cells and human basophils have been grown in liquid cultures in several laboratories. Semi-solid culture methods have been developed for rat mast cells (8), mouse mast cells (5) and human basophils (9).

The development of culture methods for murine mast cells has made it possible to identify murine mast cell growth factor

(MCGF). MCGF is routinely assayed by its ability to support the continued growth of murine mast cells in suspension culture, and purified murine   interleukin-3 (IL-3) has this activity (10). Further evidence that IL-3 is very similar, if not identical, to MCGF comes from molecular biological studies.   The amino acid sequence suggested by MCGF cDNA is identical to the NH_2 terminal sequence of IL-3 (11). However, Clark-Lewis et al. (12) reported a different NH_2-terminal sequence for P cell-stimulating activity (P cells are thought to be mast cells) than reported from IL-3 by Ihle (10). Whether this reflects differences in processing or the products of different genes remains to be determined. However, murine IL-3 does not appear to be specific for the mast cell lineage (13-15). Support for the concept that IL-3 mediates diverse biological activities comes from experiments using antiserum prepared against pure IL-3 (16).

Suda et al. in our laboratory determined that IL-3 is the component of pokeweed mitogen spleen conditioned medium (PWM-SCM) that supports the growth and differentiation of pluripotent hemopoietic precursors obtained from murine blast cell colonies (17).   IL-3 supported the development of colonies containing neutrophils, macrophages, eosinophils, mast cells, erythroid cells,   megakaryocytes and   colonies containing   various combinations of these lineages.   Furthermore, it supported the growth of multilineage colonies from single cells isolated by micromanipulation, indicating that IL-3 acts directly on multi-lineage progenitors. A characteristic of hemopoietic stem cells is that they are in a resting state referred to as G_0. Following entry into the cell cycle, progenitors of multilineage colonies proliferate at relatively constant doubling rates during the early stages of differentiation (18).   We posed the question as to whether IL-3 affects the cycling state of such hemopoietic progenitors (17). Delayed addition of IL-3 to cultures 7 days after cell plating decreased the number of multilineage colonies to half the number of colonies observed when IL-3 was added on day 0. It did not alter the proliferative and differentiation characteristics of late-emerging multilineage colonies.   New colonies continued to emerge throughout the 16-day culture period. If IL-3 had acted to trigger cells into the cell cycle, a burst of colony development would have occurred following the delayed addition.   These observations were interpreted to mean that IL-3 is necessary for the proliferation of hemopoietic cells that have entered the cell cycle.

A prevailing controversy in experimental hematology has been whether stem cell renewal and commitment to differentiation occur in a deterministic or stochastic fashion. Two decades ago, Till et al. (19) observed that the number of colony-forming units in spleen (CFU-S) varies markedly among spleen colonies, and developed a stochastic model to explain the very heterogeneous distribution of CFU-S per spleen colony.   In their birth-death model, a CFU-S randomly chooses between two fates when it divides. Each CFU-S can either form two new CFU-S or differentiate into a cell that has no colony-forming ability. They called the former event a "birth" and the latter event a "death". The distribution of the number of CFU-S per colony could be approximated by a gamma distribution as could the numbers obtained by a computer simulation of their birth-death model. Later, we (20) developed a murine culture assay which in many ways appears to be equivalent to the spleen colony assay of Till and McCulloch. In culture, hemopoietic progenitors called S cells produce blast cell colonies referred to as stem cell colonies which have exceptionally high replating capabilities. We fitted a gamma distribution to our data on the distribution of S cells per stem cell colony (21). The results of these studies supported the hypothesis that the self-renewal of murine pluripotent progenitors occurs in a stochastic fashion.

We then posed the question as to whether commitment to particular lineages occurs in a deterministic or stochastic fashion. We studied commitment through the analysis of the colony types produced by the daughter cells of individually isolated hemopoietic progenitors (22). Single progenitors were isolated by the use of a micromanipulation technique from blast cell colonies cultured from the spleens of 5-fluorouracil (5-FU)-treated mice. Eighteen to 24 hours later, the paired progenitors were separated with a micromanipulator. Six to 9 days later, the two colonies derived from the paired progenitors were individually picked and differential counts performed by using May-Grunwald-Giemsa staining. Sixty-eight pairs of non-homologous colonies and 39 pairs of homologous colonies revealing identical lineage combinations were observed. In a later study (23), we carried out sequential micromanipulation of paired progenitors. Of the total of 94 evaluable cultures, 41 cultures showed colonies revealing diverse combinations of cell lineages. Presumptive genealogic trees of the differentiation of hemopoietic progenitors constructed for the latter group of cultures suggested that monopotent progenitors may be derived from pluripotent progenitors in two ways: (1) directly during one cell division of pluripotent cells or (2) as a result of progressive lineage restriction during successive divisions of the pluripotent progenitors. Similar results have now been obtained from human paired progenitors isolated by micromanipulation (24). These observations provided experimental data in support of stochastic mechanisms of stem cell differentiation.

We observed extreme heterogeneity in the numbers of cells representing different lineages in human mixed colonies derived from single cells (3). These data suggested that the proliferative capacity of committed monopotent progenitors is randomly restricted. Most mature hemopoietic cells have lost the capacity for cell division, but murine mast cells are an exception. We recently reported that pure murine mast cell colonies have a high capacity for replating (6). The number of secondary colonies per primary colony varied widely, as shown in Fig. 1. Furthermore, the size of secondary colonies varied from a few cells to a few hundred cells. In order to test whether or not a birth–death model could be used to describe mast cell proliferation, portions of different primary colonies were replated and the number and sizes of the secondary colonies recorded after different culture periods. The size distributions for secondary colonies showed a large number of small colonies and a long spread in the distribution encompassing large

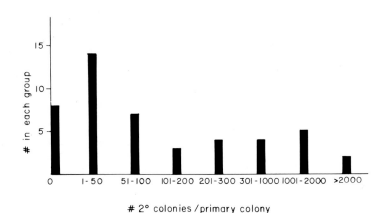

Figure 1. Distribution of the number of secondary colony-forming cells in primary mast cell colonies.

colonies. We tried fitting a simple stochastic model to these data. According to this model, at each generation a cell either disappears or divides into two proliferative cells with a probability of P_2. P_2 was assumed to be the same from generation to generation. The size distributions predicted by this model were not sufficiently heterogeneous to match our highly–skewed data. We then tested a modification of the birth–death model of Till et al. (19). This model poses the existence of two types of cells – proliferative and non-proliferative (25). Time proceeds in discrete steps corresponding to the average generation time of a proliferative cell. At each generation, all proliferative cells choose among three possibilities: (1) division into two proliferative cells; (2) division into two non-proliferative cells, and (3) disappearance. At each step, a non-proliferative cell can either do nothing or disappear. These choices are assumed to be random. The observed colony size distributions could be fitted by the distribution predicted by this stochastic model.

It is useful to compare cell culture studies to in vivo studies when that is possible. Beginning in 1978, Kitamura and colleagues have used the anemic mouse strain W/W^V which is deficient in mast cells to study mast cell development. They found that the number of mast cells in the skin, stomach, caecum and mesentery was increased to normal levels following transplantation of bone marrow cells from normal mice (26). Parabiosis of beige mice which have mast cells containing giant granules with normal mice demonstrated the presence of mast cell precursors in peripheral blood (27). These results agree with subsequent cell culture studies demonstrating circulating precursors for human mast cells and basophils (9,28,29). A more rapid in vivo assay for mast cell precursors was developed by Sonoda et al. (30). Injection of hemopoietic cells from normal mice into the skin of W/W^V mice resulted in the appearance of mast cells at the injection site. Sonoda et al. (31) compared the mast cell precursor content of in vivo and in vitro hemopoietic colonies. More than 40% of the day-12 spleen colonies and the day-14 in vitro mixed colonies contained mast cell precursors, but only a small proportion of day-7 spleen colonies or single lineage colonies obtained in culture contained mast cell precursors. Recently, the presence of mast cell precursors in the peritoneal cavity has been demonstrated (32). These results parallel our results obtained by replating pure mast cell colonies (6). Thus, both in vivo and in vitro observations suggest that some morphologically identifiable mast cells possess a relatively high proliferative capacity.

One possible interpretation of the studies of Kitamura and colleagues is that W/W^V mice are deficient in mast cell progenitors. This hypothesis was tested by Suda et al. (33) in our laboratory using cell culture assays for mast cell progenitors. The relative concentrations of mast cell progenitors in the bone marrow, spleen and peripheral blood of W/W^V mice was similar to that of normal mice. Conditioned media prepared from spleen cells of W/W^V mice were as effective as those from +/+ mice in support of mast cell colony formation. Our observation of normal numbers of circulating colony-forming cells demonstrated that mast cell progenitors can egress from the marrow to the circulation. The apparent total absence of macroscopic spleen colony formation by W/W^V marrow cells and the severe reduction in mast cells in the tissues of W/W^V mice suggests that the expression of the hemopoietic defect in W/W^V mice may require interaction of progenitors with the tissue environment.

In summary, cell culture studies as well as the transplantation studies of Kitamura have shown that mast cells belong to the hemopoietic family. The growth of mucosal–type mast cells is dependent on IL–3. According to our stochastic model of stem cell proliferation, stem cells randomly enter the cell cycle and then become dependent on IL–3 for continued proliferation. Commitment to mast cell differentiation, like that of other hemopoietic lineages, appears to be a stochastic process. A stochastic model for mast cell proliferation which allows cells to randomly lose or retain their proliferative capacity was developed. Thus, the differentiation of mast cells appears to occur through processes similar to those of other types of hemopoietic cells, except that mature mast cells retain the capacity for cell division.

REFERENCES

1. **Dvorak, A.M., H.F. Dvorak, and S.J. Galli.** 1983. Ultra-structural criteria for identification of mast cells and basophils in humans, guinea pigs and mice. Am. Rev. Respir. Dis. 128:549.
2. **Denegri, J.F., S.C. Maiman, J. Gillen, and J.W. Thomas.** 1978. In vitro growth of basophils containing the Philadelphia chromosome in the acute phase of chronic myelogenous leukemia. Br. J. Haemat. 40:351.
3. **Leary, A.G., M. Ogawa, L.C. Strauss, and C.I. Civin.** 1984. Single cell origin of multilineage colonies in culture: Evidence that differentiation of multipotent progenitors and restriction of proliferative potential of monopotent progenitors are stochastic processes. J. Clin. Invest. 74:2193.
4. **Kitamura, Y., M. Yokoyama, H. Matsuda, and T. Ohno.** 1981. Spleen colony-forming cell as common precursor for tissue mast cells and granulocytes. Nature 291:159.
5. **Nakahata, T., S.S. Spicer, J.R. Cantey, and M. Ogawa.** 1982. Clonal origin of murine mast cell colonies in methylcellu-lose culture. Blood 60:352.
6. **Pharr, P.N., T. Suda, K.L. Bergmann, L.A. Avila, and M. Ogawa.** 1984. Analysis of pure and mixed mast cell colonies. J. Cell. Physiol. 120:1.
7. **Ginsburg, H., and L. Sachs.** 1962. Formation of pure sus-pensions of mast cells in tissue culture by differentiation of lymphoid cells from the mouse thymus. J. Natl. Canc. Inst. 31:1.
8. **Zucker-Franklin, D., G. Grusky, N. Hirayama, and E. Schnipper.** 1981. The presence of mast cell precursors in rat peripheral blood. Blood 58:544.
9. **Leary, A.G., and M. Ogawa.** 1984. Identification of pure and mixed basophil colonies in culture of human peripheral blood and marrow cells. Blood 64:78.
10. **Ihle, J.M., J. Keller, S. Oroszlan, L.E. Henderson, T.D. Copeland, F. Fitch, M.B. Prystowsky, E. Goldwasser, J.W.**

Schrader, E. Palaszynski, M. Dy, and B. Lebel. 1983. Biologic properties of homogeneous interleukin 3. I. Demonstration of WEHI-3 growth factor activity, mast cell growth factor activity, P cell-stimulating factor activity, colony-stimulating factor activity and histamine-producing cell-stimulating factor activity. J. Immunol. 131:282.

11. Yokota, T., F. Lee, D. Rennick, C. Hall, N. Arai, T. Mosmann, G. Nabel, H. Cantor, and K. Arai. 1984. Isolation and characterization of a mouse cDNA clone that expresses mast-cell growth-factor activity in monkey cells. Proc. Natl. Acad. Sci. USA 81:1070.

12. Clark-Lewis, I., S.B.H. Kent, and J.W. Schrader. 1984. Purification to apparent homogeneity of a factor stimulating the growth of multiple lineages of hemopoietic cells. J. Biol. Chem. 259:7488.

13. Goldwasser, E., J.N. Ihle, M.B. Prystowsky, I. Rich, and G. Van Zant. 1983. The effect of interleukin-3 on hemopoietic precursor cells. In: Normal and Neoplastic Hematopoiesis. Edited by D.W. Golde, and P.A. Marks. New York, Alan R. Liss, Inc., p.301.

14. Greenberger, J.S., R.J. Eckner, M. Sakakeeney, P. Marks, D. Reid, G. Nabel, A. Hapel, J.N. Ihle, and K.C. Humphries. 1983. Interleukin-3-dependent hematopoietic progenitor cell lines. Fed. Proc. 42:2762.

15. Palacios, R., G. Henson, M. Steinmetz, and J.P. McKearn. 1984. Interleukin-3 supports growth of mouse pre-B cell clones in vitro. Nature 309:126.

16. Bowlin, T.L., A.N. Scott, and J.N. Ihle. 1984. Biological properties of interleukin 3. II. Serological comparison of 20 α-SDH inducing activity, colony-stimulating activity, and WEHI-3 growth factor activity using antiserum against IL-3. J. Immunol. 133:2001.

17. Suda, T., J. Suda, M. Ogawa, and J.N. Ihle. 1985. Permissive role of interleukin-3 (IL-3) in proliferation and differentiation of multipotential progenitors in culture. J. Cell. Physiol. (in press).

18. Suda, T., J. Suda, and M. Ogawa. 1983. Proliferative kinetics and differentiation of murine blast cell colonies in culture: Evidence for variable G_0 periods and constant doubling rates of early pluripotent hemopoietic progenitors. J. Cell. Physiol. 117:308.

19. Till, J.E., E.A. McCulloch, and L. Siminovitch. 1964. A stochastic model of stem cell proliferation, based on the growth of spleen colony forming cells. Proc. Natl. Acad. Sci. USA 51:29.

20. Nakahata, T., and M. Ogawa. 1982. Identification in culture of a new class of hemopoietic colony-forming units with extensive capability to self-renew and generate multipotential colonies. Proc. Natl. Acad. Sci. USA 79:3843.

21. Nakahata, T., A.J. Gross, and M. Ogawa. 1982. A stochastic

model of self-renewal and commitment to differentiation of the primitive hemopoietic stem cells in culture. J. Cell. Physiol. 113:455.

22. Suda, T., J. Suda, and M. Ogawa. 1984. Disparate differentiation in mouse hemopoietic colonies derived from paired progenitors. Proc. Natl. Acad. Sci. USA 81:2520.

23. Suda, J., T. Suda, and M. Ogawa. 1984. Analysis of differentiation of mouse hemopoietic stem cells in cultures by sequential replating of paired progenitors. Blood 64:393.

24. Leary, A.G., L.C. Strauss, C.I. Civin, and M. Ogawa. 1985. Disparate differentiation in hemopoietic colonies derived from human paired progenitors. Blood (in press).

25. Pharr, P.N., J. Nedelman, H.P. Downs, M. Ogawa, and A.J. Gross. 1985. A stochastic model for mast cell proliferation in culture. J. Cell. Physiol. (in press).

26. Kitamura, Y., S. Go, and K. Hatanaka. 1978. Decrease of mast cells in W/W^v mice and their increase by bone marrow transplantation. Blood 52:447.

27. Kitamura, Y., K. Hatanaka, M. Murakami, and H. Shibata. 1979. Presence of mast cell precursors in peripheral blood of mice demonstrated by parabiosis. Blood 53:1085.

28. Aglietta, M., G. Camussi, and W. Piacibello. 1981. Detection of basophils growing in semisolid agar culture. Exp. Hematol. 9:95.

29. Denburg, J.A., M. Richardson, S. Telizyn, and J. Bienenstock. 1983. Basophil/mast cell precursors in human peripheral blood. Blood 61:775.

30. Sonoda, T., T. Ohno, and Y. Kitamura. 1982. Concentration of mast-cell progenitors in bone marrow, spleen, and blood of mice determined by limiting dilution analysis. J. Cell. Physiol. 112:136.

31. Sonoda, T., Y. Kitamura, Y. Haku, H. Hara, and K.J. Mori. 1983. Mast cell precursors in various haematopoietic colonies of mice produced in vivo and in vitro. Br. J. Haematol. 53:611.

32. Sonoda, T., Y. Kanayama, H. Hara, C. Hayashi, M. Tadokoro, T. Yonezawa, and Y. Kitamura. 1984. Proliferation of peritoneal mast cells in the skin of W/W^v mice that genetically lack mast cells. J. Exp. Med. 160:138.

33. Suda, T., J. Suda, S.S. Spicer, and M. Ogawa. 1985. Proliferation and differentiation in culture of mast cell progenitors derived from mast cell-deficient mice of gentoype W/W^v. J. Cell. Physiol. (in press).

Mast Cell Differentiation and Heterogeneity,
edited by A. D. Befus et al.
Raven Press, New York © 1986.

Growth and Differentiation of Human Basophils, Eosinophils, and Mast Cells

*Judah A. Denburg, **Yasuo Tanno, and **John Bienenstock

*Departments of *Medicine and **Pathology, McMaster University, Hamilton, Ontario L8N 3Z5, Canada*

Until recently, there has been relative lack of information on the growth and differentiation of human basophils, eosinophils and mast cells even though biochemical, functional and hemato- logical studies have clearly demonstrated similarities among these specialized granulocytes in a variety of laboratory animals and man (1–8). Although studies in man may be crucial in under- standing the ontogeny and lineage interrelationships of baso- phils, eosinophils and mast cells and their clinical importance, most of our understanding and in vitro analysis of these cells has come from studies in rodent models such as the mouse or rat (9–14). Until recently, metachromatic cells fitting descriptions of either basophils or mast cells had not been recognized in human hemopoietic cell cultures, notwithstanding experiments in nature such as chronic myeloid leukemia in which basophilia can be a prominent and important feature (15), or systemic masto- cytosis in which a neoplastic proliferation of mast cells occurs in various organs (16).

McCarthy et al and Aglietta et al. observed clusters of metachromatic cells in semi-solid agar cultures of bone marrow from leukemic patients (17,18) and Miyoshi et al. as well as we reported the culturing in suspension of large numbers of basophils, especially from patients in blast crisis phase of chronic myeloid leukemia (19–21); however, even these leukemic cells could not be maintained in culture beyond 4 weeks (19–21). Recent observations by a number of investigators, including us, concerning differentiation of normal human peripheral blood umbilical cord blood, bone marrow or fetal liver precursors into

---

Supported by grants from the Medical Research Council of Canada, the National Cancer Institute of Canada, and Pharmacia Diagnostics. The expert technical assistance of Ms. S. Telizyn is gratefully acknowledged.
Abbreviations: G6PD, glucose 6-phosphate dehydrogenase; MBP, major basic protein; CLC, Charcot-Leyden crystal; CM, conditioned medium; IL, interleukin; DEAE, diethylaminoethyl; HPLC, high pressure liquid chromatography.

basophils or mast cells under the stimulatory influence of T-cell derived growth factor(s) (22-26) represented, in part, attempts to perpetuate lines of these cells in vitro, a goal that has remained elusive. Thus, the hope that the functional aspects of basophils and mast cells could be clarified by studying relatively homogeneous populations of cells grown in vitro has not yet been realized.

Over the last few years, we have attempted to address several basic questions regarding the growth and differentiation of basophils and mast cells. A number of different approaches have been used to culture human basophils and mast cells, including semi-solid and suspension cultures of human cord blood cultures and of leukemic cell populations (21,22,25). In addition, we have applied biochemical assays of individual, clonally-derived human hemopoietic colonies (27,28) in order to clarify progenitor-progeny relationships of these cells. In the present studies we have addressed the following questions: (i) What is (are) the lineage relationship(s) among human basophils, mast cells and eosinophils? (ii) What factor(s) regulate the differentiation of these cells?

METHODS

Cell cultures. Suspensions of density gradient separated lymphomononuclear cells, peripheral blood or human cord blood were cultured in Iscove's modified Dulbecco's medium or McCoy's 5A minimal essential medium supplemented with fetal calf serum, antibiotics, 2-mercaptoethanol, in the presence of crude or fractionated supernatants derived from a human T cell leukemia line, Mo (39), or human placental conditioned medium as described (25,29,30). Supernatants derived from a number of human transformed cell lines (squamous cell carcinoma, Colo-16 and osteogenic sarcoma, R9KL4, both kindly supplied by Dr. S.-K. Liao, McMaster University; human urinary bladder carcinoma, 5367 by Dr. B. Stadler, Bern, Switzerland) were tested for activity in cord blood cultures (above) or for interleukin (IL)-3-like activity (31) on an IL-3 dependent murine mast cell line, 32 Dcl/H4 (kindly supplied by Dr. B. Stadler, Bern, Switzerland). Methylcellulose cultures were performed as described previously (25) in small plastic petri dishes, using 0.9% methylcellulose into which cells are plated at concentrations varying from 1 x 10^4 to 1 x 10^6 per ml.

MBP and CLC assays. Analysis of cytocentrifuge preparations of suspension cultures or of single colonies for histochemical staining characteristics and immunofluorescence studies with rabbit antisera to human MBP or CLC proteins were performed by Drs. G. Gleich and S. Ackerman (Mayo Clinic, Rochester, Minn.) as described (28).

Histamine Assays. These were performed on cell suspensions or single colonies by single isotope radioenzymatic conversion method (detection limit, 0.12 ng) or by an adaptation of an automated nephelometric method (detection limit, 0.20 ng), as described (25,32).

G6PD isoenzyme studies. These were performed on individual colonies as previously described (27), by Drs. H. Messner and B. Lim of the Ontario Cancer Institute, Toronto, Ontario. The assay is sensitive to a detection limit of 30 granulocytes.

Separation of hemopoietic growth factors. Conditioned media (Mo–CM, or CM from other cell lines) were dialyzed against culture medium after ultrafiltration with Diaflo PM30 and XM50 membranes (Amicon Corp., Danvers, MA) and ammonium sulfate precipitation, heat–treated at 65°C for 30 min, and, in some instances, further subjected to a multi–step purification by DEAE–cellulose column chromatography (DE–52, Whatman), according to the method described by Yung et al. (12) (29,30). DEAE columns were eluted stepwise with buffers containing increasing concentrations (10 to 100 mM) of sodium phosphate at a flow rate 15 ml/hr at 4°C using a peristalic pump: fractions were concentrated, dialyzed, passed through a sterile 0.22μ Millipore filter (Millipore Corp., Bedford, MA) and stored at −20°C until use (29,30).

RESULTS

Peripheral blood methylcellulose cultures for basophil/mast cell progenitors. We had originally observed the presence of relatively homogeneous populations of metachromatic cells in colonies grown in the presence of Mo–CM from the peripheral blood of a patient with systemic mastocytosis (25). Such colonies contained very high levels of histamine and stained metachromatically with toluidine blue and other basic dyes; cells in some colonies stained with cyanide–resistant peroxidase. Using colony assays and morphological or biochemical (histamine) criteria to define basophil/mast cell–containing colonies (25,27,28,33,34), a low frequency of circulating basophil/mast cell progenitors was found in normals, with increased levels in chronic myeloid leukemic blood, atopic peripheral blood and near normal numbers in patients with urticaria pigmentosa (Table I). Attempts at clonally expanding individual colonies beyond the 200–500 cell stage, grown from blood of patients with systemic mastocytosis, CML or atopic conditions have been unsuccessful to date.

Biochemical analysis of granulocyte colonies. The eosinophil and basophil proteins, MBP and CLC (7,8) could be found in most histamine–positive colonies grown from atopic patient blood but not in the colonies derived from the patient with systemic mastocytosis (Table II).

Table I.   Estimated number[a] of peripheral blood basophil/mast
cell progenitors in various clinical conditions

Patient Group	Frequency/10^6 cells
Atopics (n = 53)	14.5
Chronic myeloid leukemia (n = 15)	21.6
Urticaria pigmentosa (n = 6)	1.4
Systemic Mastocytosis (n = 4)	7.6
Normal controls (n = 27)	3.0

a. Number of basophil- or histamine-positive colony-forming
   cells.
   Total number of granulocyte colonies in methylcellulose
   cultures.

The staining characteristics for CLC and MBP in these colonies
resembled those described for basophils and their content per
cell was also within the ranges described (7,8). We were unable
to predict of a given colony, given its inverted microscopic
appearance, whether it would be composed of mixed basophilic-
eosinophilic, pure basophilic or pure eosinophilic cells (28).
Moreover, there appeared to be no consistent pattern of MBP, CLC
or histamine content of single colonies (28; Figure 1); these
differentiation proteins were randomly expressed in single
colonies. All such colonies, when picked individually from the
culture dishes, could be shown to be clonally derived either by
linear plating efficiency (34) or by G6PD isoenzyme analysis
(27,28).
    Growth and differentiation of cord blood basophil and
eosinophil progenitors.   Large numbers of basophils can be
observed to grow in suspension cultures of cord blood in the
presence of PHA-CM (derived from PHA-stimulated peripheral blood
mononuclear cells) which have been separated by ion exchange
column chromatography from IL-2 (22,29,30). Further analysis in
our hands of cord blood cultures demonstrated a peak of basophils
and histamine consistently at 2 weeks in vitro in the presence
of unfractionated Mo-CM, followed by an eosinophil peak at 4
weeks and persisting up to 10 weeks in vitro (29,30). Some cultures exhibited predominant basophil growth while others showed
mostly eosinophil growth when stimulated with heat-treated, ul-
trafiltered Mo-CM (Table III). Cells of mixed basophil/eosinophil
granulation were observed in these as well as in methylcellulose
cultures (25,28,30).

TABLE II. Histamine and MBP content of single granulocyte colonies in atopic and mastocytosis patient cultures

Patient Group	Colony Type	MBP(ng)	Histamine (ng)
Atopics	Eo	<6	0.2
		<6	<0.1
		49	0.2
		<6	0.6
		15	<0.1
		16	0.22
		39	0.22
		7.5	0.4
		<6	<0.1
	GM	<6	<0.1
		<6	<0.1
		<6	<0.1
		<6	<0.1
		<6	0.5
		<6	<0.1
		<6	<0.1
		<6	<0.1
		<6	<0.1
		<6	<0.1
Mastocytosis	Eo	<6	0.54
		<6	1.0
		<6	0.22
		<6	0.44
		<6	2.4
	GM	<6	1.1
		<6	0.76
		<6	0.2

Eo = eosinophil-type
GM = neutrophil-macrophage-type
MBP = major basic protein

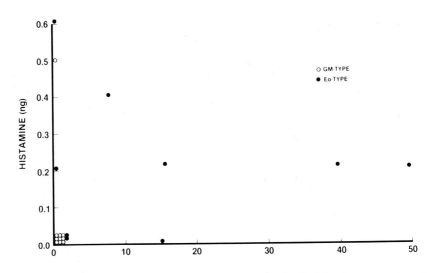

Figure 1.   Biochemical analysis of individual granulocyte
colonies. MBP (abscissa) and histamine (orginate) were measured
(ng) in individual, clonally-derived granulocyte colonies picked
from methylcellulose dishes at day 14 <u>in</u> <u>vitro</u>. Open circles,
neutrophil-macrophage (GM-type) colonies; closed circles,
eosinophil-basophil (Eo-type) colonies.

TABLE III.   Effect of heat treatment and ultrafiltration of Mo-CM
on basophil and eosinophil growth-promoting activities

	Basophils $(x10^{-3}/ml)$	Eosinophils $(x10^{-3}/ml)$
Treatment		
65°C, 30 min	78 ± 34	119 ± 51
Ultrafiltration		
>30,000 daltons	56 ± 7*	78 ± 43*
<30,000 daltons	21 ± 5	16 ± 7
Positive control[a]	60 ± 30	89 ± 34
Negative control[b]	21 ± 8	4 ± 2

a. Unfractionated Mo-CM
b. No CM
   Mean ± SEM, n = 3
* $p < 0.05$, >30,000 <u>vs</u> <30,000 daltons

Separation of basophil and eosinophil growth promoting activities. After multi-step separation of serum-containing or serum-free Mo-CM (29,30; see Methods), factors which would preferentially promote either basophil or eosinophil growth, but not both, were identified: low-salt eluates from the DEAE-cellulose column promoted basophil growth, while high-salt eluates suppressed basophil growth and enhanced eosinophil growth in cord blood suspension cultures (30). Using methylcellulose colony assays, the latter (high-salt eluate) fractions of Mo-CM could be shown to promote the growth of pure eosinophil colonies containing few or no basophils (Table IV).

TABLE IV. Granulocyte colony basophils and eosinophils in response to DEAE-fractionated T-cell[a] conditioned medium

DEAE Fraction	% Basophil-Positive Colonies	% Pure Basophil Colonies	% Pure Eosinophil Colonies
10 mM	46.2	7.7	0.0
20 mM	14.3	0.0	0.0
40 mM	25.0	0.0	0.0
100 mM	7.7	0.0	23.1

a. Mo cell line

IL-3-like activity and basophil/eosinophil growth promoting activity produced by human cell lines or tissues. Unfractionated or fractionated CM derived from a number of diverse human cell lines or tissues were tested for their ability to promote proliferation of subclones of the murine IL-3-dependent cell lines, 32Dcl/H4 (31) and WEHI-3FD (Tanno, Y and Denburg, J.A., unpublished observations). As can be seen, differences in IL-3-like activity, colony-stimulating activity, and basophil/eosinophil growth promoting activities were observed among these CM (Table V).

DISCUSSION

Although it has not been possible to date either to culture homogeneous populations of basophils, eosinophils or mast cells, or to provide direct evidence for lineage relationships among subpopulations of human metachromatic cells, the current studies provide some information regarding the ontogeny and phenotype expression of progenitors of these cells. First, we have shown that basophils and eosinophils derive from a single progenitor which gives rise to hemopoietic colonies in which phenotypic

TABLE V.   Interleukins, colony stimulating activity, and basophil
           or eosinophil growth promoting activity derived
           from human cell lines or tissues

	Cell Lines or Tissues[a]				
Activity	MO–CM	HPCM	Colo–16	5376	R9KL4
IL–1	–	–	+	–	–
IL–2	–	–	–	–	–
IL–3	–	+	+	+	±
CSA	+	+	+	+	+
BaGPA	+	+	±	NT	–
EoGPA	+	+	±	NT	–

a. defined in text
CSA   = granulocyte colony–stimulating activity in
        methylcellulose cultures
BaGPA = basophil growth–promoting activity in cord blood
        suspension or methylcellulose cultures
EoGPA = eosinophil growth–promoting activity in cord blood
        suspension or methylcellulose cultures
  NT = not tested

programs are variably expressed in methylcellulose in vitro
(27,28; Figure 1; Table II). Second, the growth and differentiation of basophils or eosinophils in suspension (29,30) or semi-
solid (Table IV) cord blood cultures appears to depend upon
separate factors present in CM derivable from a number of human
cell lines or tissues; basophil and eosinophil growth factors
appear to be at least functionally dissociable from IL-3 like
activity or from some colony-stimulating activities, since
different cell lines produce some, but not all of these activities (Table V). The kinetics of basophil and eosinophil growth
in cord blood, as well as the inversely related growth pattern of
each in both suspension and semi-solid cultures, depending on the
concentration of growth factors provided, suggest that the single
basophil/eosinophil progenitor identified may be influenced in
its phenotypic expression by microenvironmental factors. Indeed,
the presence of MBP and CLC in both basophils and eosinophils
(7,8); the demonstration of hemopoietic anomalies shared by both

these phenotypes (5,6); and the finding of "atypical eosinophils" which contain basophilic granules in some human myeloid leukemias (35) or in human promyelocytic leukemia (HL-60) cells grown under alkaline conditions associated with inversion of chromosome 16 (36), support the concept that only terminal differentiation events distinguish these two cell phenotypes. It will be of interest to test the effects of purified basophil or eosinophil growth factor(s) directly on single progenitors.

MODEL FOR
BASOPHIL/MAST CELL/EOSINOPHIL DIFFERENTIATION

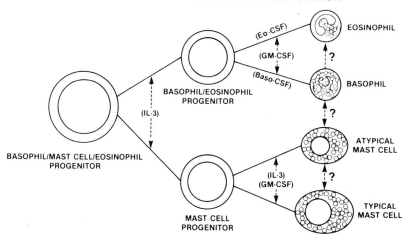

Figure 2. Schematic representation of probable (solid lines) and possible (dashed lines) differentiation pathways including putative progenitor cells and lineage-specific growth factors in the human basophil/mast cell and eosinophil lineages. A common progenitor for all three cell types may diverge into two separate lineages, basophil/eosinophil vs mast cell. A human IL-3-like molecule could act at both early and late stages of mast cell differentiation; basophil or eosinophil colony-stimulating activity (CSF) may represent a unique species of granulocyte-macrophage CSF, acting on a common basophil/eosinophil progenitor. The separate existence of subpopulations of human mast cells as well as their respective growth-factor dependence is purely speculative.

There are a number of studies demonstrating biochemical differences between human basophils and mast cells, mostly related to profiles of arachidonic acid metabolites (reviewed in 37). Mast cells may thus represent a distinct cell lineage in the human, a concept which is supported by our finding of MBP and CLC in histamine-positive colonies of CML or atopic blood but not in those from mastocytosis peripheral blood (Table II). However, it is still not at all clear whether subpopulations of mast cells and peripheral blood basophils derive from a common progenitor or from separate progenitors which respond to a common growth factor, such as IL-3-like activity (31), basophil- or eosinophil-growth promoting activity (Figure 2). More precise criteria to define human basophils and subpopulations mast cells are required to clarify this question, and the precise number of progenitors or cell lineages awaits further purification and sequencing of specific growth factors involved, as has been accomplished for murine IL-3 (38).

REFERENCES

1. **Rothwell, T.L.W.** 1975. Studies of the responses of basophil and eosinophil leucocytes and mast cells to the nematode Trichostrongylus colubriformis. I. Observations during the expulsion of first and second infections by guinea-pigs. J. Pathol. 116:51.
2. **Capron, M., J. Rousseaux, C. Mazingue, H. Bazin, and A. Capron.** 1978. Rat mast cell-eosinophil interaction in antibody-dependent eosinophil cytotoxicity to Schistosoma mansoni schistosomula. J. Immunol. 121:2518.
3. **Ogilvie, B.M., P.W. Askenase, and M.E. Rose.** 1980. Basophils and eosinophils in three strains of rats and in athymic (nude) rats following infection with the nematodes Nippostrongylus brasiliensis or Trichinella spiralis. Immunology 39:385.
4. **Brown, S.J., S.J. Galli, G.J. Gleich, and P.W. Askenase.** 1982. Ablation of immunity to Amblyomma americanum by anti-basophil serum: cooperation between basophils and eosinophils in expression of immunity to ectoparasites (ticks) in guinea pigs. J. Immunol. 129:790.
5. **Juhlin, L., and G. Michaelsson.** 1977. A new syndrome characterised by absence of eosinophils and basophils. Lancet 1:1233.
6. **Tracey, R., and H. Smith.** 1978. An inherited anomaly of human eosinophils and basophils. Blood Cells 4:291.
7. **Ackerman, S.J., G.J. Weil, and G.J. Gleich.** 1982. Formation of Charcot-Leyden crystals by human basophils. J. Exp. Med. 155:1597.
8. **Ackerman, S.J., G.M. Kephart, T.M. Habermann, P.R. Greipp, and G.J. Gleich.** 1983. Localization of eosinophil granule major basic protein in human basophils. J. Exp. Med. 158:946.
9. **Schrader, J.W.** 1981. The in vitro production and cloning of the P-cell, a bone marrow-derived null cell that expresses H-2 and Ia-antigens, has mast cell-like granules, and is regulated by a factor released by activated T cells. J. Immunol. 126:452.
10. **Nagoo, K., K. Yokoro, and S.A. Aaronson.** 1981. Continuous

lines of basophil/mast cells derived from normal mouse bone marrow. Science 212:333.

11. Razin, E., C. Cordon–Cardo, and R.A. Good. 1981. Growth of pure populations of mouse mast cells in vitro with conditioned medium derived from concanavalin A–stimulated splenocytes. Proc. Natl. Acad. Sci. (USA) 78:2559.

12. Yung, Y–P., R. Eger, G. Tertian, and M.A.S. Moore. 1981. Long–term in vitro culture of murine mast cells. III. Purification of a mast cell growth factor and its dissociation from TCGF. J. Immunol. 127:794.

13. Denburg, J.A., A.D. Befus, and J. Bienenstock. 1982. Growth and differentiation in vitro of mast cells from mesenteric lymph nodes of Nippostrongylus brasiliensis–infected rats. Immunology 41:195.

14. Haig, D.M., T.A. McKee, E.E.E. Jarrett, R. Woodbury, and H.P.R. Miller. 1982. Generation of mucosal mast cells is stimulated in vitro by factors derived from T cells of helminth–infected rats. Nature 300:188.

15. Parwaresch, M.R. 1976. In: The Human Blood Basophil. New York, Springer, p. 203.

16. Lennert, K., and M.R. Parwaresch. 1979. Mast cells and mast cell neoplasia: a review. Histopathol. 3:349.

17. McCarthy, J.H., T.E. Mandel, O.M. Garson, and D. Metcalf. 1980. The presence of mast cells in agar cultures. Exp. Hematol. 8:562.

18. Aglietta, M., G. Camussi, and W. Piacibello. 1981. Detection of basophils growing in semisolid agar culture. Exp. Hematol. 9:95.

19. Miyoshi, I., H. Uchida, T. Tsubota, Kubonishi, S. Hiraki, and K. Kitajima. 1977. Basophilic differentiation of chronic myelogenous leukaemia cells in vitro. Scand. J. Haematol. 19:321.

20. Denburg, J.A., W.E.C. Wilson, R. Goodacre, and J. Bienenstock. 1980. Basophil production in chronic myeloid leukemia: Evidence for increases in basophils and histamine in peripheral blood cultures. Br. J. Haematol. 45:13.

21. Denburg, J.A., W.E.C. Wilson, and J. Bienenstock. 1982. Basophil production in myeloproliferative disorders: Increases during acute blastic transformation of chronic myeloid leukemia. Blood 60:113.

22. Ogawa, M., T. Nakahata, A.G. Leary, A.R., Sterk, K. Ishizaka, and T. Ishizaka. 1983. Suspension culture of human mast cells/basophils from umbilical cord blood mononuclear cells. Proc. Nal. Acad. Sci. (USA) 80:4494.

23. Tadokoro, K., B.M. Stadler, and A.L. de Weck. 1983. Factor–dependent in vitro growth of human normal bone marrow–derived basophil–like cells. J. Exp. Med. 158:857.

24. Razin, E., A.B. Rifkind, C. Cordon–Cardo, and R.A. Good. 1981. Selective growth of a population of human basophil cells in vitro. Proc. Natl. Acad. Sci. (USA) 78:5793.

25. Denburg, J.A., M. Richardson, S. Telizyn, and J.

Bienenstock. 1983. Basophil/mast cell precursors in human peripheral blood. Blood 61:775.

26. Leary, A.G., and M. Ogawa. 1984. Identification of pure and mixed basophil colonies in culture of human peripheral blood and marrow cells. Blood 64:78.

27. Denburg, J.A., H. Messner, B. Lim, N. Jamal, S. Telizyn, and J. Bienenstock. 1985. Clonal origin of human basophil/mast cells from circulating multipotent hemopoietic progenitors. Exp. Hematol. 13:185.

28. Denburg, J.A., S. Telizyn, H. Messner, B. Lim, N. Jamal, S.J. Ackerman, G.J. Gleich, and J. Bienenstock. 1985. Heterogeneity of human peripheral blood eosinophil-type colonies: evidence for a common basophil-eosinophil progenitor. Blood. In press.

29. Tanno, Y., J. Bienenstock, M. Richardson, T.D.G. Lee, A.D. Befus, and J. Denburg. 1984. Human basophil and eosinophil growth in long-term cord blood cultures. Fed. Proc. 43:1935A.

30. Tanno, Y., J. Bienenstock, S. Ahlstedt, and J. Denburg. 1985. Reciprocal regulation of human basophil and eosinophil growth by separate factors in cord blood cultures. Fed. Proc. 44:587A.

31. Hirai, K., B. Stadler, and A.L. De Weck. 1985. Human IL-3-like activity is distinct from basophil-promoting activity. Int. Arch. Allergy Appl. Immunol., in press.

32. Siraganian, R.P., and M.J. Brodsky. 1976. Automated histamine analysis for in vitro allergy testing. J. Allergy Clin. Immunol. 57:525.

33. Denburg, J.A., S. Telizyn, A. Belda, J. Dolovich, and J. Bienenstock. 1985. Increased numbers of circulating basophil/mast cell progenitors in atopic patients. J. Allergy Clin. Immunol., in press.

34. Denburg, J.A., H. Otsuka, S. Telizyn, R. Rajan, D. Hitch, P. Lapp, A.D. Befus, J. Bienenstock, and J. Dolovich. 1985. Peripheral blood basophils, basophil progenitors and nasal metachromatic cells in allergic rhinitis. J. Allergy Clin. Immunol. 75:153A.

35. LeBeau, M.M., R.A. Larson, M.A. Bitter, J.W. Vardiman, H.M. Golomb, and J.D. Rowley. 1983. Association of an inversion of chromosome 16 with abnormal marrow eosinophils in acute myelomonocytic leukemia: a unique cytogenetic-clinico-pathlogical association. N. Engl. J. Med. 309:630.

36. Fischkoff, S.A., A. Pollack, G.J. Gleich, J.R. Testa, S. Misawa, and T.J. Reber. 1984. Eosinophilic differentiation of the human promyelocytic leukemia line, HL-60. J. Exp. Med. 160:179.

37. Schulman, E.S., S.W. MacGlashan Jr., R.P. Schleimer, S.P. Peters, A. Kagey-Sobotka, H.H. Newball, and L.M. Lichtenstein. 1983. Purified human basophils and mast cells: current concepts of mediator release. Eur. J. Respir. Dis. 64(Suppl. 128):53.

38. Fung, M.C., A.J. Hapel, S. Ymer, D.R. Cohen, R.M. Johnson, H.D. Campbell, and I.G. Young. 1984. Molecular cloning of cDNA for murine interleukin-3. Nature 307:233.

39. Golde, D.W., S.G. Quan, and M.J. Cline. 1978. Human T lymphocyte cell line producing colony-stimulating activity. Blood 52:1068.

Mast Cell Differentiation and Heterogeneity,
edited by A. D. Befus et al.
Raven Press, New York © 1986.

Characteristics of Human Basophilic Granulocytes Developed in *In Vitro* Culture

Teruko Ishizaka

John Hopkins University School of Medicine at the Good Samaritan Hospital, Baltimore, Maryland 21239

In IgE–mediated hypersensitivity reactions, IgE antibodies bind to mast cells and basophils through their IgE receptors, and the reaction of allergen with cell–bound IgE antibodies induces the release of a variety of preformed and newly generated mediators, such as histamine and leukotrienes (1).

Recently, several investigators succeeded in purification of human basophils (2) and human mast cells from lung and intestine (3,4,5), providing useful tools for the analysis of mediator release from these cells. However, the number of the cells obtained from these sources is limited. Possible changes in cell membranes due to purification procedures, such as extensive digestion of lung fragments, might affect the results of some experiments. With this consideration, we attempted to develop human mast cells or basophils in culture. Although we have not succeeded in developing continuous human mast cell lines from

This work was supported by National Institutes of Health, grant AI–10060, and a grant from the Watson–Hyde Foundation. This article is publication No. 590 from the O'Neill Laboratories, Good Samaritan Hospital, Baltimore, Md., USA.

Acknowledgement: I would like to express my gratitude to Drs. K. Ishizaka, D.H. Conrad, J.R. White, J.R. Niebyl, Mr. A.R. Sterk and Mrs. C.G.L. Ko, Johns Hopkins University, Baltimore, Md.; Dr. M. Ogawa, V.A. Medical Center, Medical University of South Carolina, Charleston, S.C.; Drs. A.M. Dvorak, R.L. Stevens and R.A. Lewis, Harvard Medical School, Boston, Ma.; and Dr. D.D. Metcalfe, NIDID, NIH, Bethesda, Md., for their excellent collaboration and assistance.

culture of bone marrow cells from systemic mastocytosis patients, we have succeeded in selective growth of human basophils in suspension cultures of mononuclear cells from umbilical cord blood. A large number of functionally mature basophils is available in these cultures (6). In this presentation, I would like to summarize available information on cultured basophils.

MATERIALS AND METHODS, RESULTS AND DISCUSSION

Culture Conditions and Characterization of Cultured Human Basophils. Basophil precursors and growth factor(s). Mononuclear cells or cord blood were the source of precursors of basophils. Our studies revealed that precursors of basophils in cord blood are non-adherent cells and bear neither surface immunoglobulin nor T cell markers (7). Mononuclear cells of cord blood were cultured in the presence of a fraction of culture supernatant of PHA-stimulated human T cells from which interleukin-2 had been eliminated. Growth factor(s) for basophils in the fraction is (are) heat-stable glycoprotein(s) with a molecular weight of 25,000 to 35,000.

Ultrastructural features. After 2 to 3 weeks culture of mononuclear cells, 40% to 85% of non-adherent cells in the culture contained metachromatic granules characteristic for mast cells and basophils (8), and these cells were identified as mature human basophils by electron microscopic studies carried out by Dr. A.M. Dvorak. They are polymorphonuclear cells and display cytoplasmic granules filled with particles and membranous arrays. These ultrastructural features are characteristic for mature human basophils. Furthermore, the morphology of anaphylactic degranulation of cultured cells, examined by electron microscopy, was identical to that of basophils from peripheral blood (8).

Proteoglycan content. Further evidence to support the identification of the cells as human basophils was provided by analysis of cell-associated proteoglycans by Drs. D.D. Metcalfe from NIH and R.L. Stevens from Harvard. Although they used different protocols for extraction and characterization of ^{35}S-labelled proteoglycans in the cultured cells, they obtained identical results: cultured cord blood basophils contain chondroitin 4-sulfate proteoglycan similar to those found in human leukemic basophils (9).

Basophil yield and purity. Yield of basophils in the culture greatly varied depending on cord blood employed, but is in the range of 10% to 50% of the number of mononuclear cells seeded. From approximately 10^8 mononuclear cells, which is the average number of cells obtained from 20 ml cord blood, 1 to 5 x 10^7 basophils were recovered at 2 to 3 weeks. When the purity of basophils in these cultures was lower than 60%, basophils could be purified to higher than 80% by one differential centrifugation.

Surface phenotype, histamine content and IgE receptors. Cultured human basophils are OKT 3 , OKM 1 and Ia cells and contain an average of 1.74 ± 0.82 µg histamine per 10^6 cells, which was comparable to that in the human peripheral blood basophils. Cultured basophils bear approximately 120,000 to 380,000 IgE receptors per cell. IgE receptors on cultured basophils bind human IgE with high affinity. The average equilibrium constant of the binding reaction between IgE and the receptors was 2.75 x $10^9 M^{-1}$. This value is comparable to those of the binding of rodent IgE to rodent mast cells (10).

As demonstrated with peripheral blood basophils in our previous experiments, cultured basophils bind not only human IgE, but also monoclonal mouse IgE and rat IgE with comparable high affinity. The binding of mouse IgE and rat IgE to cultured human basophils was inhibited by preincubation of the cells with human IgE but not with IgG or IgM, indicating that human and rodent IgE bind to the same receptors. IgE receptors on cultured basophils were analyzed by Dr. D.H. Conrad by SDS–polyacrylamide gel electrophoresis and characterized as glycoprotein having a molecular weight of approximately 64,000 daltons (8).

Mediator release. Although histamine content in some of the cultured basophil preparations was lower than that in basophils from the peripheral blood, all preparations of cultured basophils sensitized with human IgE released 20% to 80% of histamine upon challenge with anti-IgE. No histamine release was observed when unsensitized cultured basophils were incubated with anti-IgE. Cultured basophils could be sensitized with mouse IgE antibody as well. All of the cultured basophils, peripheral blood basophils and purified human lung mast cells, sensitized with a monoclonal mouse IgE anti–DNP antibody, released histamine upon challenge with DNP–HSA, whereas unsensitized cells failed to do so.

As expected, the reactivity of cultured basophils to various histamine releasing reagents was identical to that of basophils from peripheral blood. Cultured basophils released histamine upon exposure to calcium ionophore, chemotactic peptides, F-Met-Leu–Phe or 4 α–phorbol-12,13-didecanoate (PMA), but failed to release histamine upon challenge with compound 48/80.

BIOCHEMICAL EVENTS INVOLVED IN IGE–DEPENDENT MEDIATOR RELEASE FROM CULTURED BASOPHILS

IgE–mediated release reactions. Since histamine releasing properties of cultured basophils were identical to peripheral blood basophils, we analyzed biochemical events involved in IgE-mediated histamine release. The cells sensitized with human IgE were challenged with either the $F(ab')_2$ fragments or Fab' frag-

ments of anti-IgE.   As shown in Fig. 1, $(F(ab')_2$ fragments of
anti-IgE induced a marked enhancement in phospholipid methylation
and an increase in cAMP, which were followed by ^{45}Ca uptake and
histamine release. Phospholipid methylation reached maximum at
30 sec after challenge, while cAMP levels reached maximum at 1
min. ^{45}Ca uptake reached plateau at 3 min and maximum histamine
release was achieved by 5 min. The kinetics of these responses
are identical to those observed in human lung mast cells (4).   As
expected, the Fab' fragments of the antibody failed to induce
these responses, indicating that bridging of cell-bound IgE molecules triggered these responses.   It was also found that sensitized cultured basophils preincubated with ^{14}C-AA released ^{14}C-AA
upon challenge with anti-IgE and the kinetics of the release were
identical to that of histamine release from the cells.

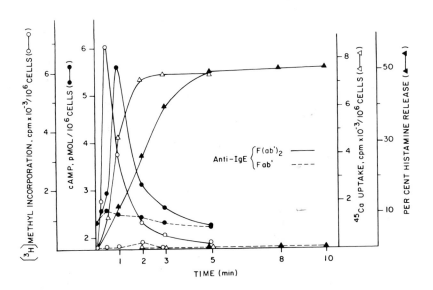

Figure 1. Kinetics of ^{3}H-methyl incorporation into
phospholipid lipids, cAMP rise, ^{45}Ca uptake and histamine release
induced by $F(ab')_2$ fragment of anti-IgE. Cultured basophils were
incubated overnight with 10 μg/ml E myeloma protein and then
challenged with an optimal concentration of $F(ab')_2$ fragments
(———) or Fab' monomer (----) of anti-IgE. Symbols as indicated.

Inhibitors of methyltransferases. As demonstrated in human
lung mast cells (4), preincubation of sensitized basophils with
inhibitors of methyltransferases, such as 3-deaza adenosine (ADO)

together with L-homocysteine thiolacetone, resulted in inhibition of not only phospholipid methylation, but also of all subsequent biochemical events, i.e. cAMP rise, ^{45}Ca uptake, histamine and ^{14}C-AA release in an identical dose response fashion. Another inhibitor of methyltransferases such as S-isobutyryl-3-deaza-adenosine (3-deaza SIBA), or 3-deaza-aristeromycin gave similar results.

In order to exclude the possibility that inhibitors of phospholipid methylation might have affected other enzymes, we examined whether the inhibiton of histamine and ^{14}C-AA release by 3-deaza ADO or 3-deaza SIBA could be reversed by the addition of the methyl donor, S-adenosyl-L-methionine (SAM). Thus, sensitized cultured basophils were preincubated with either 3-deaza SIBA or 3-deaza ADO together with L-homocysteine thiolactone, then challenged with anti-IgE in the presence or absence of SAM. The results showed that the inhibition of histamine and ^{14}C-AA released by 3-deaza SIBA and 3-deaza ADO was completely reversed by the addition of SAM, indicating that accumulation of S-adenosyl-L-homocysteine or its analogues was indeed responsible for the inhibition.

Proteolytic enzymes. As demonstrated in rodent mast cells (11), proteolytic enzyme was also involved in the IgE-mediated triggering in cultured human basophils. Diisopropylfluorophosphate (DFP), a potent inhibitor of serine esterases, inhibited all biochemical events induced by anti-IgE in an identical dose response fashion, whereas diisopropylmethylphosphate (DMP), a non-phosphorylating analogue of DFP, failed to do so (Fig. 2). It was also found that inhibitors and substrates of trypsin and chymotrypsin inhibited all of the initial biochemical events induced by anti-IgE. The results suggest an essential role of proteolytic enzyme, probably a serine esterase, in the transduction of IgE-mediated triggering signals for mediator release in cultured basophils.

Phospholipid metabolism. In rat mast cells, it has been established that bridging of IgE receptors induces alterations in phospholipid and diacylglycerol (DAG) metabolism (12, 13). Thus, we studied whether DAG is generated in sensitized cultured human basophils following stimulation with anti-IgE or specific antigen. It was found that stimulation of sensitized basophils with anti-IgE induced a selective incorporation of ^{32}P into phosphatidic acid (PA), phosphatidylinositol (PI) and phosphatidylcholine (PC). Furthermore, stimulation of cultured basophils resulted in a transient loss of PI within 30 sec after the challenge, and a concurrent accumulation of DAG and MAG (Fig. 3). We wondered if the IgE-mediated PI turnover in human basophils is biochemically connected to the activation of phospholipid methylation. In order to analyze interrelationship between these two pathways, we examined the effect of inhibitors of methyltransferases on PI turnover. As shown in Fig. 4, preincubation of basophils with 3-deaza ADO and L-homocysteine thiolactone resulted in the inhibition of phospholipid methylation,

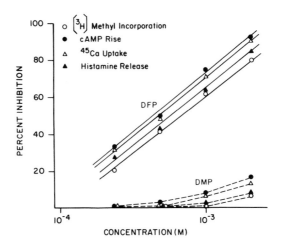

Figure 2.   Inhibition of anti–IgE induced ^{3}H–methyl incorporation into phospholipids, cAMP rise, ^{45}Ca uptake and histamine
release by DFP and DMP.   Cultured basophils sensitized with IgE
were challenged with anti–IgE in the presence of various concentrations of DFP or DMP. ^{3}H–methyl incorporation, cAMP level,
^{45}Ca uptake and histamine release were determined at 30 sec, 1
min, 3 min and 15 min, respectively. Symbols as indicated.

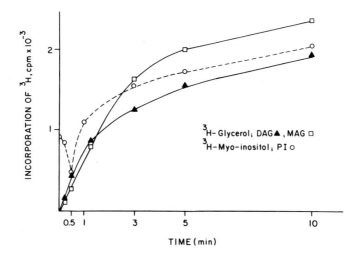

Figure 3.   Kinetics of the formation of DAG and MAG and
changes in PI induced by anti–IgE. Cultured basophils sensitized
with IgE were labelled with either ^{3}H–glycerol or ^{3}H–Myo-
inositol, then challenged with anti–IgE. Symbols as indicated.

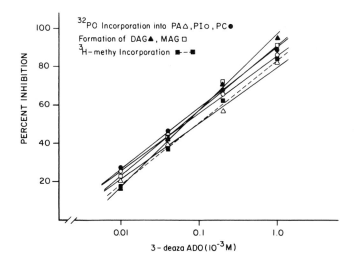

Figure 4. Inhibition of anti-IgE induced ^{3}H-methyl incorporation into phospholipids, ^{32}P incorporation into PA, PI and PC, and formation of DAG and MAG by preincubation of cultured basophils with 3-deaza ADO and L-homocysteine thiolactone (100 μM). ^{32}P incorporation, DAG and MAG formation were determined 10 min after challenge, while phospholipid methylation was determined at 30 sec after the challenge. Symbols as indicated.

^{32}P incorporation into PA, PC, PI and formation of DAG and MAG in an identical dose-response fashion. Furthermore, inhibition of ^{32}P incorporated into phospholipids and DAG formation by 3-deaza ADO was reversed by the addition of SAM to the system prior to the antigen challenge. The results suggest that activation of methyltransferases is involved in IgE-mediated PI turnover.

Further studies indicated that inhibitors and substrates of proteolytic enzymes inhibited not only phospholipid methylation, but also ^{32}P incorporation into PA, PC, PI; ^{32}P incorporation and DAG formation were comparable to that required for 50% inhibition of phospholipid methylation. Thus, it appears that activation of a putative serine esterase and methyltransferase are involved in IgE-mediated PI turnover in cultured human basophils.

Arachidonic acid metabolism. As already described, sensitized basophils release arachidonates upon challenge with anti-IgE. We wondered whether arachidonates released from cultured basophils were lipo-oxygenase or cyclo-oxygenase products. Thus, cultured human basophils sensitized with IgE were labelled with either ^{3}H-AA or ^{14}C-AA and then challenged with anti-IgE. AA metabolites released were analyzed by reversed HPLC using two different solvent systems and radioimmunoassay for leukotrienes C_4 and B_4 and prostaglandin D_2 by Dr. R.A. Lewis. The cultured basophils released 54% to 85% of radiolabelled AA upon challenge with anti–IgE. To our surprise, however, neither cyclo–oxygenase products nor lipo–oxygenase products, with the exception of small quantity of 15–HETE, was detected in the supernatant of 13 different cell preparations. Greater than 85% of radioactivity released into supernatant was identified as unmetabolized AA. Since all of the cultured basophil preparations used for the experiments were morphologically mature, contained histamine comparable to that of peripheral blood basophils, and released 22% to 48% histamine in response to anti–IgE, it appears that biochemical cascade involved in the histamine release is fully developed in the cells. Failure to generate AA metabolites indicates that enzymes essential for metabolism of AA are not fully developed, or are defective in cultured human basophils. These findings suggest that the generation of AA metabolites is not essential for histamine release. The results are in agreement with recent findings by Razin et al. that 5–lipo–oxygenation of arachidonic acid is not mandatory for IgE–mediated secretion of granule markers from bone marrow–derived murine mast cells (14).

Although these cultured basophils are not useful for studies on AA metabolism, availability of a large number of functionally active human basophils of high purity has provided us with a useful tool for analysis of the human IgE receptors, as well as membrane events for triggering and for characterization of enzymes involved in histamine release, such as proteolytic enzyme and protein kinases. Furthermore, these cultured basophils will be a valuable tool in the future for elucidation of possible relationship between basophils and mast cells in man.

REFERENCES

1. Ishizaka, T., and K. Ishizaka. 1975. Biology of immuno–globulin E. Prog. Allergy 19:61.
2. MacGlashan, D.W.Jr., and L.M. Lichstenstein. 1980. The purification of human basophils. J. Immunol. 124:2519.
3. Schulman, E.S., D.W. Jr. MacGlashan, S.P. Peters, R.P. Schleimer, H.H. Newball, and L.M. Lichtenstein. 1982. Human lung mast cells. Purification and characterization. J. Immunol. 129:2662.
4. Ishizaka, T., D.H. Conrad, E.S. Schulman, A.R. Sterk, and K. Ishizaka. 1983. Biochemical analysis of initial triggering events of IgE–mediated histamine release from human lung mast cells. J. Immunol. 130:2357.
5. Fox, C.A., A.M. Dvorak, S.P. Peters, A. Kagey–Sobotka, and L.M. Lichtenstein. 1985. Isolation and characterization of human intestinal mucosal mast cells. J. Immunol. 135:483.
6. Ogawa, M., T. Nakahata, A.G. Leary, A.R. Sterk, K. Ishizaka, and T. Ishiza. 1983. Suspension culture of mast

cells/basophils from umbilical cord blood mononuclear cells. Proc. Natl. Acad. Sci. USA 80:4494.

7. Ishizaka, T., D.H. Conrad, T.F. Huff, D.D. Metcalfe, R.L. Stevens, and R.A. Lewis. 1985. Unique features of human basophilic granulocytes developed in in vitro culture. Int. Arch. Allergy Appl. Immunol. 77:137.

8. Ishizaka, T., A.M. Dvorak, D.H. Conrad, J.R. Niebyl, J.P. Marquette, and K. Ishizaka. 1985. Morphological and immunological characterization of human basophils developed in cultures of cord blood mononuclear cells. J. Immunol. 134:532.

9. Metcalfe, D.D., C.E. Bland, and S.I. Wasserman. 1984. Biochemical and functional characterization of proteoglycans isolated from basophils of patients with chronic myelogenous leukemia. J. Immunol. 132:1943.

10. Sterk, A.R., and T. Ishizaka. 1982. Binding properties of IgE receptors on normal mouse mast cells. J. Immunol. 128:838.

11. Ishizaka, T., and K. Ishizaka. 1984. Activation of mast cells for mediator release through IgE receptors. Prog. Allergy 34:188.

12. Kennerly, D.A., T.J. Sullivan, and C.W. Parker. 1979. Activation of phospholipid metabolism during mediator release from stimulated rat mast cells. J. Immunol. 122:142.

13. Kennerly, D.A., T.J. Sullivan, P. Sylvester, and C.W. Parker. 1979. Diacylglycerol metabolism in rat mast cells. A potential role in membrane fusion and arachidonic acid release. J. Exp. Med. 150:1039.

14. Razin, E., L.C. Romeo, S. Krills, F.-T. Liu, R.A. Lewis, E.J. Corey, and K.F. Austen. 1984. An analysis of the relationship between 5-lipoxygenase product generation and secretion of preformed mediators from bone marrow-derived mast cells. J. Immunol. 133:938.

Mast Cell Differentiation and Heterogeneity,
edited by A. D. Befus et al.
Raven Press, New York © 1986.

Morphologic Expressions of Maturation and Function Can Affect the Ability to Identify Mast Cells and Basophils in Man, Guinea Pig, and Mouse

Ann M. Dvorak

Department of Pathology, Harvard Medical School; and Department of Pathology and Charles A. Dana Research Institute, Beth Israel Hospital, Boston, Massachusetts 02115

We wish to propose here the morphologic criteria for the identification of mature and immature cells and relate these to function of basophils and mast cells. We also will review the morphologic expressions of slow release reactions, of rapid release reactions, of activation, and of uptake phenomena. All of these events probably alter the baseline or control morphology of basophil and mast cells and as a result may act to create confusion in the appropriate identification of these cells.

MATERIALS AND METHODS

The ultrastructural techniques we have used include multiple tracer technologies, variable fixation, processing and embedding methods, cytochemistry, autoradiography and morphometry. All of these methods are described in detail in previous manuscripts (1).

RESULTS AND DISCUSSION

Mature Basophils. Mature human (Fig. 1), guinea pig (Fig. 2) and mouse (Fig. 3) basophils share a distinctive granulocyte morphology (1,2). Their granule substructural patterns are, however, different for each species. The shared granulocyte characteristics include polylobed nuclei with condensed chromatin, cell size range 7 to 10 microns, irregular blunt or thick surface processes, cytoplasmic glycogen, small Golgi area free of membrane-bound ribosomes, many small cytoplasmic

Supported by USPHS Grant CA 28834.
Acknowledgement. We thank Kathryn Pyne, Rita Monohan and Patricia Estrella for excellent technical assistance.

95

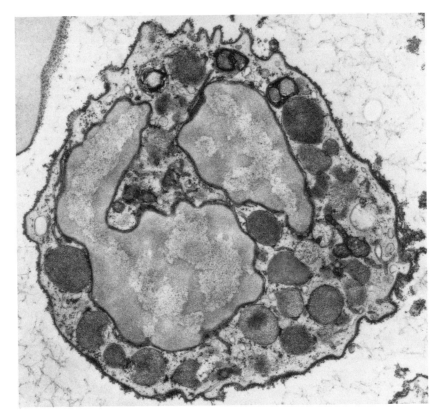

Figure 1. Mature human basophil, 23000x.

vesicles, and granules.   The granules are generally fewer in
number than is seen in mast cells of the corresponding species.
     In man and guinea pigs the granules of basophils are larger
than those of mast cells, whereas in the mouse basophil granules
may be smaller or larger than mast cell granules, depending on
the site or origin of the mast cells. The substructural patterns
seen in granules of human basophils include the more common presence of particles and the less common presence of multiple membranous arrays. In guinea pigs, basophil granular substrutural
patterns include a parallel array, a hexagonal array, and a
finely granular matrix. The granules of the mature mouse basophil are generally homogeneously dense.
     Immature Basophils. Immature human (Fig. 4), guinea pig (Fig.
5) and mouse (Fig. 6) basophils (2,3,4) also share the morphology
which is distinctive and regularly expressed by the maturing

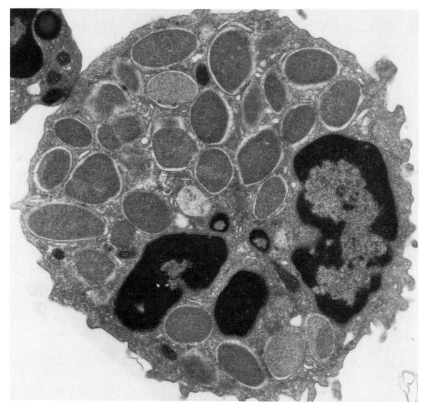

Figure 2. Mature guinea pig basophil, 14000x.

cells of all three granulocyte lineages, i.e. neutrophils, eosinophils and basophils. Thus, basophilic myelocytes are first recognizable developmentally when their specific mature and immature granules are expressed. Similarly, eosinophilic myelocytes and neutrophilic myelocytes are identifiable by the presence of mature and immature granules (eosinophils) and in neutrophils by the initial expression of their primary granules, followed by the expression of the secondary granule population.

The common shared granulocyte characteristics useful for the identification of basophilic myelocytes include the following: large cells with an eccentric, lobular nucleus, from 18 to 20 microns in size in the youngest myelocytes to 15 microns in more mature myelocytes, a cell size that is in the range of that for mast cells of these species. The nuclear chromatin is dispersed

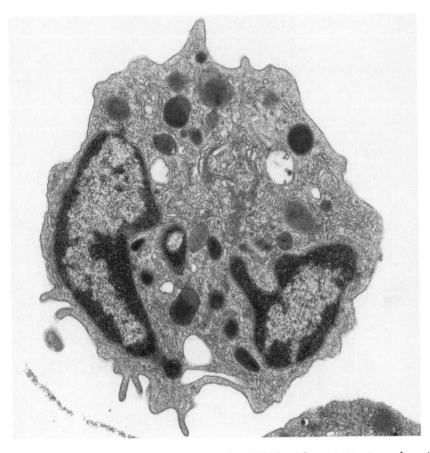

Figure 3.   Mature mouse basophil, 14000x (from A.M. Dvorak   et al., Blood 59:1279, 1982).

and the presence of this   immature, monolobed myelocyte   nucleus should not be confused with the single, eccentric nucleus of mast cells.   The surface projections are irregularly placed and blunt, like mature basophils, not mast cells.   The cytoplasmic contents include active cellular machinery for synthesis and   secretion, i.e. enlarged Golgi apparatus, and many free and membrane-bound ribosomes.   Marked distension of the cisternae of the rough endoplasmic reticulum, which is filled with a protein-like content, is characteristic of basophilic myelocytes, but not of

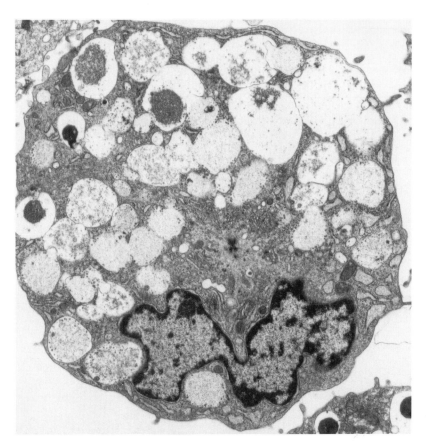

Figure 4. Human basophilic myelocyte, 9000x.

mast cells.   Basophilic myelocytes become smaller as maturation
proceeds.   This process includes the condensation of nuclear
chromatin, the segmentation of nuclei, and reduction of Golgi
structures and rough endoplasmic reticulum.   Immature granules
undergo condensation with a resultant mature granule size that is
often roughly 1/3 of the size of the immature granule of basophilic myelocytes. The immature granule of basophilic myelocytes
of all three species characteristically is large and is filled
with lightly dense, diffuse granular matrix which is surrounded
by small vesicles lying just beneath the granule membrane.   With
granule content condensation and granule size reduction, the
resultant mature granule substructural patterns appear, the
content is rendered more dense in the electron microscope, and
the vesicular content is eliminated.
    Because eosinophils are also granulocytes, eosinophilic

myelocytes are similar in general morphologic characteristics to basophilic myelocytes. Because immature granules of eosinophilic myelocytes have affinity for some metachromatic staining schedules, whereas their mature granules do not, confusion between eosinophilic myelocytes, basophilic myelocytes, and mast cells can occur (3). Ultrastructural analysis can provide appropriate distinctions that are diagnostic. For example, immature eosinophil granules are uniformly dense and generally form perfectly round dense granules that are smaller than basophilic myelocyte granules, and only rarely contain vesicles beneath their granule membranes. Mature granules, when present in myelocytes, are distinctive for the basophil (particulate

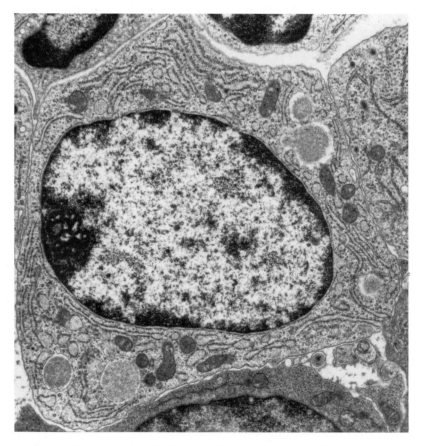

Figure 5.   Guinea pig basophilic myelocyte, 11000x.

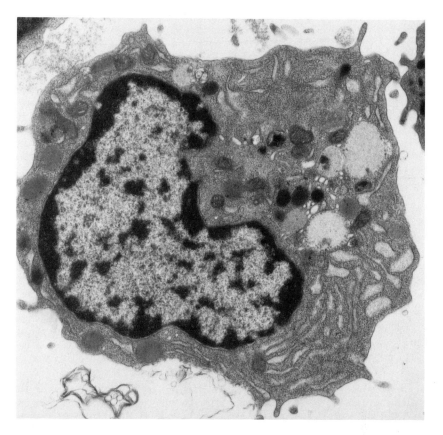

Figure 6. Mouse basophilic myelocyte, 17000x.

content in humans) or for the eosinophil (central crystal in the granule). Eosinophilic myelocytes conform to the general morphology of all myelocytes whereas mast cells do not.

Ultrastructural cytochemistry for endogenous peroxidase can be helpful in defining the eosinophilic myelocyte (Fig. 7). Eosinophilic peroxidase can be demonstrated in all immature granules, in mature granule matrix, but not in the central crystal, as well as in the cisternae of the rough endoplasmic reticulum and in the perinuclear cisterna. Golgi structures are also positive. Basophilic myelocytes (Fig. 7) do not contain endogenous peroxidase in any of the synthetic-secretory structures (3).

Mature Mast Cells. Mature human (Fig. 8), guinea pig (Fig. 9), and mouse (Fig. 10) mast cells (1,2) also share distinctive

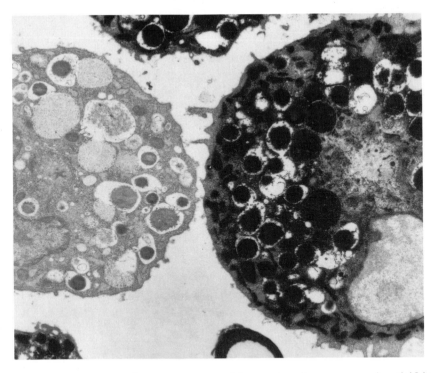

Figure 7. Endogenous peroxidase in human eosinophilic myelocyte (right). Basophilic myelocyte (left) is negative, 8500x.

morphologic features. These features are not that of mature or immature granulocytes. The common features seen include large size (14–20 microns), surface specialization of regularly placed short, narrow processes, numerous cytoplasmic granules and a single, eccentric nucleus. Like basophils, Golgi structures, and free and membrane-bound ribosomes are diminished. Unlike basophils, glycogen and numerous vesicles are absent. Lipid bodies and cytoplasmic filaments abound. Cytoplasmic granules are present in larger numbers than in basophils of the corresponding species. In man and guinea pig these granules vary remarkably in their shape and ultrastructural patterns; in mouse peritoneal mast cells, these granules are larger than granules in basophils, are round and contain extremely dense uniform material with no substructural patterns.

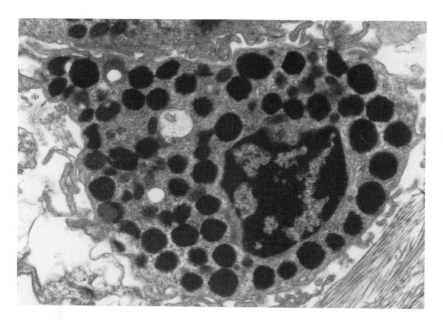

Figure 8.   Human lung mast cell in vivo, 14500x.

Mature human mast cells display a variety of substructural granule patterns (5). Granules can be filled with scrolls, crystals, particles, reticular patterns or combinations of these patterns. Individual cells can contain granules of only one such pattern or granules of all varieties. We have found that isolated human lung (5,6) and gut mast cells (7) predominantly contain granules of the scroll type and numerous lipid bodies, as observed in vivo (1). By contrast, human dermal mast cells observed in vivo generally display granules filled with crystals and no lipid bodies (1). In general, lipid bodies are increased in mast cells found in areas of disease and are less frequent in mast cells remote from areas of obvious disease. We feel that such lipid body variation reflects mast cell functional expression.

Like human mast cells, those of guinea pigs also have granules filled with scrolls, crystals or particles whereas those of mice are uniformly dense. That human mast cells can be completely filled with granules of the particle type is important to know, since this granule type is morphologically identical to that of most human basophil granules. Therefore, morphologic rules other than granule content alone must be applied for the absolute identification of human mast cells and basophils (3).

Immature Mast Cells. Mast cell morphologic maturational sequences are just now becoming clear and more information is currently available for mouse mast cells (1) than for human mast

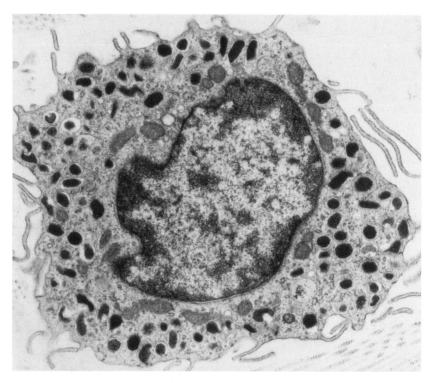

Figure 9. Guinea pig skin mast cell in vivo, 8500x (from A.M. Dvorak et al. (2).

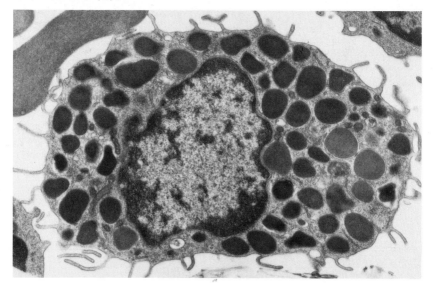

Figure 10. Mouse peritoneal mast cell in vivo, 12000x.

cells; none is yet available for mast cells of guinea pigs.
Clones and mast cell lines have recently been derived from mouse
fetal liver, spleen and bone marrow (1).   These mast cells   are
immature and provide an opportunity to evaluate maturational
events induced by butyrate (1).   Immature mast cells are cells
with a single oval or lobular nucleus (Fig. 11). Their surface
specialization includes   regularly-spaced narrow processes,   the
cytoplasm is filled with   large granule chambers, and while many
of the immature granules are completely empty, others are filled
with vesicles,   and some   display a mixture of very dense progranules and vesicles.   Individual progranules are packaged in
active Golgi zones prior to being added to granule contents.
Later maturational events include the condensation and homogenization of granule contents such that granules are completely
filled with dense homogeneous material (Fig. 12).   These mature
granules are smaller than immature granules.

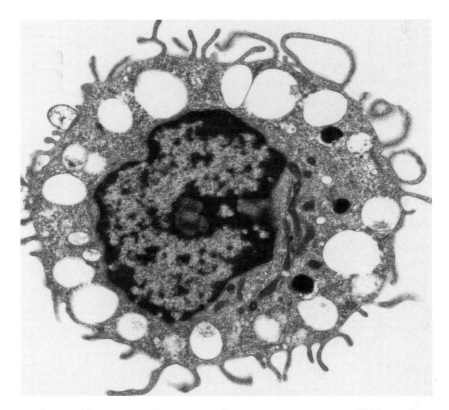

Figure 11.   Immature mouse bone marrow mast   cell in vitro,
13000x.

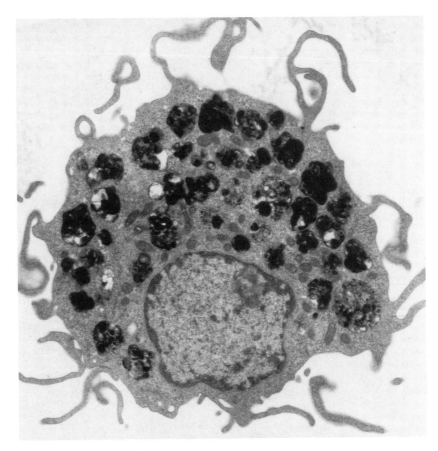

Figure 12. Butyrate-induced maturation of immature mouse bone marrow mast cells in vivo, 11000x.

Morphologic Expressions of Function which can Influence the
Identification of Basophils and Mast Cells

Slow Release Reactions. Human basophils (Fig. 13) and mast cells (Fig. 14) can release granule matrix materials slowly (days). The morphologic expression of this process is the progressive loss of granule materials which are transported by vesicles to the cell's surface. This results in the existence of cells with varying numbers of empty granule chambers, partially-filled granules and completely filled granules. Cells displaying this piecemeal degranulation are readily found in vivo in a wide variety of neoplastic and inflammatory processes (1).

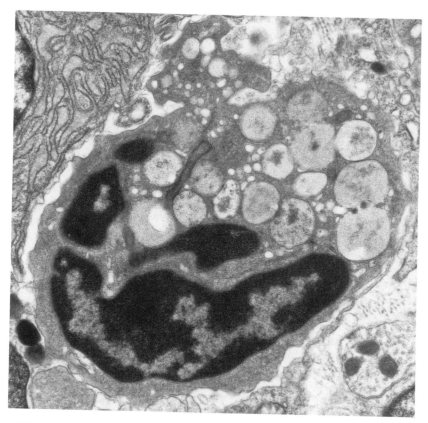

Figure 13. Human basophil in vivo (ileum) shows piecemeal degranulation, 17000x.

Rapid Release Reactions.   Human basophils undergo rapid extrusion of their particle–filled cytoplasmic granules singly through multiple openings in the plasma membrane (Fig. 15) when appropriately stimulated to do so (1,8). Human mast cells (Fig. 16), by contrast, when similarly stimulated undergo complex intracytoplasmic events which result in the formation of numerous degranulation channels and solubilization of granule matrix materials prior to the development of multiple openings to the exterior of these degranulation channels (1,5,6,8).
    Guinea pig basophils (Fig. 17), when appropriately stimulated, release all of their granules into a centrally–enlarging degranulation sac. This sac then opens to the exterior through a single pore and all granules are then extruded through this opening. These completely degranulated basophils can then rebuild new granules (1).

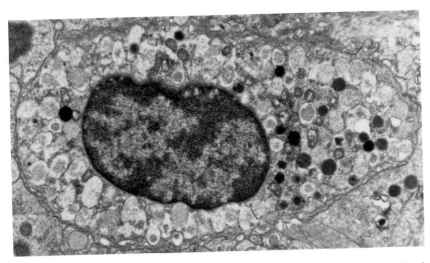

Figure 14. Human mast cell <u>in vivo</u> (ileum) shows morphologic expression of slow release of granule materials, 8500x.

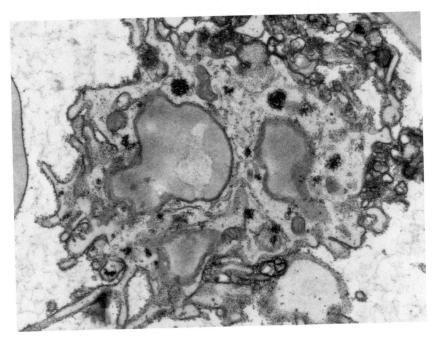

Figure 15. Human peripheral blood basophil undergoes ana-phylactic degranulation in response to histamine-releasing activity <u>in vitro</u>, 19000x.

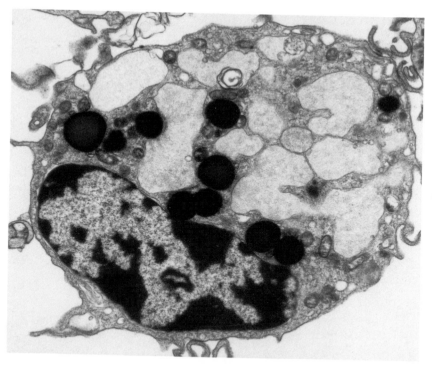

Figure 16.   Human lung mast cell undergoes anaphylactic degranulation in response to anti-IgE in vitro, 14500x.

Immature mouse mast cells (Fig. 18), when stimulated, extrude granule matrix materials (dense progranules and clear vesicles) through multiple membrane openings to the cells' exterior.

These rapid release events in three species have morphologic similarities and differences. The similarities include the following: (1) multiple membrane openings (human basophils and mast cells, immature mouse mast cells); (2) formation of new intracellular spaces (channels in human mast cells, sacs in guinea pig basophils; (3) extrusion of recognizable granule materials (human and guinea pig basophils, immature mouse mast cells). The differences include the following: (1) intracytoplasmic (channel) solubilization of granules prior to release in guinea pig basophils. Knowledge of these morphologic expressions of function may be useful to further substantiate the identity of putative basophils or mast cells.

Lipid Body Formation.   Lipid bodies are non-membrane-bound spherical organelles found in many types of cells. They increase in mast cells (Fig. 19) and basophils of humans when these cells

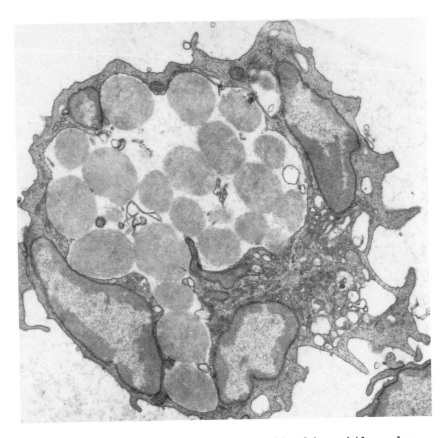

Figure 17.   Guinea pig peripheral blood basophil undergoes anaphylactic degranulation in response to Concanavalin A, 12500x.

are found in a wide variety of disease processes. For this reason we consider that their presence in increased numbers may represent a morphological expression of cellular activation. Recently, using an ultrastructural autoradiographic approach, we have implicated these previously neglected organelles in the arachidonic acid (AA) metabolic pathways. In addition to label- ling these structures with ^{3}H AA in human mast cells (5,9), we have labelled them in macrophages of humans, guinea pigs and mice (9), as well as in eosinophils (10) and neutrophils (11) of man.

Endocytosis. Guinea pig basophils have the capacity to internalize exogenous horseradish peroxidase (HRP) tracer in smooth pinocytotic vesicles which then fuse with granules and

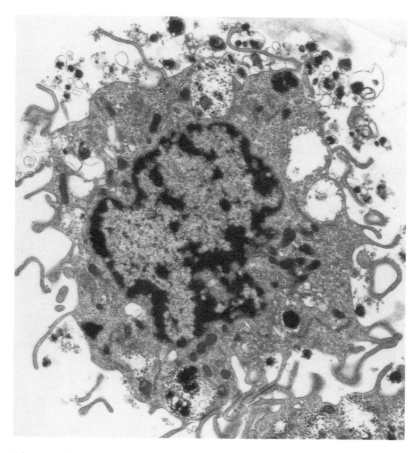

<u>Figure 18</u>. Immature mouse bone marrow mast cell undergoes anaphylactic degranulation <u>in vitro</u>, 12000x.

release HRP into them. Such vesicular transport and granule storage has been demonstrated <u>in vivo</u> and <u>in vitro</u> (1). Moreover, this granule-bound HRP can be released from granules and transported by vesicles to be released from the cell (1). We recently have shown that guinea pig basophils can also internalize and store eosinophil peroxidase in their granules by utilizing a similar uptake mechanism (12). In this event, EPO is tightly bound to basophil granules and washout experiments failed to release such bound EPO.

Human basophils which develop in basophil-growth factor-containing cultures also internalize eosinophil peroxidase by

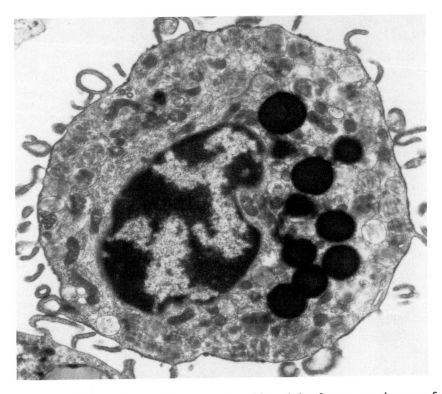

Figure 19.   Human   lung mast cell with large numbers of cytoplasmic lipid bodies, 15000x.

vesicles and   store it   in individual granules   in the   cytoplasm (3). This potential for sequestration of eosinophil peroxidase by human basophils may be the mechanism responsible for reports of eosinophil products in human basophils (13) and may serve to eliminate potentially toxic products from inflammatory reactions rich in eosinophils.

Immature mouse mast cells also internalize eosinophil peroxidase by an endocytic system of vesicles, vacuoles and tubules and sequester eosinophil peroxidase in immature granules (12).

CONCLUDING REMARKS

Morphologic expressions of maturation and function can affect the ability to identify mast cells and basophils in man, guinea pig, and mice. The accurate identification of mast cells and basophils will depend on application of morphologic guidelines which include these altered morphologic expressions. These data, in concert with various biochemical, functional, and immunological data, will provide the best possible scheme for the classification of these cells in vivo and in new in vitro experimental models. This multifaceted approach should allow more rapid evaluation of the role(s) of these cells in health and disease.

REFERENCES

1. **Galli, S.J., A.M. Dvorak, and H.F. Dvorak.** 1984. Basophils and mast cells: Morphologic insights into their biology, secretory patterns and function. In: Progress in Allergy: Mast Cell Activation and Mediator Release, Vol. 34. Edited by K. Ishizaka. Basel, Switzerland, S. Karger, p. 1.

2. **Dvorak, A.M., H.F. Dvorak, and S.J. Galli.** 1983. Ultrastructural criteria for identification of mast cells and basophils in humans, guinea pigs and mice. Am. Rev. Resp. Dis. (Suppl): Comparative Biology of the Lung 128:549.

3. **Dvorak, A.M., T. Ishizaka, and S.J. Galli.** 1985. Ultrastructure of human basophils developing in vitro. Evidence for the acquisition of peroxidase by basophils, and for different effects of human and murine growth factors on human basophil and eosinophil maturation. Lab. Invest. 53:57.

4. **Dvorak, A.M., and R.A. Monahan.** 1985. Guinea pig bone marrow basophilopoiesis. J. Exp. Pathol. 2:(in press).

5. **Dvorak, A.M., I. Hammel, E.S. Schulman, S.P. Peters, D.W.Jr. MacGlashan, R.P. Schleimer, N.H. Newball, K. Pyne, H.F. Dvorak, L.M. Lichtenstein, and S.J. Galli.** 1984. Differences in the behavior of cytoplasmic granules and lipid bodies during human lung mast cell degranulation. J. Cell. Biol. 99:1678.

6. **Dvorak, A.M., E.S. Schulman, S.P. Peters, D.W.Jr. MacGlashan, H.H. Newball, R.P. Schleimer, and L.M. Lichtenstein.** 1985. Immunoglobulin—E—mediated degranulation of isolated human lung mast cells. Lab. Invest. 53:(in press).

7. **Fox, C.C., A.M. Dvorak, S.P. Peters, A. Kagey—Sobotka, and L.M. Lichtenstein.** 1985. Isolation and characterisation of human intestinal mucosal mast cells. J. Immunol. 135:483.

8. **Dvorak, A.M., S.J. Galli, E.S. Schulman, L.M. Lichtenstein, and H.F. Dvorak.** 1983. Basophil and mast cell degranulation: Ultrastructural analysis of mechanisms of mediator release. Fed. Proc. 42:2510.

9. **Dvorak, A.M., H.F. Dvorak, S.P. Peters, E.S. Schulman, D.W.Jr. MacGlashan, K. Pyne, V.S. Harvey, S.J. Galli, and L.M. Lichtenstein.** 1984. Lipid bodies: Cytoplasmic

organelles important to arachidonate metabolism in macro-phages and mast cells. J. Immunol. 132:1586.

10. **Weller, P., and A.M. Dvorak.** 1985. Arachidonic acid incor-poration by cytoplasmic lipid bodies of human eosinophils. Blood 65:1269.

11. **Galli, S.J., A.M. Dvorak, S.J. Peters, E.S. Schulman, D.W.Jr. MacGlashan, T. Isomura, K. Pyne, V.S. Harvey, I. Hammel, L.M. Lichtenstein, and H.F. Dvorak.** 1985. Lipid bodies: Widely distributed cytoplasmic structures that represent preferential non-membrane repositories of exogenous ^{3}H-arachidonic acid incorporated by mast cells, macrophages and other cell types. In: Prostaglandins, Leukotrienes and Lipoxyns: Biochemistry, Mechanisms of Action and Clinical Applications. Edited by Bailey. New York, Plenum Press (in press).

12. **Dvorak, A.M., S.J. Klebanoff, W.R. Henderson, R.A. Monahan, K. Pyne, and S.J. Galli.** 1985. Vesicular uptake of eosinophil peroxidase by guinea pig basophils and by cloned mouse mast cells and granule-containing lymphoid cells. Am. J. Pathol. 118:425.

13. **Ackerman, S.J., G.J. Weil, and G.J. Gleich.** 1982. Formation of Charcot-Leydon crystals by human basophils. J. Exp. Med. 155:1597.

Mast Cell Differentiation and Heterogeneity,
edited by A. D. Befus et al.
Raven Press, New York © 1986.

Fibroblasts Are Required for Mast Cell Granule Synthesis

S. Davidson, A. Kinarty, *R. Coleman, A. Reshef, and H. Ginsburg

*The Rappaport Family Institute for Research in Medical Science, Departments of Immunology and *Biological Structure, Faculty of Medicine, Technion, Israel Institute of Technology, Haifa, Israel*

Interest in mucosal mast cells (MMC) of various types has grown considerably in the past few years. These cells extensively proliferate in helminth infection.

Clonal growth of MMC in cultures of lymphoid cells was first demonstrated by Ginsburg (1,2) and by Ginsburg and Lagunoff (3). Subsequently, it was shown that this growth was the outcome of a factor released from antigen-sensitive or mitogen-stimulated T-cells (4-7). Reports of many studies on the in vitro cultivation of mast cells disclose an immature form with a small number of granules which appear structurally incomplete and with low levels of histamine content, usually in the range of 0.1-0.7 $\mu g/10^6$ cells. This immaturity state has made the cultures unsuitable for studying degranulation. By growing mast cells on embryonic skin fibroblast monolayers, average values of histamine between 3-7 $\mu g/10^6$ cells were obtained. A 54 day old culture initiated from mesenteric lymph nodes of Schistosoma mansoni infective mice yielded values of as high as 18 μg histamine per 10^6 mast cells (Table V). These mast cells, fully packed with granules (Fig. 1), were totally degranulated by monoclonal IgE molecules and the specific antigen releasing 95% of the histamine. Regeneration of the mast cells following total degranulation was demonstrated; by 9-11 h the histamine content was replensiehd (8,9). A study on the interation of two monoclonal IgE specificities, the receptors and the antigens in relation to degranulation will be reported elsewhere (Rofolovitz and Ginsburg, in preparation).

When mast cells are allowed to develop in cultures in the absence of fibroblasts they appear vacuolated. They are devoid of metachromatic granules and lack histamine (10). However, when these cells (termed vacuolated mast cells) are plated on fibroblasts, both granule and histamine synthesis is triggered.

Supported by NIH grant A116255 and by a grant from the Israel National Council for Research and Development.

MATERIALS AND METHODS AND RESULTS

Cultures of Vacuolated Mast Cells. Populations of vacuolated mast cells were obtained by the following procedures: (i) culture of mesenteric lymph node cells from horse serum injected mice (10) or infection of mice with 50–300 cercaria of Schistosoma mansoni and (ii) culture of the mesenteric lymph node cells 40–50 days later in the presence of 10 μg protein of the cercaria antigen (10,11). Four to 8d medium from these cultures can be shown to contain the growth factor designated as "mast cell stimulating factor" (MSF). This preparation is particularly valuable since it selectively stimulates colonies of mast cells in cultures of lymph node cells of nude mice. In more recent experiments, MSF or mouse TCGF (supernatant from Con A – spleen culture) were added to cultures prepared from both immunized or untreated normal or congenic nude mice. The combination of both immunization and addition of either growth factor was a substantial improvement in terms of mast cell number, maturation and survival of the mast cell population. In cultures of lymph node cells form nude mice and MSF in methylcellulose semi-solid medium, the mast cells that developed in the colonies were entirely devoid of granules. No other cell types were present in these cultures. Populations of vacuolated mast cells were obtained after 7 to 13 days incubation in test tubes of 10^7 to 2×10^7 mesenteric lymph node cells in 2 ml medium containing 15% heat inactivated horse serum and 50% MSF or 4% to 10% mouse TCGF (10,11). Mast cells were morphologically evaluated and counted in cytocentrifuge prints after staining with toluidine blue. Both vacuolated and granulated mast cells were easily distinguished from macrophages or from fibroblasts after staining for acid phosphatase or May–Grunwald–Giemsa, respectively (Fig. 1).

An electron micrograph of the commonest cell type seen in cultures of vacuolated mast cells is represented in Figure 2a; 5 days after plating, vacuolated mast cells appear on fibroblast monolayers (Figure 2b).

Attempts to Induce Granule Synthesis with Products of Fibroblasts. The following preparations were added to 6 day old cultures of vacuolated mast cells: (a) supernatants (neat or concentrated) from a variety of fibroblast cultures; (b) extracts of fibroblasts; (c) plating on fibroblasts fixed with glutaraldehyde or methanol (data not shown); (d) extracellular matrix of fibroblast monolayers prepared as described by Voldavski et al. (12) or of bovine endothelial cells provided by Dr. Voldavski. None of these preparations triggered granule synthesis (Table I), the mast cells in these cultures remained totally vacuolated, and no histamine was detected.

Separation of Vacuolated Mast Cells and Fibroblasts by a Porous Membrane. Since only viable fibroblasts triggered granule synthesis in vacuolated mast cells, experiments were carried out

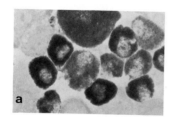
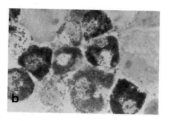
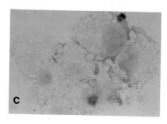

Figure 1. Cytocentrifuge prints of mast cell cultures. Mesenteric lymph node cells from Schistosoma mansoni-infected BALB/c mice. (Toluidine blue x 1350).

a) Same as culture shown in Table V, 54 days old. MSF was omitted at the 42nd day. The culture contained 0.525×10^6 mast cells and 18.66 μg histamine per 10^6 mast cells.

b) Same as (a), but maintained with MSF from the 25th till the 54th day. The culture contained 1.525×10^6 mast cells and 11.6 μg histamine per 10^6 mast cells (Table V).

c) Vacuolated mast cells: 13-day old culture maintained without fibroblasts.

to test whether contact between the vacuolated mast cells and the fibroblasts is necessry for granular maturation. A Marbrook culture apparatus was constructed following the model of Eipert et al. (13), consisting of an assembly of two compartments separated by a membrane filter (Gelman Instrument Co.) of 0.45 μm pore size and 130 μm thickness. Wells of 16mm diameter were placed in a plate of 24 wells (Costar, Cat. No. 3524) into which glass tubes (8mm outer diameter) were inserted. Into each of the 16mm wells, 10^5 fibroblasts were plated while 0.75×10^6 vacuolated mast cells were plated into both compartments. Results are shown in Table II. In the glass tubes the mast cells remained vacuolated, while in the cells adherent to the monolayer synthesized granular content. Of the free floating mast cells, only a fraction (15-25%) were granulated. In a separate study electron micrographs were made from large numbers of mast cells adhered to fibroblasts in order to find any type membrane junction between the two cell types; none was discernible.

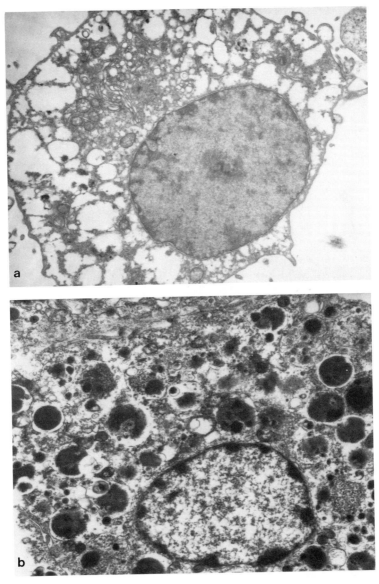

Figure 2.   Electron micrographs of mast cells in culture.

   a) Vacuolated mast cell from 14-day-old culture of mesenteric
lymph node cells of BALB/c mice infected with S. mansoni. 15 x
10⁶ cells in 2 ml plus 10 μg protein of cercaria antigen were
maintained in test tube. Pellet was fixed with 0.5% glutaraldehyde in cacodylate buffer.   Note well developed Golgi zone and
membrane-limited vacuoles. (x9,000)
   b) Same as (a), 5 days after plating a fibroblast monolayer
(18 days in culture). (x8200).

TABLE I. Growth of vacuolated mast cells in the presence of culture supernates[a] and extracts of fibroblasts

Factor and % added	No. Vacuolated mast cells ($\times 10^6$/plate)	No. granulated mast cells ($\times 10^6$/plate)	Total cell harvest ($\times 10^6$/plate)	Histamine ug/plate
Supernate from fibro-blasts 10% (v/v)	0.153	0	0.60 ± 0.07	0.02
Supernate from fibro-blasts 50% (v/v)	0.121	0	0.65 ± 0.12	0.05
Supernate[b] from fibro-blasts 50% (v/v) (concentrated X50)	0.375	0	0.49 ± 0.01	ND
Extracts prepared from fibroblasts 50% (v/v)	0.143	0	0.84 ± 0.04	0.06
Monolayer fixed with glutaraldehyde 1% (v/v)	0.153	0	0.92 ± 0.07	0.02
No factor added	0.161	0	1.01 ± 0.07	0.02
Plating on fibroblast monolayer	0	0.338	1.74 ± 0.18	2.00

a. Balb/c mice were infected with 300 cercariae of S. mansoni 2 months before sacrifice. 10^7 mesenteric lymph node cells in 2 ml medium plus 10 ug protein of cercaria antigen were grown in culture tubes for 6 days, then the preparation to be tested was added and incubated for further 5 days. Extracts were prepared from fibroblasts by triple freezing and thawing of monolayers containing 0.5×10^6 cells per plate. Supernatant from fibroblast monolayers was collected 48 hrs after adding fresh medium. The vacuolated mast cells were also plated on the same monolayers.
b. Supernatant from fibroblasts was concentrated 1×50 by dehydration.

TABLE II. Required Contact Between Vacuolated Mast Cells and Fibroblasts for Granule Synthesis[b]

Cells	Days in Marbrook's Culture	Compartment					
		with fibroblasts				without fibroblasts	
		Non-adherent fraction		Adherent fraction			
		No. per well	%	No. per well	%	No. per well	%
Vacuolated mast cells		20839	55.57	0	0	77600	80.00
Granulated mast cells	8	5775	15.40	105875	14.75	0	0
Total cell harvest		37500	–	726666	–	97000	–
Vacuolated mast cells		14189	94.59	0	0	43906	85.00
Granulated mast cells	11	811	5.41	117361	11.93	0	0
Total cell harvest		15000	–	983333	–	51666[b]	–
Vacuolated mast cells		22500	75.00	0	0	51995	86.66
Granulated mast cells	13	7500	25.00	113812	15.38	0	
Total cell harvest		30000	–	740000	–	60000[b]	

a. Balb/c mice were immunized by two intraperitoneal injections, one week apart, of 0.2 ml horse serum. Five days later 15×10^6 mesenteric lymph node cells per tube were plated in 1 ml medium + 15% horse serum and 4% (v/v) of mouse TCGF (ConA-conditioned T-cell growth factor). Eleven days later, the cell suspension was harvested and distributed into two compartments in fibroblast Marbrook's culture in 17 mm wells, each containing 0.75×10^6 cells. At intervals adherent cells were recovered by trypsin treatment.

b. Beside mast cells, lymphocytes, macrophages and unidentified cells were also present.

TABLE III. Incubation of Vacuolated Mast Cells[a] with Varying Numbers of Fibroblasts

Time after plating on fibroblasts (days)	No. fibroblasts plated[b] (x 10^6)	Histamine (ug/plate)	No. of granulated mast cells (x 10^6)	%	No. of intermediate mast cells[c] (x 10^6)	%	Total no. of mast cells[e] (x 10^6)	% of mast cells in culture	Total cell harvest (x 10^6)
5	0	0	0	0	0	0	0.920	33	2.83 ± 0.12
	0.062	0.40	0.030	3	0.039	3	0.950	46	2.06 ± 0.29
	0.125	0.48	0.170	23	0.167	23	0.732	46	1.59 ± 0.21
	0.250	0.54	0.293	27	0.286	27	1.107	64	1.69 ± 0.30
	0.500	0.83	0.479	83	0.251	43	0.578	46	1.26 ± 0.20
	1.000	1.67	0.368	85	0.099	23	0.430	28	1.56 ± 0.07
	2.000	1.66	0.560	97	0.136	24	0.576	16	3.71 ± 0.34
8[d+]	0.062	0.47	0.304	36	0.304	36	0.840	59	1.43 ± 0.21
	0.125	0.46	0.314	75	0.217	52	0.420	42	1.01 ± 0.41
	0.250	–	0.524	94	0.209	37	0.560	44	1.78 ± 0.28
	0.500	0.71	0.499	90	0.196	36	0.551	29	1.43 ± 0.28
	1.000	1.70	0.666	99	0.051	8	0.676	32	2.10 ± 0.68
	2.000	0.84	0.445	100	0.021	5	0.445	20	2.25 ± 0.21
14	0.05	0.165	0.333	77	0.333	76	0.441	46	0.95 ± 0.08
	0.10	0.590	0.182	98	0.149	80	0.186	40	0.47 ± 0.12
	0.50	1.550	0.893	100	0.139	16	0.893	47	1.92 ± 0.08
	1.00	1.750	1.270	100	0.118	9	1.270	46	2.78 ± 0.34

a. Culture of vacuolated mast cells was established by incubating 1.5 x 10^7 mesenteric lymph node cells of Balb/c mice in test tubes in 2 ml medium containing 10% mouse TCGF. 11 days later the developed population of vacuolated mast cells was diluted twice and plated on the fibroblast monolayers.
b. 3 days after the fibroblasts were seeded they were x-irradiated with 2000 rads before mast cell plating.
c. From the 8th day onward the cultures with fibroblasts were in well advanced degenerate state.
d. Represents the lightly granulated immature mast cell type in the granulated mast cell compartment.
e. Vacuolated and granulated mast cells.

There is a clear correlation with the fibroblast concentration in terms of a number of mast cells containing granules and, as well, in the degree of maturity. Thus, after 5 and 8 days the percentage of granulated mast cells is directly proportional to fibroblast concentration. At the 14th day, almost all the mast cells developed granules; however, at low fibroblast concentrations most of the mast cells were of the intermediate, immature type.

Growth of Vacuolated Mast Cells on Varying Numbers of Fibroblasts. Five to 14 days after vacuolated mast cells were plated on fibroblasts they were harvested, free floating cells were spun down, and adherent cells were harvested by trypsin treatment. The cell suspension thus obtained was combined with the cell pellet; half was frozen for determination of cellular histamine content, while in the other half the cells were counted and cytocentrifuge preparations were made. Differential counts were carried out after staining with toluidine blue, acid phosphatase and May–Grunwald–Giemsa. The mast cells were then classified arbitrarily and subjectively by degree of metachromasia. Three types were discernible: (i) vacuolated mast cells and two kinds of granulated mast cells; (ii) an intermediate type, which was only slightly granulated; (iii) a more mature type (10). Results are summarized in Table III.

TABLE IV. Growth of Vacuolated Mast Cells on Peritoneal Exudate Cells

MONOLAYER	No. vacuolated mast cells	%[b]	No. of granular mast cells	%	Total No. of mast cells	%[c]	No. of cells removed by trypsin
Peritoneal exudate 0.2 x 10[6] cells	26,250	12.50	76,293	74.40	102,543	48.83	210,000
Peritoneal exudate 1 x 10[6] cells	960	0.60	21,440	95.71	22,400	14	160,000
Peritoneal[a] macrophages after trypsin treatment 1 x 10[6] cells	58,752	100.00	0	0	58,752	34.56	170,000
Fibroblasts 1 x 10[6] cells	9,830	1.45	666,070	98.55	675,890	32.18	2,100,000

a. 12 day old peritoneal macrophage cultures were treated with trypsin to remove fibroblasts which grow readily from the peritoneal cavity. The vacuolated mast cells were then plated.
b. percentage from total mast cells.
c. percentage from total cells removed by trypsin (the macrophages resist trypsin).

Attempts to Trigger Granule Synthesis with Cell Line and Peritoneal Macrophages. The next question was whether the same synthesis-triggering effect could be produced by cell types other than fibroblasts. Accordingly, vacuolated mast cells were plated on peritoneal macrophages and on the fibrosarcoma A9 HT. In these cultures fibroblast populations also developed among the macrophages. When these cells were removed with trypsin, the weak triggering effect vanished altogether (Table IV). The established line did not trigger granule synthesis and the mast cells remained vacuolated.

Protocol of Long-Term Culture of Fully Mature Mast Cells. Follow-up of a selected culture serves to demonstrate the maturation and persistence of fully mature mast cells when both MSF and fibroblasts are combined. The culture was initiated from mesenteric lymph nodes of mice that had been infected with S. mansoni, in the absence of cercaria antigen. Addition of cercaria antigen markedly increases mast cell number (data not shown) in such cultures. On the 25th day, subcultures were prepared by making four new plates from each older plate; to half of each, MSF was added. Seventeen days later, on the 42nd day of culture, marked differences were noted. With MSF, macrophage growth was arrested, while on day 53 there was twice as much histamine in the plates that were replenished with MSF. On day 54, a three-fold increase of histamine is seen, with maximal 18.66 μg per 10[6] mast cells (Table V). Cultures maintained in the presence of MSF from day 25 *in vitro* contained three times more histamine than cultures from which MSF was omitted, in which macrophage numbers rose markedly. Mast cells of 54 day old cultures are shown in Figure 1: plates that were kept without MSF from the start of the culture became confluently populated with macrophages, as previously shown (14).

TABLE V. Proliferation and Maturation of Mast Cells and Macrophages in Cultures of Lymph Node Cells from <u>Schistosoma mansoni</u>-infected Mice:[a] Effect of MSF on Late Cultures

Time in culture (days)	Treatment	Time of treatment (days)	Histamine in cells (ug/plate)	Histamine in cells (ug/10^6 mast cells)	Mast cells (no/plate)	Mast cells (%)	Macrophages (no/plate)	Macrophages (%)	Total cells (no/plate x 10^6)
5	Plating on	3	0.01	-	-	-	-	-	-
8	fibroblast	3	0.015	-	-	-	-	-	-
12	monolayers	3	ND	-	19,000	1.96	-	-	0.95
	Passage, cultures diluted 4-fold:								
42	MSF + 10% added	25	4.40	6.33	695,000	42.00	18,683	1.19	1.57
42	Without MSF	25	0.75	9.75	76,900	8.01	242,000	25.25	0.96
53	MSF continues	42	6.55	-	-	-	-	-	-
53	MSF discontinued	42	3.35	-	-	-	-	-	-
54	Fresh medium (MSF continued)	53 (42)	17.75	11.60	1,525,000	64.17	1,960	0.04	2.40
54	Fresh medium (MSF discontinued)	53	9.80	18.66	535,000	36.94	301,000	21.5	1.42
54	Monolayer only	-	0	-	0	-	0	-	0.52

a. DBA/2 female mice, 5 months old, were injected with 50 cercaria organisms. Four months later, cultures were made from mesenteric lymph nodes. 10^7 cells per test were incubated for 3 days and then on skin fibroblast monolayers.

MSF = mast-cell-stimulating factor, medium of 4-day culture of mesenteric lymph nodes from <u>S. mansoni</u>-infected mice.
Note macrophage proliferation after MSF is omitted. Note increase in histamine after medium change.

DISCUSSION

The present work and the studies previously described (10,11) on both mast cells and granular NK-mucus secreting cells (15) show that an extraneous factor dependent on cell-to-cell contact is required for the synthesis of granular contents; the nature of this trigger has yet to be resolved. Observations made on vacuolated mast cells in vitro and in particular on granular NK cells in colonies that develop on non-triggering monolayers such as embryonic skin (11) suggest that all the components required for granule maturation are present (e.g. glycogen is stored in the cells), but a factor is required to initiate this maturation process.

The present work points to the requirement for fibroblasts for macromolecular packaging in mast cells. Peritoneal macrophages and an established fibrosarcoma cell line did not trigger granule synthesis. This does not rule out the possibility that other types of phagocytic cells, as well as other non-fibroblastoid cells are capable of triggering granule maturation in mast cells. It was difficult to exclude possible fibroblast growth that may originate from suspensions of vacuolated mast cells plated on any cell monolayer since fibroblasts are most frequently contaminant of spleen, bone marrow and peritoneal exudate cell suspensions. In cultures of bone marrow in methylcellulose, semi-solid medium in the presence of mouse TCGF and eythropoietin, fibroblasts are common components of many mixed colonies, which also were found to contain granulated mast cells. On the other hand, in semi-solid cultures of lymph node cells of nude mice in the presence of MSF, only colonies of mast cells developed without other hemopoietic lineages observed. Mast cells in such colonies are entirely devoid of granules (unpublished observations).

It thus appears that when mast cells are stimulated to grow, the amount of histamine and metachromasia may be a reflection of the degree of fibroblast and other cellular contamination in the culture.

In a previous work (16) we have shown that mast cells developed in cultures of lymph node cells from mice immunized with horse serum (horse serum is a constituent of the culture medium), and maintained on fibroblast monolayers, contained large amounts of sulfated proteoglycans. Sixty to 70% of this was heparin and 30%-40% chondroitin sulfate and dermatan sulfate. Since the monolayers were not irradiated (the control monolayers were entirely devoid of mast cells) (16), possible contamination of connective tissue mast cells could not be excluded. Several workers have shown both in vitro and in vivo that mucosal mast cells are distinct by containing chondroitin sulfate and lack heparin. Therefore, we decided to repeat our previous study using x-irradiated fibroblast monolayers. The heparin and chondroitin sulfate E determination was as performed by Razin et al. (17). Results of three different sets of cultures were identical to those reported by Bland et al. (16). Thus, it appears the fibroblasts trigger the synthesis of heparin in mast cells.

The cultures described in the present and the previous work (16) were maintained under conditions avoiding any stimuli for degranulation. Therefore, the mast cells, during the prolonged maintenance, could accumulate a large number of heparin-containing granules. In vivo, the possibility that a population of mucosal mast cells, if developed in the mucosa, is subjected to constant degranulation stimuli, must also be taken into account. In such a case the mast cells may appear immature and may lack heparin. This question can be resolved if we addition-ally study in vitro the kinetics of proteoglycan synthesis during multiple degranulation events.

REFERENCES

1. **Ginsburg, H.** 1963. The in vitro differentiation and culture of normal mast cells from the mouse thymus. Ann. N.Y. Acad. Sci. 103:20.

2. **Ginsburg, H., and L. Sachs.** 1963. Formation of pure suspension of mast cells in tissue culture by differentia-tion of lymphoid cells from mouse thymus. J. Natl. Cancer Inst. 31:1.

3. **Ginsburg, H., and D. Lagunoff.** 1967. The in vitro differ-entiation of mast cells. Culture of cells from immunized mouse lymph nodes and thoracic duct lymph on fibroblast monolayers. J. Cell. Biol. 35:685.

4. **Ginsburg, H., E.C. Olson, T. Huff, H. Okudaira, and T. Ishizaka.** 1981. Enhancement of mast cell differentiation in vitro by T cell factor(s). Int. Archs. Allergy Appl.

Immunol. 66:447.

5. Schrader, J.W., and I. Clark–Lewis. 1982. A T–cell–derived factor stimulating multipotential hemopoietic stem cells: Molecular weight and distinction from T–cell growth factor and T–cell–derived granulocyte-macrophage colony-stimulating factor. J. Immunol. 129:30.

6. Galli, S.J., A.M. Dvorak, J.A. Marcum, T. Ishizaka, G. Nabel, H. Der Simonian, K. Pyne, J.M. Goldin, R.D. Rosenberg, H. Cantor, and H.F. Dvorak. 1982. Mast cell clones: a model for the analysis of cellular maturation. J. Cell. Biol. 95:435.

7. Yung, Y.P., S.Y. Wang, and M.A.S. Moore. 1983. Characterization of mast cell precursors by physical means: dissociation from T–cells and T–cell precursors. J. Immunol. 130:2843.

8. Ginsburg, H., Y. Nir, I. Hammel, R. Eren, B.–A. Weissman, and Y. Naot. 1978. Differentiation and activity of mast cells following immunization in cultures of lymph node cells. Immunology 35:485.

9. Ginsburg, H., D. Ben–Shahar, I. Hammel, and Ben–Daid, E. 1979. Degranulation capacity of mast cells grown in cell culture in the presence of histamine releaser. Nature (London) 280;151.

10. Davidson, S., A. Mansur, R. Gallily, M. Smolarsky, and H. Ginsburg. 1983. Mast cell differentiation depends on T cells and granule synthesis on fibroblasts. Immunology 48:439.

11. Ginsburg, H., E. Ben–David, A. Kinarty, M. Rofolovitch, E. Chriqui, and S. Davidson. 1983. Murine interleukin–2 generates glycogen-rich mucus-secreting NK cells. Immunology 49:371.

12. Voldavski, I., G.M. Lui, and D. Gospodarowicz. 1980. Morphological appearance, growth behaviour and migratory activity of human tumor cells maintained on extracellular matrix versus plastic. Cell 19:607.

13. Eipert, E.F., L. Adorini, and J. Couderc. 1978. A miniaturized in vitro diffusion culture system. J. Immunol. Meth. 22:283.

14. Ginsburg, H., D. Ben–Shahar, and E. Ben–David. 1982. Mast cell growth on fibroblast monolayers. Two cell entities. Immunology 45:371.

15. Ginsburg, H., T. Yehuda–Cohen, R. Colemen, Z. Lapidot, Y. Hecht, and A. Kinarty. 1985. Granular NK-cells develop into mucus secreting cells. Immunol. Lett. (in press).

16. Bland, E.C., H. Ginsburg, E. Silbert, and D.D. Metcalfe. 1982. Mouse heparin proteoglycan: synthesis by mast cell fibroblast monolayers during lymphocyte-dependent mast cell proliferation. J. Biol. Chem. 257:8661.

17. Razin, E., J.N. Ihle, D. Seldin, J.–M. Mencia–Huerta, H.R. Katz, P.A. Leblanc, A. Heinz, J.P. Caulfield, K.F. Austen, and R.L. Stevens. 1984. Interleukin 3: a differentiation and growth factor for the mouse mast cell that contains chondroitin sulfate E proteoglycan. J. Immunol. 132:1479.

Mast Cell Differentiation and Heterogeneity,
edited by A. D. Befus et al.
Raven Press, New York © 1986.

Factors Influencing the Differentiation of Murine Intestinal Mast Cells

*D. Guy-Grand and **P. Vassalli

*INSERM U. 132, Groupe d'Immunologie et de Rhumathologie Pediatriques, Hopital Necker-Enfants Malades, 75730 Paris, France; and **Department of Pathology, University of Geneva, 1211 Geneva 4, Switzerland

In normal mice, intestinal lamina propria mast cells (IMC), which require the use of special fixatives to be detected on histological sections (1), are nearly absent. In contrast, they are observed in large number in the intestinal lamina propria of nematode-infested mice; however, when nude mice are infested, their intestinal lamina propria lacks mast cells (1). This strongly suggests that IMC differentiate under the influence of factors released by antigenically stimulated T-cells. Indeed, cultured mouse mast cell lines, which look like IMC by their cytological and biochemical properties (2) can differentiate in culture under the influence of products released by stimulated T lymphocytes or by cultured myelomonocytic leukemic cells (WEHI-3B). The factor responsible for this differentiation may correspond to a multispecific CSF (3) and has received different names, among them interleukin 3 (IL-3) (4).

In the present work, we studied, with the help of several experimental models (engraftment of WEHI tumor cells, Nippostrongylus brasiliensis (Nb) infestation, or graft-versus-host reactions (GVHR)), conditions for differentiation of IMC or related mast cells, including the role of IL-3 in epithelial lesions of the intestinal mucosa which accompany IMC accumulation in vivo.

Abbreviations used in this paper: BM, bone marrow; CSF, colony stimulating factor; GVHR, graft versus host reaction; IL-3, interleukin-3; MLN, mesenteric lymph nodes; IMC, intestinal lamina propria mast cells; IMC-P, intestinal lamina propria mast cell precursors; IE, intestinal epithelial; Nb, Nippostrongylus brasiliensis; PLN, popliteal lymph nodes; PP, Peyer's patches; TD, thoracic duct, ^{3}H-TdR, ^{3}H-thymidine.

MATERIALS AND METHODS

Animals and Experimental Models.   Nb infestation in mice was studied as previously described (5). WBB6 F_1 strain mice, W/W^v, were purchased from Jackson Laboratories (Bar Harbor, Maine). GVHR were studied in F_1 hybrid C3H-DBA mice with C3H lymph node lymphocytes, using either newborn mice, irradiated adult mice (900 rad) injected with 10^7 lymphocytes, or non-irradiated adult mice injected twice with 5 x 10^7 lymphocytes as previously described (6).

Preparation of Cell Suspensions, Treatment of Cells and Tissues, Culture and Studies of Supernatant Activities.   Lympho-cytes were depleted of Lyt 2^+ or L3T4$^+$ cells as previously described (5) (GK 1.5 anti-L3T4 hybridoma was a kind gift of Dr. F. Fitch, University of Chicago, USA).   IL-3 was detected by a proliferation assay using either bone marrow (BM) mast cells grown in the presence of WEHI-3B supernatant, or an IL-3-dependent mast cell line (kind gift of T.M. Dexter, Manchester, England).

Autoradiographs of Intestinal Sections.   Mice received 1 μCi per gram of body weight of ^{3}H-thymidine (^{3}H-TdR) (CEA Saclay, France) 26 h before sacrifice and intestine (at 5 cm from the pylorus) was oriented on filter paper and fixed in Carnoy's fluid.   Sections were dipped in K5 emulsion (Ilford, Essex, England), exposed for 1 to 3 months, and stained by methyl green pyronin.

RESULTS

Intestinal Lamina Propria Mast Cell Precursors (IMC-P): Frequency in the Gut of Normal, Nude, and W/W^v Mice and Their Traffic. As we have previously shown (5), while IMC are extremely rare in normal mice, IMC-P are, in contrast, present in high number.   In vitro studies, with mast cells obtained from different organs in the presence of IL-3 under conditions of limiting dilution (Table I), revealed that the frequency of IMC-P in lymphoid cell populations from intestinal mucosa was close to or even higher than the frequency of BM mast cell precursors, while frequencies of presumably related mast cell precursors were very low in other lymphoid organs and in the thoracic duct (TD) lymph.   IMC-P occur with similar frequencies in normal adult mice, newborn mice, germ-free mice and in nude mice; in contrast, their frequency is very low in the intestine of newborn and adult W/W^v mice which have a normal frequency of mast cell precursors in BM.

TABLE I. IMC-P and related mast cell precursor frequencies[a] in various organs of mice

	BM	Spleen	Intestinal Mucosa	PP	MLN	TDL	PLN
Normal C57BL/6	900	34	700	25[b]	6	3	<1.5
Newborn C57BL/6			1500				
Nude	450	50	1700				
Germ-free	800	140	2000				
Nb-infested C57BL/6:							
d 9-10	900	350	10500	65[b,c]	120[c]	330[c]	<1.5
d 15		80	1500		20	12	
Normal W/W^v	2000	33	40				
Newborn W/W^v			120				
Nb-infested W/W^v:							
d 10	105		15		26		
Normal Balb/c	800	70	3000		6		<1.5
Balb/c with:							
WEHI i.p.	5000	1250	5000				<1.5
WEHI s.c.					36		

a Average of two to six experiments. Number of precursors per 10^6 cells.

b Always contaminated by IMC-P present in the adjacent epithelium.

c On day 9, the total cell number obtained from MLN and from 25h TD drainage showed a two-fold increase, while PP showed a two-fold cell decrease.

d Normal intestinal cells comprise LP cells and IE cells from the crypts. Pure IE cells from villous epithelium contain about 100 IMC-P per 10^6 cells; IMC-P frequency in villous IE cells cannot be tested in Nb-infested mice because under these conditions IE cell recovery from the villi is made difficult and poorly reproducible due to the thick layer of mucus covering the gut.

IMC-P appear to be the progeny of BM mast cell precursors and do not arise in the intestine, since undifferentiated hematopoietic stem cells (CFU-S) are absent from the intestinal mucosa.

IMC–P, coming from the BM, appear to home directly from the blood to the intestinal mucosa rather than passing via TD lymph–like T-cells (7) (since TD cannulation for 2 days does not decrease the frequency of IMC–P) or via the spleen (since neither splenectomy nor selective irradiation of the spleen for 5 days with ^{32}P polyvinylchloride modified their frequency). It appears that circulating IMC–P are attracted from the blood to the intestinal mucosa by local factors independent of antigenic stimulation or T lymphocytes. Otherwise, their frequency in the intestine would not be normal in newborn, germ–free and nude mice.

IMC and Related Mast Cells Proliferate and Differentiate Under the Influence of IL–3. In several pathological conditions, it can be shown that proliferation and differentiation of IMC–P is under the influence of IL–3. The evidence for this is three–fold:

1) Mice bearing a WEHI–3B tumor: In adult Balb/c mice bearing a WEHI–3B tumor, the blood contains detectable levels of IL–3 and the frequency of IMC–P and related mast cells is increased 6– to 20–fold in BM, spleen, mesenteric lymph nodes (MLN) and intestinal mucosa. In addition, mature mast cells appear under these conditions in BM, spleen red pulp and intestinal mucosa. IMC are located intraepithelially within the crypts and at the bottom of villi; no IMC are observed in Peyer's patches (PP). There are no mast cells in any of the other mucosae nor in the peripheral lymph nodes (PLN). It is interesting to note that the number of dermal mast cells is also markedly increased in WEHI–3B tumor–bearing mice.

2) Nb mice: In Nb–infested mice, the blood does not contain detectable levels of IL–3 and mast cell precursor frequency is normal in BM; but, as long as the nematode is present in the gut (day 8 to 10 after infestation), it can be shown that local IL–3 release takes place in gut–associated lymphoid tissue (including MLN, TD lymph and cells isolated from the mucosa). Indeed, lymphocytes from these sources can be induced to secrete IL–3 under the influence of specific nematode antigen (Table II).

The IL–3 secreting lymphocytes in Nb–infested mice are of the Thy 1$^+$, L3T4$^+$ phenotype (as demonstrated by velocity sedimentation at unit gravity). IL–3 secreting lymphocytes are, like intestinal T–lymphocytes observed in other circumstances (7), the progeny of immature blasts arising in antigenically stimulated PP and reaching the gut mucosa through a traffic cycle via the TD lymph and the blood. Indeed, 48 h cannulation of the TD from day 8 to 10 after infestation, interrupting T-cell traffic, leads to a 10-fold decrease in the capacity of intestinal mucosal T-cells to secrete IL–3 under in vitro Nb antigenic stimulation. These specific, IL–3-releasing lymphocytes are short–lived, disappearing shortly after cessation of antigenic stimulation in vivo following worm expulsion (around day 15).

TABLE II.   <u>Lymphocyte release of IL-3 under the influence of antigen</u>[a]

Mouse strain and condition	Spleen	Gut Mucosa	PP	MLN	TDL	PLN
<u>N. brasiliensis-infested</u> <u>C57BL/6</u>						
d 8 to 10	80	143	0	100	157	0
<u>N. brasiliensis-infested</u> <u>W/W^v</u>						
d 10	23	50		51		
C3H DBA with GVHR						
total gut lymphocytes		31				
Lyt 2$^+$ gut T-cells		22			15	
L3T4$^+$ gut T-cells		56			70	

a. Supernatants of 48 h culture of 10^6 cells, stimulated with specific antigen (Nb worm extracts of Thy-1-treated and irradiated spleen cells of the appropriate H-2 MHC). IL-3 was assayed either on cultured murine mast cells or on an IL-3-dependent mast cell line (see materials and methods). Results are expressed as percentage of ^{3}H-TdR incorporation induced by an optimal amount of WEHI-3 supernatant and are given at the dilution allowing the maximum effect (1/2 to 1/10) (some inhibition is often observed with undiluted supernatants).

Local IL-3 release from specifically stimulated intestinal T-lymphocytes appears to be responsible for the proliferation and the differentiation of IMC-P, as evidenced by the striking increase in the frequency of IMC-P (Table I), particularly in intestinal mucosa and TD. In the former, immature IMC proliferate, displaying few granules but bearing IgE receptors (they are identical to the earliest form of identifiable MC observed in IL-3-stimulated cultures). In the TD, no such immature mast cells are detectable. After parasite expulsion ("self-cure"), IMC-P frequency decreases but fully mature IMC with large granules appear and remain conspicuous for a few days on histological sections (Table III). There are three conditions in which Nb-infested mice do not display the normal appearance of mature IMC: (i) in mice whose TD is cannulated from day 8 to 10 (before "self-cure") and whose intestine is thus depleted of antigen-specific T-cells, the number of IMC decreases 5-fold; (ii) in mice bearing fetal intestinal grafts, the grafts are populated, like the animal's own intestine, with T-cells arising from host PP (7), but no IMC are observed after Nb infestation.

Since worms are never found in these grafts and since IMC–P can be detected in the grafts, it appears that the missing element required for local IMC appearance is worm antigen; (iii) finally, in Nb–infested W/W^v mice, intestinal mucosal T–cells are able to secrete IL–3 under antigenic stimulation, but the very low IMC–P frequency does not increase, and no IMC are found on tissue sections (Tables I and II).

TABLE III. Number of IMC detectable on tissue sections and acceleration of the epithelial renewal in pathological conditions

Mouse strain and condition	IMC[a]	Index of Epithelial Renewal[b]
Normal (adult or newborn)	0	5
Nb–infested		
d 8	0[c]	1.6
d 10	73	1
d 13	136	
d 15	16	
Fl with GVHR		
adult	up to 12	up to 1.6
newborn	up to 37	up to 5
irradiated adult d 6	0[e]	2.5
WEHI–3 tumor bearing[d]		
adult	146	2.3
newborn	133	3.9
irradiated adult	0[e]	1.7

a. Intestinal sections, performed at 5 cm from the pylorus, were fixed in Ruitenberg's fluid (1), stained by toluidine blue and IMC were numbered in 20 villous crypt units (1).
b. Extent of the ^{3}H–TdR labelling along the villous, estimated with a micrometer in pathological conditions/control conditions.
c. IMC–P, highly proliferating and poorly granulated, may be isolated from the intestine but are not yet apparent on tissue sections.
d. Control mice were injected with P815 tumor.
e. These mice have no IMC–P as a result of irradiation.

3) <u>Graft versus host reactions:</u> IL-3 secretion is also demonstrable in the course of GVHR. In adult healthy lethally irradiated mice, 6 days after the injection of semi-allogeneic, unselected LN lymphocytes or of lymphocytes depleted of Lyt 2$^+$ or L3T4$^+$ cells (see Materials and Methods), lymphocytes collected from TD or isolated from intestinal mucosa are able to secrete IL-3 under specific allogeneic stimulation (Table II). In tissue sections, no IMC are detectable, but in these irradiated mice, no IMC-P survive (data not shown). In newborn mice with GVHR, donor T lymphocytes (mainly of the Lyt 2$^+$ phenotype) are found in the intestinal mucosa. In non-irradiated adult mice, gut infiltration by T-cells is observed after massive injection of semi-allogeneic lymphocytes. In these newborn or adult mice, provided that host cells (which contain IMC-P) are not destroyed by the GVHR, mature IMC are observed on tissue sections in variable numbers depending upon the conditions and time of observation.

<u>Evidence for an Association Between IL-3 Release and Acceleration of Intestinal Epithelial Renewal.</u> Villous epithelial cells proliferate in the crypts and move to the tops of the villi in 2 or 3 days, as assessed by autoradiography performed at various times after injection of ^{3}H-TdR. Rate of intestinal epithelial renewal may thus be assessed by the extent of labelling of the villi 26 hours after ^{3}H-TdR injection (9). Pathological conditions can be compared to normal conditions in these experiments and thus an index of acceleration may be determined in this manner (Table III). In GVHR and in Nb-infected mice, an acceleration of intestinal epithelial renewal is observed concomitantly with the appearance of IMC. That this acceleration may be the result of IL-3 release is supported by the observation that a WEHI-3B tumor graft, which leads to the maturation of IMC-P <u>in</u> <u>situ</u> (5) (in non-irradiated adult and newborn mice) also results in an acceleration of epithelial renewal, without other gut modifications (such as increase either of goblet cells or of Ia-positive epithelial cells). Grafts of non-IL-3 releasing tumors, such as P815 cells, are not accompanied by acceleration of intestinal epithelial renewal.

DISCUSSION

The bone marrow origin and the nature of IMC-P have been previously discussed (5) and our conclusions are in agreement with those of Crapper and Schrader (10) and Schrader et al. (11). The peculiarities of IMC-P traffic deserve some comments. Under normal conditions, mouse TD contains very few IMC-P, indicating that these cells probably directly home to the intestinal mucosa from the blood without undergoing a cycle of amplification involving intestinal mucosa, MLN and TD before returning to the intestine. This cycle has been shown to exist for intestinal mucosal T-cells or for IgA-bearing blasts which are the progenitors of intestinal mucosal IgA plasma cells, both cell types stimulated by antigenic stimulation in the PP (7,8). Since IMC-P

and related mast cell precursors are much more frequent among the
lymphoid cells isolated from intestinal mucosa than among cells
from any other lymphoid or hematopoietic tissue with the exception of BM, it appears that IMC-P released into the circulation
may be attracted specifically to the intestine by unknown
factors, some of which may be released by intestinal mucosa
itself. This possibility is reinforced by the observations made
in mice bearing a WEHI-3B tumor, a condition in which there is a
detectable blood level of IL-3, which presumably allows for the
maturation of IMC from IMC-P in situ: large numbers of IMC are
observed only in those areas of intestinal mucosa where there are
also increased numbers of IMC-P. This postulated "gut factor"
allowing the localization of IMC-P may be the factor (or one of
the factors) missing in W/W^v mice, since in these mice IMC-P are
present in normal amounts in the bone marrow but are very low in
frequency in intestinal mucosa; for example, IMC do not appear
after Nb infestation of W/W^v mice.

In Nb-infested mice, IMC-P are not only induced to differentiate locally into IMC, but IMC-P or related mast cell precursors
may also be induced to recirculate through the TD to the blood,
presumably allowing further homing all along the intestinal
tract, a process of dissemination to the mucosa similar to that
observed for T blasts (7) and IgA bearing blasts (8). Transfer
experiments performed in rats have indeed shown that TD blasts
are able to migrate to the intestinal mucosa where they mature
into fully differentiated IMC (5). The presence in normal rats
as well as in newborn and nude rats of IMC, which are not
observed in mice under these conditions, suggests that rat
intestinal mucosa may contain more of the postulated "gut factor"
which favours homing of IMC-P and their differentiation in situ.

However, a major factor responsible for IMC-P proliferation
and subsequent differentiation is IL-3 released by antigen-
stimulated T lymphocytes. The present and previous experiments
show that T lymphocytes of L3T4$^+$ but also of Lyt 2$^+$ phenotype,
obtained from intestinal mucosa of Nb-infested mice or from mice
undergoing GVHR, produce in vitro large amounts of IL-3 in
response to stimulation by Nb antigen or by cells of the host MHC
haplotype. Immature, dividing IMC can be isolated from the
intestine at the beginning of the Nb expulsion or of the GVHR,
i.e. at a time when the local encounter between specific T-cells
and the antigen they recognize induce a strong release of IL-3.
Following the appearance of these immature IMC forms, IMC mature
and become conspicuous on tissue sections. They then disappear,
when IL-3 secretion stops, presumably as a result of renewal of
antigenic stimulation (after Nb expulsion) or suppressive effects
of IMC-P which derive from the host (unpublished data). The role
of IL-3 in producing activated intestinal T-cells or of released
IL-3 in stimulating the proliferation and differentiation of
IMC-P is clearly demonstrated by the failure of IMC to appear
when one of the elements of this chain of events is missing: a)
IMC-P in W/W^v mice; b) specific antigen activated T-cells when

the TD of infested mice is cannulated shortly before the self-cure; c) specific antigen which is absent from the fetal intestinal grafts in Nb infestation.

Finally, it seems likely that at the level of the intestine, IL-3 is responsible not only for maturation of IMC-P but also for acceleration of epithelial renewal, leading to villous atrophy, a major alteration of the intestinal wall in numerous disorders, and in particular, in intestinal GVHR. The preliminary data presented here show that both IMC proliferation and acceleration of epithelial renewal appear concomitantly under the influence of factor(s) secreted by the WEHI tumor. Whether only IL-3 or other factors are involved, the nature of the target cells (i.e. epithelial cells of the crypts), and the possible influence of potential antagonist factors (such as gamma interferon) will be the subjects of further studies.

REFERENCES

1. **Ruitenberg, E.J., and A. Elgersma.** 1976. Absence of intestinal mast cell response in congenitally athymic mice during Trichinella spiralis infection. Nature 264:258.
2. **Sredni, B., M.M. Friedman, C.E. Bland, and D.D. Metcalfe.** 1983. Ultrastructural, biochemical and functional characteristics of histamine-containing cell clones from mouse bone marrow: tentative indentification as mucosal mast cells. J. Immunol. 131:915.
3. **Watson, J.D., and R.L. Prestige.** 1983. Interleukin 3 and colony-stimulating factor. Immunol. Today 4:278.
4. **Ihle, J.N., J. Keller, S. Oroszlan, L.E. Henderson, T.D. Copeland, F. Fitch, M.B. Prystowsky, E. Goldwasser, J.W. Schrader, E. Palaszynski, M. Dy, and B. Lebel.** 1983. Biologic properties of homogeneous interleukin 3. I. Demonstration of Wehi-3 growth factor activity, mast cell growth factor activity, P cell-stimulating factor activity, colony-stimulating factor activity, and histamine-producing cell-stimulating factor activity. J. Immunol. 131:282.
5. **Guy-Grand, D., M. Dy, G. Luffau, and P. Vassalli.** 1984. Gut mucosal mast cells. Origin, traffic and differentiation. J. Exp. Med. 160:12.
6. **Rolink, A.G., and E. Gleichmann.** 1983. Allosuppressor and allohelper-T cells in acute and chronic GVH disease. J. Exp. Med. 158:546.
7. **Guy-Grand, D., C. Griscelli, and P. Vassali.** 1978. The mouse gut T lymphocyte, a novel type of T cell. Nature, origin, and traffic in mice in normal and graft-versus-host conditions. J. Exp. Med. 148:1661.
8. **Guy-Grand, D., C. Griscelli, and P. Vassalli.** 1974. The gut-associated lymphoid system: nature and properties of the large dividing cells. Eur. J. Immunol. 4:435.
9. **Guy-Grand, D., and P. Vassalli.** 1983. Gut mucosal lymphocyte subpopulations and mast cells. In: Regulation of the

Immune Response. 8th Int. Convoc. Immunol. Edited by P.L. Ogra, and D.M. Jacobs. Buffalo, N.Y., Karger, Basel, p.122.

10. **Crapper, R.M., and J.W. Schrader.** 1983. Frequency of mast cell precursors in normal tissues determined by an in vitro assay: antigen induces parallel increases in the frequency of P cell precursors and mast cells. J. Immunol. 131:923.

11. **Schrader, J.W., R. Scollay, and F. Battye.** 1983. Intra-mucosal lymphocytes of the gut: Lyt-2 and Thy-1 phenotype of the granulated cells and evidence for the presence of both T cells and mast cell precursors. J. Immunol. 130:558.

Mast Cell Differentiation and Heterogeneity,
edited by A. D. Befus et al.
Raven Press, New York © 1986.

Probable Transdifferentiation Between Connective Tissue and Mucosal Mast Cells

Yukihiko Kitamura, Toru Nakano, Takashi Sonoda,
Yoshio Kanayama, Teiichi Yamamura, and Hidekazu Asai

*Institute for Cancer Research and Departments of Medicine and Dermatology,
Kita-ku, Osaka 530; and Shizuoka Laboratory Animal Center, Hamamatsu-city,
Shizuoka 435, Japan*

Morphological and functional differences among rodent peritoneal, dermal or other non-intestinal and intestinal mast cells have been described (1,2), and similarities between rat or murine intestinal and $\underline{in}$ $\underline{vitro}$ cultured mast cells have been claimed (3,4). If these claims are admitted, mast cells may be divided into two subgroups; the first subgroup consists of mast cells such as peritoneal and dermal mast cells, and the second subgroup includes intestinal and cultured mast cells. Since purification of murine intestinal mast cells is difficult, we used cultured mast cells as representative of intestinal mast cells. Although both $\underline{in}$ $\underline{vivo}$-derived dermal or peritoneal and cultured mast cells are the progeny of multipotential hemato-poietic stem cells (5-8), the proliferative state of these two types of cells appears to be different. Division of the former is rare (9) but that of the latter is frequent (10,11). In spite of their relative dormancy, some peritoneal or dermal mast cells have extensive proliferative potentiality. When morphologically identifiable peritoneal mast cells are injected into the skin of genetically mast cell-deficient (WB X C57BL/6)F_1-W/W^v (hereafter WBB6F_1-W/W^v) mice, about 5% of the cells form clusters containing 400 to 8,000 mast cells (12). However, this does not necessarily mean that differentiated peritoneal or dermal mast cells divide extensively; in fact, morphologically identifiable mast cells disappear soon after injection, and mast-cell clusters develop rather abruptly 15 to 20 days thereafter. The result suggests the de-differentiation of peritoneal or dermal mast cells and their re-differentiation after proliferation (13). Since even dormant peritoneal mast cells may form mast-cell clusters, there

This work was supported by the Ministry of Education, Science and Culture, the Ministry of Health and Welfare, the Mitsubishi Foundation, the Princess Takamatsu Cancer Research Fund, and the Japanese Foundation for Multidisciplinary Treatment of Cancer.

is a possibility that cultured mast cells may proliferate and differentiate in tissues of WBB6F$_1$-W/W^v mice. Furthermore, there is also a possibility that murine dermal or peritoneal mast cells may change to that type of mast cell found in vitro or vice versa, through de-differentiation, proliferation and re-differentiation.

MATERIALS AND METHODS AND RESULTS

Morphological Identification of Peritoneal, Dermal and Intestinal Mast Cells. Cultured cells and peritoneal cells were spun in a cytocentrifuge and fixed with Carnoy's fluid. Skin pieces and stomachs were also fixed with Carnoy's fluid, and paraffin sections were made with the routine method. Cytocentrifuge preparations and tissue sections stained were with either alcian blue, safranin or berberine sulfate (14). Cultured mast cells and mast cells in the mucosa of glandular stomachs were stained with alcian blue but not with safranin or berberine sulfate, whereas most of the mast cells in the peritoneal cavity and skin of WBB6F$_1$-+/+ mice stained with berberine sulfate and some of them with safranin. The effect of either chondroitinase ABC or heparinase digestion on the staining with berberine sulfate was examined to confirm whether berberine sulfate specifically stained heparin. The number of peritoneal mast cells stained with berberine sulfate was significantly reduced by heparinase digestion but not by chondroitinase ABC digestion. Thus, we considered that at least some of the berberine sulfate-positive material in peritoneal mast cells was heparin, as reported by Enerback (14).

Since Dvorak et al. (15) and Galli et al. (11) showed that the electron microscopical features of cultured mast cells were different from those of other in vivo-derived basophils or mast cells in several species, we also examined by electron microscopy the morphological change of cultured mast cells after transplantation into the peritoneal cavity of WBB6F$_1$-W/W^v mice.

In Vitro Culture of Mast Cells. Bone marrow cells of WBB6F$_1$-+/+ mice were cultured at 10^6 cells per ml in α-medium supplemented with 2-mercaptoethanol (20^{-4}M), horse serum (20% vol/vol), together with medium conditioned by pokeweed mitogen-stimulated spleen cells (10% vol/vol) (10). Culture flasks were incubated at 37°C in a humidifed atmosphere flushed with 5% CO_2 in air. Cells were fed by replacement of half the medium at 7-day intervals.

Cultured cells were harvested at the time of the half medium change. The concentration of spleen colony-forming cells and mast-cell precursors were measured. For assay of mast-cell precursors, various numbers of cultured cells were directly injected into the skin of WBB6F$_1$-W/W^v mice. By determining the proportion of injection sites at which mast-cell clusters appeared, the concentration of mast-cell precursors was calculated by limiting dilution analysis (16). The concentration of spleen colony-forming cells decreased rapidly, and actually none were detectable 3 weeks after the initiation of the culture. In contrast, the concentration of mast-cell precursors increased rapidly in 3 weeks after the initiation of the culture. On the 21st day of the culture, more than 95% of the cells had the morphology of immature mast cells, and about 2% of the cells had an ability to form mast-cell clusters in the skin of $WBB6F_1-W/W^v$ mice.

In the next experiment, cultured cells were harvested 4 weeks after the initiation of the culture and injected into the skin of $WBB6F_1-W/W^v$ mice. When the recipient $WBB6F_1-W/W^v$ mice were killed 10 weeks after the injection, some mast cells in the cluster stained with safranin.

Intraperitoneal Injection of Cultured Mast Cells. Cultured mast cells of $WBB6F_1-+/+$ mouse origin were harvested 4 weeks after initiation of the culture, and 10^6 cells were injected i.p. into $WBB6F_1-W/W^v$ mice. Peritoneal cells were recovered at various times after i.p. injection, and cytocentrifuge specimens were prepared. Mast cells which stained with berberine sulfate appeared 1 week after the injection. At day 21, most of the mast cells recovered from the peritoneal cavity stained with berberine sulfate. Mast cells which stained with safranin developed 15 weeks after the i.p. injection. The electron microscopical features of mast cells recovered from the peritonal cavity of the $WBB6F_1-W/W^v$ mice 5 weeks after the injection of cultured mast cells closely resembled those of peritoneal mast cells recovered from non-treated $WBB6F_1-W/W^v$ mice. Moreover, the content of histamine increased significantly after i.p. transplantation.

Intravenous Injection of Cultured and Peritoneal Mast Cells. Cultured mast cells (10^6) of $WBB6F_1-+/+$ mouse origin were harvested 4 weeks after initiation of cultures, and injected i.v. into $WBB6F_1-W/W^v$ recipients. The $WBB6F_1-W/W^v$ mice were killed 10 weeks after the injection. Mast cells appeared either in the skin or in the mucosa of the glandular stomach. When each tissue was stained with alcian blue-safranin, some mast cells in the skin stained with safranin, but mast cells in the gastric mucosa stained only with alcian blue.

Peritoneal mast cells (10^5) of $WBB6F_1-+/+$ mice were also injected i.v. in $WBB6F_1-W/W^v$ recipients; the W/W^v mice were killed 10 weeks after the injection. Safranin-positive mast cells appeared in the peritoneal cavity, whereas mast cells in the gastric mucosa stained only with alcian blue.

Direct Injection of Peritoneal Mast Cells Into the Stomach. Peritoneal mast cells (10^3) from $WBB6F_1-+/+$ mice were directly injected into the wall of the glandular stomach of $WBB6F_1-W/W^v$ mice. Recipients were killed 5 weeks after the injection and stomach tissue was stained with alcian blue-safranin. Mast cells appeared both in the muscular and mucosal layers. Some mast cells in the muscular layer stained with safranin, whereas all mast cells in the mucosal layer stained only with alcian blue.

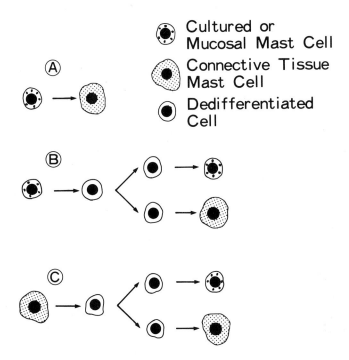

Figure 1 A scheme for probable <u>transdifferentiation</u> of murine mast cells. A. A cultured mast cell may transform into a dermal or peritoneal mast cell. B. A cultured mast cell may differentiate into intestinal, dermal or peritoneal mast cell through probable <u>de-differentiation</u>. C. A dermal or peritoneal mast cell may differentiate into intestinal, dermal or peritoneal mast cell through probable <u>de-differentiation</u>.

DISCUSSION

Mast cells staining with safranin appeared after the injection of cultured murine mast cells into the skin of the peritoneal cavity of WBB6F$_1$-W/W^v mice. The electron microscopic features of mast cells which developed in the peritoneal cavity were comparable to those of normal peritoneal mast cells recovered from WBB6F$_1$-+/+ mice. Moreover, histamine content per cell increased significantly after i.p. injection of cultured mast cells. The present results are not consistent with the results of Crapper et al., who also transferred cultured murine mast cells into the skin of W^f/W^f mice. They reported that safranin-positive mast cells did not appear at the injection sites and that the transferred mast cells were maintained only in the hosts bearing WEHI-3B tumors (4). Currently, we cannot explain the discrepancy between the results of Crapper et al. and those reported here. Our results suggest that cultured mast cells themselves (Fig. 1A) or their de-differentiated progeny (Fig. 1B) may differentiate into dermal mast cells. The first possibility, as shown in Fig. 1A, has been proposed by Galli et al. (11). Since mast cells which stained only with alcian blue appeared in the gastric mucosa of $WBB6F_1-W/W^v$ mice after i.v. injection of cultured mast cells, at least a subpopulation of cultured murine mast cells seems to have the potential to differentiate into either dermal or gastric mast cells (Fig. 1B).

When peritoneal mast cells of $WBB6F_1-+/+$ mouse origin were injected i.v. or directly into the glandular stomach of $WBB6F_1-W/W^v$ mice, mast cells staining only with alcian blue developed in the mucosa. This result suggests that some murine peritoneal mast cells may change to gastric mast cells, probably after de-differentiation and proliferation (Fig. 1C).

The present observations seem to suggest that each of the murine dermal, peritoneal and intestinal (gastric) mast cells may transform to any of the subtypes as evidenced by staining properties. There is a possibility that the phenotypes may be dependent on microenvironmental influences, which influence final differentiation pathways.

REFERENCES

1. **Enerback, L.** 1981. The gut mucosal mast cells. Monogr. Allergy 17:222.
2. **Bienenstock, J., A.D. Befus, F. Pearce, J. Denburg, and R. Goodacre.** 1982. Mast cell heterogeneity: derivation and function, with emphasis on the intestine. J. Allergy Clin. Immunol. 70:407.
3. **Haig, D.M., T.A. McKee, E.E.E. Jarrett, R. Woodbury, and H.R.P. Miller.** 1982. Generation of mucosal mast cells is stimulated in vitro by factors derived from T cells of helminth-infected rats. Nature 300:188.
4. **Crapper, R.M., W.R. Thomas, and J.W. Schrader.** 1984. In vivo transfer of persisting (P) cells; further evidence for the identity with T-dependent mast cells. J. Immunol. 133:2174.
5. **Kitamura, Y., M. Yokoyama, H. Matsuda, T. Ohno, and K.J. Mori.** 1981. Spleen colony-forming cell as common precursor for tissue mast cells and granulocytes. Nature 291:159.
6. **Sonoda, T., Y. Kitamura, Y. Haku, H. Hara, and K.J. Mori.** 1983. Mast-cell precursors in various haematopoietic colonies of mice produced in vivo and in vitro. Br. J. Haematol. 53:611.

7. Schrader, J.W., S.J. Lewis, I. Clark-Lewis, and J.G. Culvenor. 1981. The persisting (P) cell: histamine content, regulation by a T-cell-derived factor, origin from a bone marrow precursor, and relationship to mast cells. Proc. Natl. Acad. Sci. USA 78:323.

8. Nakahata, T., S.S. Spicer, J.R. Cantey, and M. Ogawa. 1982. Clonal assay of mouse mast cell colonies in methylcellulose culture. Blood 60:352.

9. Padawer, J. 1974. Mast cells: extended life span and lack of granule turnover under normal in vivo conditions. Exp. Mol. Pathol. 20:269.

10. Hasthorpe, S. 1980. A hemopoietic cell line dependent upon a factor in pokeweed mitogen-stimulated spleen cell conditioning medium. J. Cell. Physiol. 105:379.

11. Galli, S.J., A.M. Dvorak, and H.F. Dvorak. 1984. Basophils and mast cells: morphologic insights into their biology, secretory patterns, and function. Prog. Allergy 34:1.

12. Sonoda, T., Y. Kanayama, H. Hara, C. Hayashi, M. Tadokoro, T. Yonezawa, and Y. Kitamura. 1984. Proliferation of peritoneal mast cells in the skin of W/W^v mice that genetically lack mast cells. J. Exp. Med. 160:138.

13. Kitamura, Y., T. Nakano, and Y. Kanayama. 1985. Probable dedifferentiation of mast cells in mouse connective tissues. Curr. Top. Dev. Biol. (in press).

14. Enerback, L. 1974. Berberine sulfate binding to mast cell polyanions. Histochemistry 42:301.

15. Dvorak, A.M., G. Nabel, K. Pyne, H. Cantor, H., H.F. Dvorak, and S.J. Galli. 1982. Ultrastructural identification of the mouse basophil. Blood 59:1279.

16. Sonoda, T., T. Ohno, and Y. Kitamura. 1982. Concentration of mast-cell progenitors in bone marrow, spleen, and blood of mice determined by limiting dilution analysis. J. Cell. Physiol. 112:136.

Mast Cell Differentiation and Heterogeneity,
edited by A. D. Befus et al.
Raven Press, New York © 1986.

A Comparison of Some Properties of Mast Cells and Other Granulated Cells

*Graham Mayrhofer and **Rosanne Pitts

*Department of Microbiology and Immunology, The University of Adelaide, Adelaide 5000, South Australia; and **The Clinical Immunology Research Unit, Princess Margaret Hospital, Subiaco 6008, Western Australia*

The intention of this paper is to review briefly some of the relationships that have been suggested to exist between mast cells (MC) and several other cell-types that have basophilic cytoplasmic granules as a common feature. Discussion will include the MC that are encountered at various sites in the body, the intestinal granular intraepithelial lymphocytes (IEL) and the large granular lymphocytes (LGL) found in blood and lymphoid tissues. It will centre mainly on cells that have differentiated in vivo, because anatomical location is a key characteristic for the identification of these various cells.

Clearly the best classification of granular cells would be one based on function, but at present too little is known about most of them to make this possible. Another method could be to group cells together on the basis of shared characteristics such as morphology, histamine content, glycosaminoglycan composition, surface receptors or markers, etc. However, we do not yet know the diagnostic probity of these various characteristics and neither do we know which are consistently expressed throughout the greater part of the cell's differentiation. The following discussion deals with tantalizing similarities and differences between several types of granular cells that nevertheless fall short of either proving or disproving relationships between them.

Abbreviations: MC, mast cells; LGL, large granular lymphocytes, IEL, intraepithelial lymphocytes; MFAA, methanol–formalin–acetic acid; CAE, chloroacetate esterase; NE, non-specific esterase; AP, acid phosphatase; EF, formalin; PBS, phosphate–buffered saline; MCA, methylcholanthrene; DNCB, dinitrochlorobenzene; FACS, fluorescence activated cell sorter.

MATERIALS AND METHODS

Animals. Female WAG or PVG/c rats were housed in clean conventional conditions. Nude mice and heterozygous littermates were of BALB/c background and were maintained in clean barrier conditions. Nude rats and B rats were described elsewhere (1).

Tissue Preparation. Sections for immunoperoxidase studies were either from frozen unfixed gut or from paraffin blocks fixed for 1 h in cold Carnoy. Paraffin-embedded material fixed in methanol-formalin-acetic acid (MFAA) (2) was used to demonstrate chloroacetate esterase (CAE) and frozen sections of tissue fixed in 4% paraformaldehyde were used to detect non-specific esterase (NE) and acid phosphatase (AP). To simultaneously identify specific granules staining with alcian blue as well as AP and NE, frozen sections were prepared from tissue fixed with ethanol containing 10% formalin (EF).

The methods to detect AP, NE and CAE were essentially as described by Lojda, Grossrau and Schiebler (3). Staining with alcian blue and safranin O was as described by Mayrhofer (2).

Histochemistry

Immunological Reagents and Immunochemistry. The indirect immunoperoxidase technique has been described elsewhere (4). Primary antibodies were culture supernatants of IgG-secreting mouse anti-rat hybridomas. ^{125}I-labelled purified rat myeloma IR2 IgE was used to measure binding of IgE to MC. Cells were incubated with radio-iodinated myeloma protein (2 μg/ml) for 1 h on ice in phosphate-buffered saline (PBS) containing 5% fetal calf serum, followed by counting of the washed cells in a gamma counter.

Isolation and Examination of Cells. IEL and lymphocytes were isolated as previously described (1). Peripheral blood mononuclear cells were prepared by isopyknic sedimentation on Hypaque-Ficoll, density 1.090. Histochemical studies were performed on frozen sections of cell pellets fixed with cold EF. Peritoneal mast cells were purified on a Percoll step gradient (5). The surface phenotype of isolated IEL was investigated using flow cytometry (1).

Contact Dermatitis. Contact dermatitis was induced on the shaved flanks of mice by twice weekly painting with either 0.5% 20-methyl cholanthrene (MCA) in benzene or 0.1% dinitrochlorobenzene (DNCB) in acetone. The contralateral flank was painted with solvent. At intervals, mice were sacrificed and strips of skin from the treated areas were fixed overnight at room temperature in MFAA. Paraffin sections were stained with alcian blue and safranin O. MC were counted between the epidermal basement membrane and the subdermal fat layer.

RESULTS AND DISCUSSION

Mast Cells and T Lymphocytes. It is difficult to discuss the lineage of MC without considering their possible relationship to T lymphocytes (6,7). For those rodent MC present in connective tissues (skin, serosal surfaces, peritoneum, bone marrow), there is no obligate requirement for a thymus during development (8,9) although its presence may affect numbers and maturity of these MC (10). The case for rodent mast cells at mucosal sites is less clear. Although the hyperplasia following nematode infestations appears to be T cell–dependent (11,12,13), the numbers of MC in the intestinal mucosa of thymus–deficient rodents are normal (10), suggesting that this dependence is not obligate. In vitro studies on murine MC show that these cells also arise from non–T cells, although they require T-cell derived factors for their growth (14); but their precise nature is uncertain. Thus, murine MC in connective tissues are bone marrow-derived and probably non–thymus–processed, but a possible T cell lineage for MC at mucosal sites cannot yet be completely discarded.

MC and Granular IEL in Intestinal Mucosa. It has been speculated that these cells share a common lineage (7,14). The suggested precursor relationship of intestinal granular IEL to MC is based on certain shared properties (16). The results presented here compare the surface markers and histochemistry of the two cell types.

In the normal rat, studies on isolated IEL or frozen sections establish the following surface antigen phenotype: LCA^+, $W3/13^+$, $OX8^+$, Thy 1.1^-. All express LCA, most express both W3/13 and OX8, but small subpopulations express either only OX8 or only W3/13 (Figs. 1 and 2a). W3/13 and OX8 antigens can be detected in Carnoy-fixed tissue by immunoperoxidase. Most granular and non-granular IEL were found to express each antigen but in each case, significant proportions of both morphological subsets were unstained. In contrast, intestinal MC were all $OX8^-$ and $W3/13^-$. Under conditions where peritoneal MC bound large amounts of ^{125}I-IgE, binding to either IEL or to lymph node cells was insignificant. Histochemical studies on sections showed that the granules of both intestinal MC and granular IEL were positive for CAE. In frozen sections fixed with either paraformaldehyde or with EF the granules of IEL were NE positive but AP negative. Simultaneous staining with alcian blue showed that these MC were negative for both AP and NE. Finally, the critical electrolytic concentration at which staining of IEL is inhibited is considerably lower than in intestinal MC (2).

There are therefore considerable differences between intestinal MC and granular IEL in terms of the surface antigens, granule enzymes, IgE Fc receptors and the granule glycosaminoglycan. If it is assumed that intestinal MC, like other rodent MC (see below), are also Thy 1.1^+, this is a further difference. However,

(a)   (b)

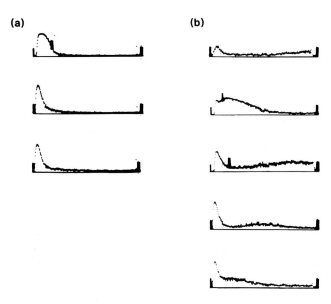

Figure 1. Fluorescence profiles obtained with a fluorescence—activated cell sorter (FACS) from isolated rat IEL labelled
by an indirect technique employing mouse monoclonal primary antibodies. From top to bottom in each case: a. antibodies that do
not bind:- W6/32 (anti-human HLA), OX12 (anti-rat kappa chain)
and OX7 (anti-Thy-1.1); b. antibodies that bind:- OX1 (leukocyte-
common antigen, LCA), W3/13 (rat thymocytes, peripheral T cells,
some myeloid cells, plasma cells), OX8 (suppressor and cytotoxic
T cells, NK cells), W3/25 (helper T cells, some macrophages) and
OX19 (rat thymocytes and peripheral T cells). Vertical axis, log
cell number in each channel; horizontal axis, fluorescence
intensity. 1×10^{5} cells analysed.

intestinal MC and granular IEL do share certain characteristics.
The possession of CAE is of interest, although Huntley et al.
(17) have shown that rat intestinal MC have the RMCPII serine
protease, whereas granular IEL do not.   Finally, although rat
intestinal MC are W3/13$^-$, some non-intestinal MC are W3/13$^+$ (see
below) while granular IEL can be either W3/13$^+$ or W3/13$^-$.   Thus,
while the evidence is strongly on the side of separate lineages,
it remains possible that a subpopulation of W3/13$^-$ granular IEL
could differentiate into intestinal MC.
    Relationship of IEL to T Cells. Studies of IEL in nude mice
raised the possibility that they were thymus-dependent (18).

Guy-Grand et al. (7) suggested that IEL were a type of T lymphocyte but later studies in thymus-deficient mice and rats suggested that only the non-granular IEL were thymus-dependent (15).
Functional studies on IEL have produced conflicting conclusions.
On the one hand, their natural cytotoxicity (19,20), the close
similarity of their surface antigen phenotype (see below) to NK
cells in the rat (21) and relative radioresistance (1) have
suggested a non-T cell lineage. On the other hand, some authors
have demonstrated certain T cell functions (22) by these cells.
    Recent evidence supports the non-T cell nature of IEL. The
phenotype of IEL from normal rats is shown in Fig. 2a and
expression of W3/13 and OX8 antigens are described above.   MRC
OX19 antibody is a more specific pan T cell marker than W3/13
(23) and this labels only a small subpopulation of IEL which
encompasses all W3/25$^+$ cells and a proportion of the W3/13$^+$, OX8$^+$
cells. The OX19$^+$ subpopulation is very small in B rats and is
absent from young nude rats (Fig. 2b,c). Its presence in normal
rats probably represents contamination from lymphoid follicles or
the lamina propria because: (a) OX19$^+$ IEL are rare in frozen
sections of normal gut; and, (b) granular IEL are absent from
OX19$^+$ cells separated on the fluorescence activated cell sorter
(FACS).
    Both granular and non-granular IEL can arise from bone marrow
in the absence of the thymus (1). The apparent thymus-dependence
of non-granular IEL (15) might be due to inductive effects of T
lymphocytes on IEL maturation, the excess of granular cells in
thymus-deficient animals perhaps representing a low turnover
state and a "shift to the right" in maturation. A maturation
shift might also explain the W3/13$^+$, OX8$^-$ phenotype of IEL in
nude rats (Fig. 2c). At the other extreme, the intestinal epithelium of fetal rats becomes colonized with OX8$^+$, W3/13$^-$
lymphoid cells that are mainly non-granular between 18 and 20
days gestation (Fig. 3). These findings in nude and fetal rats,
and the small OX8$^+$, W3/13$^-$; OX8$^-$, W3/13$^+$ and OX8$^-$, W3/13$^-$ subpopulations in normal rats, suggest the following maturation
sequence:

OX8$^+$, W3/13$^-$ (non-granular) → OX8$^+$, W3/13$^+$ (granular and non-
                                                    granular)
→ OX8$^-$, W3/13$^+$ (predominantly granular) → OX8$^-$, W3/13$^-$.

    Interrelationships of MC at Different Anatomical Sites in
Rodents. Differences between rat intestinal MC and peritoneal or
other non-intestinal MC have been discussed in a number of
reviews (16, 24). Two broad possibilities exist concerning their
lineage - either they diverge at a relatively early point in

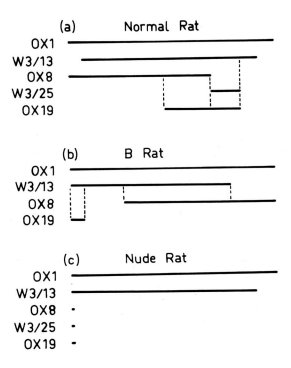

Figure 2. Bar chart representing subpopulations of rat IEL. Surface antigen phenotypes were deduced from fluorescence profiles from the FACS using IEL labelled either singly or with appropriate combinations of monoclonal antibodies. One hundred percent of IEL were labelled by OX1 and the sizes of subpopulations labelled by other antigens are represented to scale. Subpopulations labelled by more than one antibody are delineated by vertical broken lines. a) IEL from normal rats. b) IEL from B rats (adult-thymectomized,irradiated, bone marrow-reconstituted). W3/25 antibody omitted, possibly equivalent to the OX19[+] subpopulation. c) IEL from approximately 6 week-old nude rat.

their differentiation from stem cells or they diverge late in differentiation, perhaps influenced by local inductive factors. In general, with respect to rodent non-intestinal MC, differentiation of the mature cell occurs at peripheral sites under direction from tissue-derived factors (25). Rat intestinal MC are believed to be produced locally (26) and precursors of these cells have been demonstrated in murine intestinal mucosa (27). It is not known whether there are intrinsic inductive influences from the mucosa itself, but hyperplasia can be induced by local T-dependent immune responses (see above). In contrast, non-intestinal (e.g. dermal, peritoneal, serosal) MC are widely believed to be T-independent.

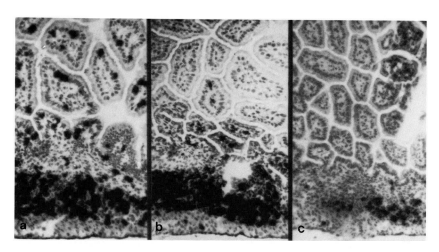

Figure 3. Localization of surface antigens on IEL in the proximal small intestine of a 21-day gestation rat fetus. a. OX8 antibody. Labelled cells in the villi have mainly an intra-epithelial distribution and OX8+ presumptive T cells are also present in the rudimentary Peyer's patch. b. W3/13 antibody. W3/13+ T lymphocytes are already present in the Peyer's patch but the IEL are W3/13 negative. c. OX19 antibody also labels T lymphocytes in the Peyer's patch, but the IEL are unlabelled.

We have examined several aspects of the physiology of MC in the gut mucosa. In normal rats, an injection of tritiated thymidine (^{3}HTdR) labelled approximately 1% of the cells in autoradiographs of tissue taken 1 hour later (Fig. 4). These cells are assumed to have been labelled in the mucosa itself. In the succeeding 12-48 hours, the proportion of labelled cells doubled and dilution of the label suggested that further cell division had occurred. After 2 to 4 days, the proportion of labelled cells rose to over 5% but most were only lightly labelled. This second increase could reflect an influx of cells from another tissue or could have been generated from previously non-granular resident cells. When ^{3}HTdR was infused intravenous-

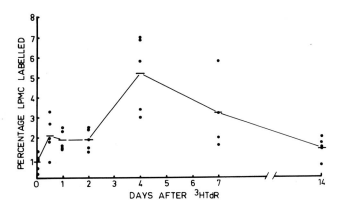

Figure 4. Kinetics of labelling of lamina propria mast cells (LPMC) in the jejuna of normal rats following a single intravenous injection of tritiated thymidine (1 μCi/gram body weight). Labelling of LPMC was determined by autoradiography on MFAA-fixed paraffin sections stained by the alcian blue-safranin O method. Each point represents the proportion of labelled LPMC in an individual rat and means for each group are indicated by horizontal bars.

ly for 4 days, approximately 8% of jejunal and ileal MC were labelled, suggesting a turnover of about 2% of the population per day (Table I). In contrast, tongue MC had a very low labelling index. However, about 2% of skin MC were labelled, some half lying immediately below the epidermis. In this area a high proportion of the MC are immature as judged by their affinity for alcian blue.

In rats infested with Nippostrongylus brasiliensis, the percentage of lamina propria MC in S phase is increased several fold (Fig. 5), and nearly all MC are labelled by continuous infusion of ^{3}HTdR between 10 and 14 days after infestation (Table I). Most of these cells arose by cell division either locally within the lamina propria or within the gut-associated lymphoid tissues, as shown by pulse labelling at 12 days post-infestation during temporary occlusion of the superior mesenteric artery and its anastomoses, then sampling the intestine at selected times after labelling (Table II).

Thus, rat intestinal MC turn over considerably more rapidly than MC at other sites, although in skin there appears to be a subpopulation of the latter that are less mature and which divide. Most of the increase in intestinal MC numbers caused by

Table I.   Labelling of MC_2 at different anatomic sites by tritiated thymidine (^{3}HTdR) in normal and Nippostrongylus-infested rats after 4 days of continuous intravenous administration.[1]

Site sampled	Normal Rats[+]		Infested Rats	
	Number of MC scored	Percent labelled	Number of MC scored	Percent labelled
Upper jejunum[c]	435	6	1003	71
	447	8	1005	64
Jejunum[d]	378	12	1001	85
	338	5	1010	84
Ileum	154	9	1004	70
	479	7	1004	55
Tongue[e]	711	0.1	1393	0.2
	753	0.3	1413	1.2
Skin	925	2.2	886	2.0
	662	1.8	465	5.8

[a]   All rats received a loading dose of ^{3}HTdR by intraperitoneal injection (1 μCi/gram body weight) and then received a continuous intravenous infusion (1 μCi/gram body weight/24 hours) for the next 4 days. This spanned days 10–14 post-infestation in infested rats.
[b]   2 normal and 2 infested rats used.
[c]   Sampled 4 cm from the ligament of Treitz.
[d]   Sampled 8 cm from the ligament of Treitz, near the site of maximal worm burden.
[e]   Submucosa and muscle. Most labelled cells submucosal.

a nematode infestation is due to cell division and most of this appears to occur locally in the gut or its associated lymphoid tissues. The findings complement the work of Schrader et al. (27), who have demonstrated that murine gut mucosa is rich in MC precursors.

Rat bone marrow, peritoneal and intestinal MC have been investigated for surface antigens by the use of monoclonal antibodies (Table III). The most interesting antigen is that defined by W3/13 antibody. It was strongly expressed by all bone marrow MC and was absent from intestinal MC. However, peritoneal MC were heterogeneous, with the majority either negative or only weakly W3/13[+]. Among non-intestinal MC, expression of W3/13 antigen may reflect maturity, but whether its absence from intestinal MC should be taken as evidence in favour of a different lineage is not clear.

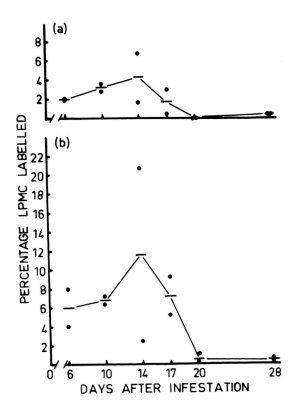

Figure 5. The proportion of LPMC in S phase in the jejuna of rats at various times after infestation with 1000 Nippostrongylus brasiliensis stage 3 larvae. Each rat received 1 μCi/gram body weight of tritiated thymidine intravenously 1 h before sacrifice. Animals were sacrificed in pairs and the horizontal bar repre-sents the mean percentage labelled LPMC (determined as in Fig. 4) in a. jejunum 4 cm from the ligament of Treitz and b. jejunum 8 cm from the ligament of Treitz, near the region of maximum worm burden.

Finally, we question whether thymus-dependency is really a diagnostic criterion restricted to intestinal MC. Cramer and Simpson (28) showed that a hyperplasia of MC occurred in the skins of mice after repeated local application of methylcholan-threne (MCA). This effect could be related to either its carcin-ogenic or its skin-sensitizing activities (29). We have studied the effect of MCA and the thymus-dependent skin sensitizing agent

TABLE II. Origin of labelled MC in the jejunal mucosae of rats infested with Nippostrongylus brasiliensis determined by vascular isolation of the small intestine during administration of tritiated thymidine (^{3}HTdR)[a]

Days post-infestation	Vascular isolation	Number MC scored	Percent labelled	Percent lightly labelled	Percent too heavily labelled
14	+	139	2.2	100	0
	+	384	4.2	94	6
	−	577	6.9	15	85
16	+	530	3.8	100	0
		520	3.1	100	0
	−	505	10.5	74	26
18	+	520	0.4	100	0
	+	525	3.4	100	0
	−	490	27.6	77	23

a. Animals received a single intravenous dose of ^{3}HTdR (1 μCi/ gram body weight on day 12 after infestation. Fifteen min later they received an intravenous dose of "cold" thymidine (3 mg). In some animals the vascular anastomoses with the stomach and large intestine had been tied off, and during the labelling period the superior mesenteric artery was temporarily occluded. They were then sacrificed either 2, 4 or 6 days after isotope injection. Jejunum was sampled for autoradiography 8 cm distal to the ligament of Treitz.

dinitrochlorobenzene (DNCB) on the local MC population. Compared to treatments with solvent alone, both compounds produced MC hyperplasia in the skins of phenotypically normal +/+ or Nu/+ BALB/c mice while in Nu/Nu mice, neither chemical had any effect on MC numbers. The results for DNCB are shown in Fig. 6. A thymus-dependent immune response can therefore lead to local hyperplasia of dermal MC, and T cells may also affect maturation of similar MC (10).

The Relationship of IEL to LGL. It is worth considering briefly the relationship of IEL (and MC) to LGL and NK cells. This is relevant because some granulated cells that have been produced in vitro have exhibited NK-like activity (30) whilst

TABLE III. Surface   antigen phenotypes of rat small   intestinal,
            bone marrow and peritoneal cavity MC defined by mouse
            monoclonal antibodies.

Mouse monoclonal	Small[b] intestine	MC Bone[c] marrow	Peritoneal[c] cavity
OX1[d]	−	±	±
OX7	?	+	+
OX8	−	−	−
OX19	−	−	−
W3/13[e]	−	+(−)	−(+)
W3/25	−	−	−

[a] See Fig. 1 for specificities.
[b] Indirect immunoperoxidase, paraffin sections of Carnoy-fixed
     tissue.   Preserves OX7, OX8 and W3/13 antigens well.   May
     reduce sensitivity for OX1, OX19 and W3/25.
[c] Indirect immunoperoxidase, ethanol-fixed smears.
[d] Staining of lamina propria too intense to distinguish MC.
[e] Most bone marrow MC strongly stained, majority of peritoneal
     cavity MC weakly stained or negative.

others have the broader cytotoxic activity attributed to MC (27).
In man (31) and rat (32), NK activity is contained in part within
the LGL subpopulation of leukocytes.   The distribution of NK
activity and LGL is characteristic and follows the pattern:
peripheral blood > spleen > LN > thymus.
     We have found a population of cells in rat lymphoid prepara-
tions that contains alcian blue-positive granules.   It has the
same tissue distribution and size as LGL scored in Giemsa-stained
smears.   A similar proportion of cells also contains granules
that are positive for NE, AP and CAE.   Thus LGL, and by inference
some NK cells, have granules containing glycosaminoglycan plus
some enzymes that are found in IEL. The exception is AP, which
is either absent or present in small amounts in granular IEL.
The functional evidence that IEL have NK activity (20) also
increases the similarity and suggests that LGL and granular IEL
may be quite closely related.   The presence of glycosaminoglycan
and CAE in LGL also invites either confusion or comparison with
MC.
     Other similarities between LGL, NK cells and IEL are a) the
sharing of markers with T lymphocytes and b) heterogeneous

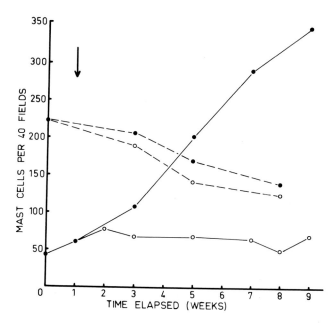

Figure 6. The effect of repeated exposure to dinitrochloro-benzene (DNCB) on the dermal mast cell populations in Balb/c Nu/+ mice (solid lines) and in Nu/Nu littermates (broken lines).   One flank of each mouse was painted twice-weekly with DNCB (0.15 in acetone)   and   the   other   with   acetone alone.   Applications commenced at the time indicated by the vertical arrow.   Points represent the means of total mast cells counted in 40 consecutive microscopic fields (x 25 objetive) of the dermis in sections of skin from the four mice in each group.   Solid symbols, DNCB: open symbols, acetone alone.

expression of markers.   With respect to LGL and NK cells, a strong case has been made in man that these are not T cells, although they express certain T cell markers during the middle portion of their differentiation (33).   In mice, a proportion of NK cells expresses Thy 1 antigen, at a level considerably below that found on T cells (34).   Rat NK cells express W3/13 and OX8 antigens, but with a heterogeneity uncharacteristic of T cells (21).

IEL in the rat have a phenotype similar to either cytotoxic T cells or NK cells.   The heterogeneous expression of W3/13 and OX8 antigens, the absence of OX19 antigen, the granular morphology and the exhibition of natural cytotoxic activity (20) suggest a closer relatedness to NK cells.   IEL in man have both similarities to and differences from LGL and NK cells (33,35) and it will be interesting to see whether they indeed represent a different lineage or simply a local variant in the differentiation pathway of the latter cells.

CONCLUSION

Rodent MC in skin, bone marrow and peritoneal cavity, as well as probably intestinal MC, can be separated from the T lymphocyte lineage. Although there are differences between them, it is not clear whether these are fundamental or whether they reflect local inductive influences on terminal maturation. Until more is understood about lineage and functions, it seems that a classification based on anatomical considerations has the virtue of being understood by most workers. It provides a clue to identification of granular cells in vivo which is lacking for those who work in vitro. This may be particularly important when considering any immature MC as they otherwise resemble intestinal MC in their staining characteristics.

The distinction between granular IEL or LGL from T cells is more controversial, although the weight of evidence now favours separate lineages. IEL and LGL have close similarities. The evidence in support of a relationship between granular IEL and MC at mucosal sites has never been compelling and is now less impressive. However, the distinction between granular cells of the IEL or LGL type from MC in vitro presents a real problem. Again, the advantage of in vivo studies is that cells are contained in defined compartments. As with MC, the definitive classification of "granulated lymphocytes" also awaits more detailed information on their lineages and functions.

REFERENCES

1. **Mayrhofer, G., and R.J. Whatley.** 1983. Granular intra-epithelial lymphocytes of the rat small intestine. I. Isolation, presence in T lymphocyte-deficient rats and bone marrow origin. Int. Arch. Allergy Appl. Immunol. 71:317.
2. **Mayrhofer, G.** 1980. Fixation and staining of granules in mucosal mast cells and intraepithelial lymphocytes in the rat jejunum, with special reference to the relationship between the acid glycosaminoglycans in the two cell types. Histochem. J. 12:513.
3. **Lojda, Z., R. Grossrau, and T.H. Schiebler.** 1979. Enzyme Histochemistry. A Laboratory Manual. Springer-Verlag.
4. **Mayrhofer, G., C.W. Pugh, and A.N. Barclay.** 1983. The distribution, ontogeny and origin in the rat of the Ia-positive cells with dendritic morphology and of Ia antigen in

epithelia, with special reference to the intestine. Eur. J. Immunol. 13:112.

5. **Enerback, L., and I. Svensson.** 1980. Isolation of rat peritoneal mast cells by centrifugation on density gradients of Percoll. J. Immunol. Methods 39:135.

6. **Burnet, F.M.** 1975. Possible identification of mast cells as specialized post-mitotic cells. Med. Hypoth. 1:3.

7. **Guy–Grand, D., C. Griscelli, and P. Vassalli.** 1978. The mouse gut T lymphocyte, a novel type of T cell. Nature, origin, and traffic in normal and graft–versus–host conditions. J. Exp. Med. 148:1661.

8. **Vicklicky, V., P. Sima, and H. Pritchard.** 1973. On the origin of mast cells in adult life. Folio Biol. (Praha.) 19:247.

9. **Kitamura, Y., M. Shimada, S. Go, H. Matsuda, K. Hatanaka, and M. Seki.** 1979. Distribution of mast-cell precursors in haemopoietic and lymphopoietic tissues of mice. J. Exp. Med. 150:482.

10. **Mayrhofer, G., and H. Bazin.** 1981. Nature of the thymus dependency of mucosal mast cells. III. Mucosal mast cells in nude mice and nude rats, in B rats and in a child with the Di George syndrome. Int. Arch. Allergy Appl. Immunol. 64:320.

11. **Ruitenberg, E.J. and A. Elgersma.** 1976. Absence of intestinal mast cell response in congenitally athymic mice during Trichinella spiralis infection. Nature 264:258.

12. **Mayrhofer, G., and R. Fisher.** 1979. Mast cells in severely T-cell depleted rats and the response to infestation with Nippostrongylus brasiliensis. Immunology 37:145.

13. **Mayrhofer, G.** 1979. The nature of the thymus dependency of mucosal mast cells. II. The effect of thymectomy and of depleting recirculating lymphocytes on the response to Nippostrongylus brasiliensis. Cell. Immunol. 47:312.

14. **Clark–Lewis, I., and J.W. Schrader.** 1981. P cell-stimulating factor: biochemical characterization of a new T cell-derived factor. J. Immunol. 127:1941.

15. **Mayrhofer, G.** 1980. Thymus–dependent and thymus–independent subpopulations of intestinal intraepithelial lymphocytes: A granular subpopulation of probable bone marrow origin and relationship to mucosal mast cells. Blood 55:532.

16. **Mayrhofer, G.** 1984. Physiology of the intestinal immune system. In: Local Immune Responses of the Intestinal Tract. Edited by Newby, T.J. and C.R. Stokes, p.1, CRC Press.

17. **Huntley, J.F., B. McGorum, G.F.J. Newlands, and H.R.P. Miller.** 1984. Granulated intraepithelial lymphocytes: their relationship to mucosal mast cells and globule leucocytes in the rat. Immunology 53:525.

18. **Parrott, D.M.V., and M.A.B. de Sousa.** 1974. B cell stimulation in nude (Nu/Nu) mice. In: Proceedings of the First International Workshop on Nude Mice. Edited by Rygaard, J., and C.O. Poulsen. Stuttgart., Fisher, p. 61.

19. **Arnaud–Battandier, F., B.M. Bundy, M. O'Neill, J. Bienenstock, and D.L. Nelson.** 1973 Cytotoxic activities of gut mucosal lymphoid cells in guinea pigs. J. Immunol. 121:1059.

20. **Flexman, J.P., G.R. Shellam, and G. Mayrhofer.** 1983. Natural cytotoxicity, responsiveness to interferon and morphology of intraepithelial lymphocytes from the intestine of the rat. Immunology 48:733.

21. **Cantrell, D.A., R.A. Robbins, C.G. Brooks, and R.W. Baldwin.** 1982. Phenotype of rat natural killer cells defined by monoclonal antibodies marking rat lymphocyte subsets. Immunology 45:97.

22. **Dillon, S.B., and T.T. MacDonald.** 1984. Functional properties of lymphocytes isolated from murine small intestinal epithelium. Immunology 52:501.

23. **Dallman, M.J., D.W. Mason, and M. Webb.** 1982. The roles of host and donor cells in the rejection of skin allografts by T cell–deprived rats injected with syngeneic T cells. Eur. J. Immunol. 12:511.

24. **Jarrett, E.E.E., and D.M. Haig.** 1984. Mucosal mast cells in vivo and in vitro. Immunol. Today 5:115.

25. **Hatanaka, K., Y. Kitamura, and Y. Nishimura.** 1979. Local development of mast cells from bone marrow–derived precursors in the skin of mice. Blood 53:142.

26. **Miller, H.R.P., and W.F.H. Jarrett.** 1971. Immune reactions in mucous membranes. I. Intestinal mast cell response during helminth expulsion in the rat. Immunology 20:277.

27. **Schrader, J.W., R. Scollay, and F. Battye.** 1983. Intramucosal lymphocytes of the gut: Lyt–2 and Thy–1 phenotype of the granulated cells and evidence for the presence of both T cells and mast cell precursors. J. Immunol. 130:558.

28. **Cramer, W., and W.L. Simpson.** 1944. Mast cells in experimental skin carcinogenesis. Cancer Res. 4:601.

29. **Old, L.J., B. Benacerraf, and E. Carswell.** 1963. Contact sensitivity to carcinogenic polycyclic hydrocarbons. Nature 198:1215.

30. **Galli, S.J., A.M. Dvorak, T. Ishizaka, G. Nabel, H. Der Simonian, H. Cantor, and H.F. Dvorak.** 1982. A clonal cell with NK function resembles basophils by ultrastructure and expresses IgE receptors. Nature 298:288.

31. **Saksela, E., T. Timonen, A. Ranki, and P. Hayry.** 1979. Morphological and functional characterization of isolated effector cells responsible for human natural killer activity to fetal fibroblasts and to cultured cell line targets. Immunol. Rev. 44:71.

32. **Reynolds, C.W., T. Timonen, and R.B. Herberman.** 1981. Natural killer (NK) cell activity in the rat. I. Isolation and characterization of the effector cells. J. Immunol. 127:282.

33. **Horwitz, D.A., and A.C. Bakke.** 1984. An Fc receptor–bearing, third population of human mononuclear cells with cytotoxic

and regulatory function. Immunol. Today 5:148.

34. **Herberman, R.B., and H.T. Holder.** 1978. Natural cell-mediated immunity. Adv. Cancer Res. 27:305.

35. **Cerf–Bensussan, N., E.E. Schneeberger, and A.K. Bhan.** 1983. Immunohistologic and immunoelectron microscopic characterization of the mucosal lymphocytes of human small intestine by the use of monoclonal antibodies. J. Immunol. 130:2615.

Mast Cell Differentiation and Heterogeneity,
edited by A. D. Befus et al.
Raven Press, New York © 1986.

A Growth Factor for Human Mast Cells

Beda M. Stadler, Koichi Hirai, *Thomas Schaffner,
and Alain L. de Weck

*Institute of Clinical Immunology and *Institute of Pathology,
University of Bern, Bern, Switzerland*

Murine mast cell growth factor has recently been shown to be identical with a number of other synonymous factors, including P cell stimulating factor, WEHI-3 derived growth factor, histamine inducing factor or interleukin 3 (1). The sequence of murine interleukin 3 is well established and the gene has been cloned (2). Nevertheless, a human counterpart for interleukin 3 has not yet been characterized. Therefore, we have utilized IL-3-dependent cell lines of mouse origin to study whether human tissue-derived activities could be found to promote their growth and thus potentially develop an assay for a human IL-3-like factor. As recently published, a subclone of the murine 32Dcl line (3) proliferated in the presence of a growth factor derived from human lectin-stimulated peripheral blood mononuclear cells (4). This growth factor was different from a growth factor for human basophil-like cells, which we termed basophil promoting activity (BaPA) (5), in terms of its molecular weight and isoelectric points. In the present studies, we have investigated whether this human growth factor that induced the growth of IL-3-dependent murine cells is identical to human mast cell growth factor.

MATERIALS AND METHODS

Factor Production. Human mononuclear cells were isolated by Ficoll-Hypaque sedimentation. 2–4 x 10^6 cells/ml were cultured in RPMI 1640 and stimulated with PHA 10 μg/ml (Difco, Detroit,

Supported by the Swiss National Science Foundation, Grant No. 3.500-0.83
Abbreviations: BaPA, basophil promoting activity; IL-2, interleukin 2; IL-3, interleukin 3; ^{3}H-TdR, ^{3}H-thymidine.

MI, USA). Supernatants (PHA-LCM) were harvested by centrifugation and stored at -20°C until use.

Biochemical Procedures. Conditioned medium from mononuclear
cell cultures was concentrated on membranes with a molecular
cut-off of 10,000 daltons. Concentrates were precipitated by
addition of 80% solid ammonium sulfate. Pellets were dialysed
against 10 mM Tris-HCl starting buffer pH 8.0 and applied on DEAE
Sephacel (Pharmacia, Uppsala, Sweden) equilibrated in starting
buffer. The DEAE column was washed with starting buffer supplemented with 60 mM NaCl. Active materials were eluted stepwise
(150 mM NaCl). Eluted activity was applied on a phenyl-Sepharose
CL-4B column and subsequently eluted with 40% ethylene glycol.
Peak fractions were concentrated by Amicon filtration (YM5) and
loaded on a 95 x 5cm AcA 54 column; finally, eluted material was
used as a source of IL-3-like activity.

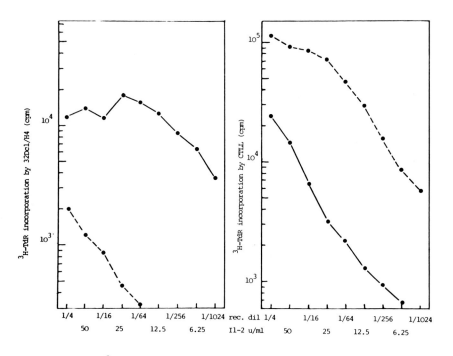

Figure 1. ^{3}H-TdR incorporation by 32Dcl/H4 cells (left panel)
or CTLL cells (right panel) in the presence of semi-purified IL-
3-like activity or 100 U/ml recombinant IL-2 (broken line).

Bone Marrow Cultures. Ficoll-Hypaque isolated bone marrow cells from normal human donors undergoing orthopedic surgery were cultured in suspension in McCoy 5A at a density of $0.33 - 1 \times 10^6$ cells/ml. Cultures were supplied with various concentrations of semi-purified PHA-LCM (above), as a source of IL-3-like activity.

RESULTS

Human IL-2 and IL-3-like Activity Assays on 32Dcl/H4 Cells. It has recently been shown that the 32Dcl line also responds to growth factors other than murine IL-3, e.g. IL-2 (6). Figure 1 shows the response, as measured by ^{3}H-TdR incorporation, of a subclone of the 32Dcl line to both human recombinant IL-2 (Cetus Corp., Emeryville, CA, USA) and to semi-purified IL-3-like activity from human PHA-LCM. The subclone of 32Dcl actually responded to a small degree to recombinant IL-2 (100 U/ml). However, as shown on the right-hand side of Figure 1, the murine IL-2-dependent cell line CTLL showed a greater response to

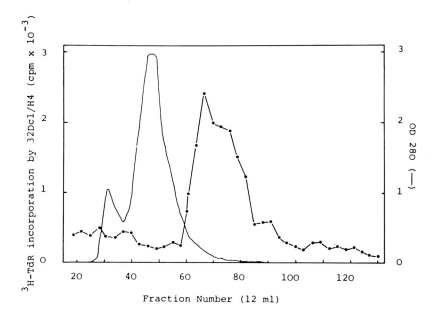

Figure 2. AcA 54 gel filtration (90 x 5cm) of PHA stimulated human mononuclear cell supernatant concentrated 30 times by diafiltration and subsequently precipitated by ammonium sulfate then chromatographed by DEAE ion exchange followed by phenyl-Sepharose chromatography, as described in Materials and Methods. Dotted line represents ^{3}H-TdR incorporation by 32Dcl/H4 cells in the presence of 1/20 dilution of each fraction.

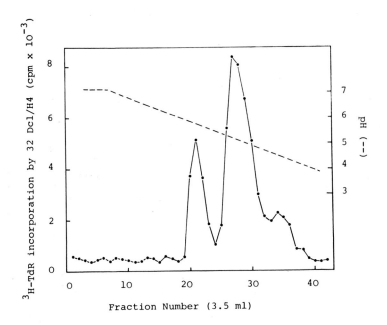

Figure 3.　Chromatofocusing of IL-3-like activity purified　as described in Figure 2.　Each fraction was tested at a dilution of 1/32 in the microassay using the 32Dcl/H4 cell line ( ).

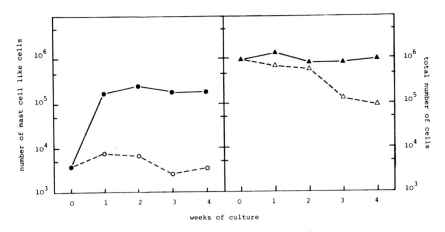

Figure 4.　Liquid cultures of human bone marrow cells　(broken lines) and in　the presence　of semi-purified IL-3-like　activity (closed lines).

recombinant IL-2 than to IL-3-like activity. Thus, 32Dcl/H4 cells seemed to respond to both growth factors, but the dose response curves for each differed. Crude conditioned medium from 3 day PHA stimulated human mononuclear cells contained usually less than 10 U/ml IL-2, and only minimal interference of IL-2 in the IL-3 assay was observed.

Assay for Semi-purified Human IL-3-like Activity Using 32Dcl/H4 Cells. We used the 32Dcl/H4 line to assay for purified human IL-3-like activity. The factor was produced by PHA stimulation of human mononuclear cells. Batches of 20-30 liters of human mononuclear cell supernatants were concentrated and then subjected to DEAE chromatography. Subsequently, active materials were passed over phenyl-Sepharose columns and the effluent activity was concentrated and subjected to AcA 54 gel filtration. Figure 2 gives a typical example of the optical density and ^{3}H-TdR incorporation of material from PHA-LCM separated after this step. Figure 3 shows the isoelectric points of the IL-3-like activity eluted from gel filtration. The human IL-3-like activity had isoelectric points between pH 6.0 to 4.2.

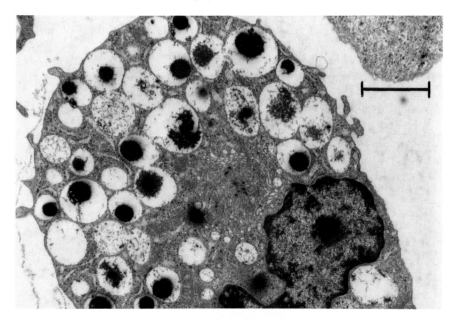

Figure 5. Electron micrograph of mast cell-like cell-derived from cultured human bone marrow in the presence of semi-purified human IL-3-like activity. Epoxy-embedded glutaraldehyde/osmium fixed cell suspension contrasted with Sato's potassium permanganate-lead staining sequence. Bar represents 2 μm. A typical mononuclear cell is shown containing numerous electron-dense inclusions and vacuoles containing fluffy granular material.

Normal Human Bone Marrow Cultures in Presence of Human
IL-3-like Activity. Such semi-purified material was subsequently
used for the stimulation of normal human bone marrow cultures.
The cells were cultured with a source of IL-3-like activity
corresponding to approximately 200% of the activity originally
present in crude PHA-LCM.   In these bone marrow cultures in the
presence of PHA-LCM-derived IL-3-like activity, the number of
metachromatic cells resembling mast cells under light microscopy
increased usually during the first 2 weeks of culture.   Figure 4
shows a typical experiment where only in the presence of IL-3-
like activity did mast cell-like cells appear in culture.   These
mast cell-like cells stained positively with alcian blue, and
histamine content was below 0.2 pg/cell. The percentage of mast
cell-like cells in culture ranged between 40-60%.   Figure 5
represents a typical mast cell-like cell by electron microscopy.
Granule contents were usually homogenous in the centre but not
completely filled with electron dense material as observed in
tissue mast cells.

DISCUSSION

The murine 32Dcl/H4 cell line seems to be a suitable cell line
for the assessment of human IL-3-like activity.   As shown by
others, this cell line is not absolutely specific for IL-3 (6).
However, under appropriate conditions, the marginal effect of
IL-2 can be excluded. In highly purified preparations containing
IL-3-like activity we were able to demonstrate that 99% of the
IL-2 was effectively removed as determined by the CTLL cell line.
Nevertheless, when the 32Dcl cell line was used to monitor the
purification of a factor in human lectin-stimulated mononuclear
cell supernatants (PHA-LCM), we obtained a material that stimulated the growth of mast cells in human bone marrow suspension
cultures. A high percentage of these cultured cells resembled
mast cells on light microscopy; electron microscopical evaluation
showed similarity to cultured mast cells of mouse origin.   The
stimulation of IL-3-dependent growth of murine cell lines and the
in vitro production of mast cell-like cells represent two
criteria for this human factor in PHA-LCM to be considered
equivalent to murine interleukin 3. The assumption is made that
interleukin 3 has, in the human, the same broad spectrum as it
has in the mouse, where IL-3 seems to be identical with a mast
cell growth factor (1).   It has already been shown for other
growth factors such as interleukin 2 that only a limited species
restriction exists.   Human interleukin 2 can very easily and
precisely be assessed by mouse IL-2-dependent cells (CTLL) (7).
Thus, it might be possible that the human interleukin 3-like
activity can be assessed by a mouse IL-3-dependent cell line. So
far we have no explanation for the finding that human recombinant
IL-2 also has an effect on the mouse IL-3-dependent cell line,
even though this effect is not very pronounced.   In these
experiments 100 units of recombinant IL-2 were used, an amount

which is approximately 5–10 times more than what is found in crude supernatants of lectin–stimulated mononuclear cells used for the production of the IL–3–like activity.

The cultured bone marrow cells have been termed mast cell–like because of their resemblance to mast cells on light microscopy, using May–Grunwald–Giemsa staining. Additionally, cells stain positively when stained by alcian blue and have low histamine content. A resemblance to mast cells is also shown on the electron microscope pictures. While these cultured cells do not show the complete range of mast cell ultrastructural morphology, they resemble mast cells more closely than other cell types.

REFERENCES

1. Ihle, J., J. Keller, S. Oroszan, L.E. Henderson, T.D. Copeland, F. Fitch, M.B. Prytoswky, E. Golwasser, J.W. Schrader, E. Palazynski, M. Dy, and B. Lebel. 1983. Biological properties of homogeneous interleukin 3. I. Demonstration of Wehi–3 growth factor activity, mast cell growth factor activity, P cell stimulating factor activity, colony stimulating factor activity, and histamine–producing cell–stimulating factor activity. J. Immunol. 131:282.
2. Fung, M.C., A.J. Hapel, S. Ymer, D.R. Cohen, R.M. Johnson, H.D. Campbell, and I.G. Young. 1984. Molecular cloning of cDNA for murine interleukin 3. Nature 307:233.
3. Greenberger, J.S., R.J. Eckner, M. Sakakeeny, P. Marks, D. Reid, G. Nabel, A.J. Hapel, J.N. Ihle, and K.C. Humphries. 1983. Interleukin 3–dependent hematopoietic progenitor cell lines. Fed. Proc. 42:2762.
4. Stadler, B.M., K. Hirai, K. Tadokoro, and A.L. de Weck. 1985. Distinction of the human basophil promoting activity from human interleukin 3. Int. Arch. Allergy Appl. Immunol. 77:151.
5. Tadokoro, K., B.M. Stadler, and A.L. de Weck. 1983. Factor-dependent in vitro growth of human normal bone marrow-derived basophil like cells. J. Exp. Med. 158:857.
6. Hapel, A.J., H.S. Warren, and D.A. Human. 1984. Different colony–stimulating factors are detected by the "interleukin-3"–dependent cell lines FDC–Pl and 32Dcl–23. Blood 64:786.
7. Stadler, B.M., S.F. Dougherty, J.J. Farrar, and J.J. Oppenheim. 1981. Relationship of cell cycle to recovery of IL–2 activity from human mononuclear cells, human and mouse T cell lines. J. Immunol. 127:1936.

Mast Cell Differentiation and Heterogeneity,
edited by A. D. Befus et al.
Raven Press, New York © 1986.

Mast Cell Heterogeneity: Can Variation in Mast Cell Phenotype Be Explained Without Postulating the Existence of Distinct Mast Cell Lineages?

Stephen J. Galli

Department of Pathology, Harvard Medical School; and Department of Pathology and Charles A. Dana Research Institute, Beth Israel Hospital, Boston, Massachusetts 02215

In order to discuss variations in mast cell phenotypes clearly, it is important to define some of the important terms related to this area. In distinguishing between mast cells and basophils, "mast cells" will designate cells which ordinarily mature outside of the bone marrow or circulation, generally in the connective tissues or serosal cavities (in rodents), which bear high affinity IgE receptors, and which possess prominent cytoplasmic granules that stain metachromatically with certain basic dyes. In those cases where they have been analyzed, mast cell cytoplasmic granules have been found to contain sulfated glycosaminoglycans/proteoglycans and vasoactive amines (1,2). Basophils also bear high affinity IgE receptors and contain prominent cytoplasmic granules incorporating sulfated glycosaminoglycans and histamine; however, basophils have the kinetics and natural history of granulocytes: they ordinarily mature in the bone marrow, circulate in the blood, and retain certain characteristic ultrastructural features even after migrating into the tissues during inflammatory responses (1). Identifiable mast cells appear to retain at least latent or limited proliferative capacity, even in such sites as the peritoneal cavity (3) or skin

---

Supported by U.S.P.H.S. Grants AI20292 and CA28834.

Abbreviations: CTMC, connective tissue–type mast cell; IL–3, interleukin 3; MMC, mucosal mast cell.

(4). By contrast, there is no convincing evidence that mature basophils, whether in the circulation or in the tissues, express mitotic capability.

With regard to the definition of heterogeneity (see 5), cells within a given population can be said to exhibit "heterogeneity" once a certain minimum (but generally undefined) level of variation in one or more of their characteristics has been demonstrated; however, this is purely a descriptive term. For example, a particular cell type can be said to exhibit "heterogeneity" with respect to size, ultrastructural appearance, mediator content, or any combination of these, and/or other characteristics. Yet the occurrence of such "heterogeneity" may reflect the operation of many different underlying mechanisms.

The terms "differentiation" and "maturation" are problematic. Differentiation may be defined as "specialization; the acquisition or the possession of character or function different from that of the original type" (6). By contrast, "maturation" can be defined as "the process of achieving full development or growth" (6). It would seem reasonable to conclude from these definitions that most or all processes of cellular "differentiation" can be subsumed within the broader process of "maturation", but that certain maturational events (e.g. the storage of increasing amounts of histamine by "maturing" mast cells), may not properly be considered examples of "differentiation". To avoid confusion, in this paper the process by which mature mast cells develop from their precursors will be termed "maturation/differentiation".

Variation in the morphological, biochemical and/or functional characteristics of mast cells derived from different anatomical locations (or even involving mast cells from the same organ or site) has been reported in several mammalian species (1,2,7–24); thus, they exhibit heterogeneity. The distinction in the literature between rodent "connective-tissue-type: and "mucosal-type" mast cell subsets is a classical example of "heterogeneity".

Maximow was probably the first to recognize that certain mast cells in the rat intestinal mucosa were atypical in that their staining characteristics differed from those of mast cells observed in other anatomical sites (7). Enerback greatly extended these observations and defined conditions of fixation and histochemical staining which discriminated between such atypical or mucosal mast cells (MMC) and the "connective-tissue-type" mast cells (CTMC) of the skin, peritoneal cavity, and other sites in the rat (14). In addition to differing in morphological and histochemical characteristics, MMC and CTMC, if this distinction can be applied to species other than the rat, appear to differ in many other aspects of natural history, biochemistry, function and role in inflammation and immunity (1,2,7–24).

The possible basis for phenotypic heterogeneity of the mast cell includes: (1) the existence of distinct mast cell lineages; (2) the process of cellular maturation/differentiation; (3) the functional status of the cell; and (4) the influence of microenvironmental factors.

Distinct mast cell lineages. What precisely constitutes a cellular lineage is a moot point, although two classical examples immediately come to mind: granulocytes and lymphocytes.

Granulocytes. The sequence of branch points in the generation of neutrophils, eosinophils and basophils from their less differentiated precursors, it will be agreed, leads to the expression of distinct cell types. Two important sets of observations support this opinion: first, the three types of granulocytes exhibit clear qualitative and/or quantitative differences in multiple characteristics, including morphology, biochemistry and function; second, the commitment of these cells to their own particular phenotypic pattern appears to be irrevocable. We know of no set of factors which can influence eosinophils to become neutrophils, or neutrophils to become basophils. At the molecular level, this irrevocable phenotypic commitment must reflect a stereotyped and, under physiological conditions, apparently irreversible alteration in the pattern of gene expression which occurs during the maturation/differentiation of the three different granulocytes from their common precursors.

Similar molecular events may also underlie the generation of lymphocyte subsets from their less differentiated precursors. But one distinction between granulocyte and lymphocyte development which has particular bearing upon the controversy about mast cell subsets should be noted. In contrast to the differentiation of the three granulocyte lineages, which may be readily monitored using morphological approaches, conventional morphology is considerably less informative about the process of lymphocyte differentiation. It may be very difficult to distinguish some B-cells from some T-cells by their morphological features, even at the ultrastructural level, despite the profound differences in their natural history and patterns of gene expression.

In view of what is known about the differentiation of granulocytes and lymphocytes, what points can be made about the issue of mast cell subsets? Arguing from the natural history of granulocytes, it might be required that proof of the existence of distinct mast cell subsets include not only demonstration of predictable differences in phenotype among the putative subsets, but also evidence of irrevocable commitment to the expression of such differences. In other words, the demonstration of differences in morphology, histochemistry, biochemistry and/or function may not be sufficient to identify distinct mast cell lineages, unless these differences are irreversible. The natural history of lymphocyte differentiation teaches a different lesson: cells which appear quite similar may exhibit striking differences in gene expression and function. Thus, the existence of distinct mast cell lineages cannot be excluded simply because the population of mast cells examined appears morphologically homogeneous.

Maturation/differentiation. It may not be simple to establish the existence of distinct mast cell subsets analogous to the different classes of granulocytes or lymphocytes. However, there are alternative mechanisms which may account for at least some of the phenotypic diversity observed in mast cell populations in vivo. For example, consider the striking changes in phenotype which can accompany maturation/differentiation within a single cell lineage. Again, two examples come to mind. Circulating monocytes and tissue or peritoneal exudate macrophages exhibit many quantitative and qualitative differences in phenotype, yet the latter populations clearly are derived from the former (25).

B cells exhibit what is arguably an even more complex natural history, which includes striking alterations in their mitotic potential, morphology, and gene expression (26). Immature B cells may resemble small T cells by morphology, and both of these cell types appear strikingly different from mature plasma cells. Yet, despite the differences in their appearance, immature B cells give rise to plasma cells. The natural history of the B cell also illustrates how cells in the same lineage can express seemingly fundamental differences in important gene products. In the development of a clone of B cells capable of secreting IgG antibodies, the cells pass through a stage during which they synthesize and express IgM of the same idiotype (26). This remarkable "isotype switch" is now fully accepted as a critical and characteristic feature of B cell maturation/differentiation. But if the mechanisms accounting for this "change of phenotype" were not appreciated, it might be tempting to speculate that B cells synthesizing IgM must represent a "lineage" distinct from those synthesizing IgG.

How do these examples influence an understanding of mast cell heterogeneity? Kitamura and his colleagues (27) and Zucker-Franklin et al. (28) already have pointed out certain similarities between the natural history of monocytes/macrophages and mast cells: both are derived from the bone marrow, and both circulate as forms which differ morphologically from the corresponding mature populations in the tissues. Another similarity is apparent: like murine mast cells, the derivatives of circulating monocytes may express distinct phenotypes in different anatomical locations (e.g. alveolar macrophages, Kupffer cells, etc.) (25). To what extent these distinct derivatives of circulating monocytes are irrevocably committed to their particular pattern of phenotypic expression remains to be determined. But at least some of the properties of these cells may be strikingly altered in response to microenvironmental factors (25). For example, under certain circumstances tissue macrophages may transform into "epithelioid" cells, or form different varieties of multinucleated giant cells.

Murine mast cells may also exhibit striking phenotypic changes during their maturation/differentiation in the tissues and serous cavities, including changes in the cells' mitotic capability and content of biogenic amines, and alterations of cytoplasmic granule size, number and histochemistry (29–32). Thus, morphological and histochemical studies indicate that the immature CTMC of murine rodents exhibit a high proliferative potential, contain low levels of vasoactive amines, and synthesize sulfated glycosaminoglycans other than heparin (31). These characteristics also appear to apply to murine MMC (1,2,10,11,14,15,17,19,21,24), raising the possibility that some of the features of MMC may reflect their stage of maturation/differentiation rather than their membership in a distinct mast cell lineage.

Functional characteristics. Profound alterations in mast cell ultrastructure accompany the stimulus-induced release of the cells' stored and newly generated mediators (1,33-35). While rapid, "anaphylactic" patterns of mast cell degranulation have been the most intensively studied (1,33-35), ultrastructural investigations suggest that mast cells may also be able to release granule-associated mediators in a slower, more gradual fashion (reviewed in 1). Both "anaphylactic" and "piecemeal" patterns of mast cell degranulation can alter the staining characteristics of the cell, probably because of changes in the physicochemical properties of granule-associated proteoglycans and other granule constituents. Furthermore, mast cells stimulated to undergo degranulation can synthesize a new complement of cytoplasmic granules (reviewed in 1), and these newly synthesized granules may differ in appearance from those in mature, unstimulated cells. Many of the features of cellular immaturity, degranulation or regranulation are most easily recognized and evaluated at the ultrastructural level. As a result, it may be prudent to examine the ultrastructure of populations of mast cells which exhibit "heterogeneity" by routine light microscopy or histochemistry, in order to determine whether any of the distinctive light microscopic features of the cells might reflect recent or ongoing stimulation of secretion, or regranulation.

Microenvironmental factors. At least some variation in mast cell phenotype, perhaps including the distinction between CTMC and MMC, might reflect the influence of microenvironmental factors. One way the microenvironment may affect mast cell phenotype is by influencing the cells' proliferation (since immature mast cells exhibit different features from mature mast cells) and/or by influencing the rate or extent of maturation/differentiation. An obvious candidate to act by this mechanism is the molecule bearing the names IL-3 (36,37), mast cell growth factor (38-43), and P cell stimulating factor (44,45). The effects of these molecules on mast cell proliferation have been discussed primarily in the context of "mucosal" (or "T-cell-dependent") mast cells. But modest proliferation of cutaneous mast cells has been observed in association with T-cell-dependent reactions in human skin (4), and preliminary observations suggest that the same may be true in mice (SJ Galli, unpublished data). Whether such "connective tissue-type" mast cell proliferation reflects the influence of IL-3 remains to be determined.

Many other mechanisms by which the microenvironment might influence mast cell phenotype could be proposed. For example, factors which influence mast cell function can alter the cells' morphology and mediator content, and the concentration of these factors may vary as a function of anatomical site or proximity to local immunological or pathological processes. Such agents include IgE and antigen, anaphylatoxins, and certain basic peptides, among a host of other possibilities (reviewed in 1,2,15).

The agents already mentioned and/or other factors might affect mast cell phenotype by influencing the cells' stage of maturation /differentiation. This general mechanism is compatible with at least two different models of mast cell heterogeneity. In the first, local factors might affect mast cell phenotype by promoting irreversible changes in patterns of gene expression, resulting in vectors of mast cell differentiation which vary acording to anatomical site. In this scheme, mast cell heterogeneity would reflect the locally regulated radiation of a single mast cell precursor into distinct, committed lineages (e.g. CTMC, MMC). Alternatively, mast cell heterogeneity might reflect the influence of local factors whose net effect is to position that particular mast cell population at a certain point along a single vector of maturation/differentiation. In this model, the various phenotypes exhibited by mast cells in different anatomical sites would not require irrevocable commitment to a distinct branch of maturation/differentiation. It might be possible, for example, to define conditions which promote the maturation/differentiation of MMC (or their progeny) to CTMC.

Discussions of the effect of the microenvironment on mast cell phenotype generally focus on how local factors might influence patterns of mast cell gene expression. However, one other mechanism which can alter mast cell phenotype deserves brief mention. It has been recognized for some time that mast cells can take up a variety of substances from their immediate environment and store them in their cytoplasmic granules (reviewed in 46). This process clearly may produce a certain amount of phenotypic "heterogeneity". One example immediately comes to mind. There has been some controversy about whether basophils or mast cells contain peroxidase activity (reviewed in 47,48). In collaboration with Seymour Klebanoff, Bill Henderson and Terry Ishizaka, we recently provided one possible explanation for variable expression of peroxidase activity by basophils and mast cells (47,48). We found that freshly isolated guinea pig peripheral blood basophils (47), cloned mouse mast cells maintained in vitro (47), and human basophils generated in vitro (48), lacked peroxidase activity in Golgi structures and rough endoplasmic reticulum, indicating that the cells were not synthesizing the enzyme. But all three cell types were able to take up eosinophil peroxidase within endocytotic vesicles. The peroxidase-positive vesicles then fused with the cytoplasmic granules, which eventually became markedly positive for the enzyme activity by ultrastructural cytochemistry. These experiments reveal a mechanism by which mast cell (or basophil) phenotype can be altered in the apparent absence of a significant change in the cells' pattern of gene expression.

One final body of information relevant to the issue of mast

cell heteroeneity in murine rodents should be considered: data derived from studies of mast cells generated in vitro. In 1981, several groups, including our own, reported that cells with many of the features of mast cells developed in cultures of normal mouse hematopoietic cells maintained in media containing macromolecules derived from mitogen-activated T cells, cloned Ly 1^+2^- inducer T-cells, or WEHI-3B tumor cells (38,44,49-51). Hasthorpe had reported a similar finding in 1980 (52), although her mast cell-like cell line (which contained c-type virus particles by electron microscopy) was derived from the spleen of a mouse previously injected with cell-free supernatant from Friend-virus producing erythroleukemia cells. All of these observations represented extensions of seminal work by Haim Ginsburg (51,52) and Terry Ishizaka (53,54) and their associates, who grew mast cells in vitro from cells in mouse or rat lymphoid tissue.

It was apparent from the beginning that the mouse mast cells grown in suspension cultures differed in certain important respects from the best characterized cell type available for detailed comparative studies: mature peritoneal mast cells (reviewed in 1). The cultured cells appeared immature by ultrastructure (1,38,57), contained low levels of histamine (1,38,49-51,57-59), and expressed fewer surface receptors for IgE immunoglobulin than did mature peritoneal mast cells (57). In addition, we (57) and others (59-61) found that the cultured mast cells incorporated $Na_2{}^{35}SO_4$ into granule-associated chondroitin sulfates. These were characterized in detail by Ehud Razin, Rick Stevens and Frank Austen and their associates, who showed that the molecules included an over-sulfated chondroitin designated chondroitin sulfate E (60,61). In contrast to the cultured cells, normal mouse peritoneal mast cells synthesized heparin (57,60). Finally, cultured mast cells and peritoneal mast cells were shown to differ in their surface antigen expression (62) and in the products of arachidonic acid oxidation they elaborated upon stimulation (63).

We (1,38,57) and several other investigators (18,59,61,64) noted that cultured mouse or rat mast cells expressed similarities to "T-cell-dependent" or "mucosal" mast cells. "Mucosal" mast cells contain low levels of histamine (14,17) and microspectrophotometric evidence suggests that their granules contain a poorly sulfated glycosaminoglycan similar to chondroitin sulfate (10). Cultured mast cells and MMC also exhibit similar patterns of responsiveness to certain secretagogues (59) and, in the rat, contain a similar or identical granule-associated protease (18). Furthermore, the growth of both cultured mast cells and MMC may be regulated by similar signals. In normal mice and rats, striking proliferation of MMC occurs during T-cell-dependent responses to certain intestinal parasites (for reviews, see 1, 13-15,19,21-23,64). By contrast, athymic nude mice lack such a response (8). The nude mouse intestinal mucosa contains MMC precursors (22), however, and MMC proliferation does occur if nude mice are reconstituted with T-cells prior to

intestinal infection (8).   Cultured mast cells proliferate   in
response   to   mast   cell   growth   factor   (IL-3),   which   can   be
elaborated by   cloned Ly 1^+2^- inducer T-cells   (38-40,57).   In
addition, Crapper   et al   (65)   reported that   the   survival   of
cultured mast cells injected into the skin of (B10pd x DBA/2)F_1 -
W^f/W^f mast   cell   deficient   mice   required that   the   mice   also
receive a subcutaneous inoculum of the WEHI-3B tumor as a   source
of mast cell growth factor.

Although some investigators have suggested that cultured   mast
cells may be committed to express the MMC phenotype, we (1,38,57)
and Yung and Moore (43) have argued that an alternative   hypothesis cannot be excluded: that many of the properties   of cultured
mast cells might reflect their immaturity. But attempts to induce
further maturation of mast cells in suspension culture met   with
only limited success. The inducing agent sodium butyrate   caused
a marked inhibition of cultured mast cell proliferation, resulted
in increased storage   of histamine and chondroitin sulfate,   and
favoured partial maturation of cytoplasmic   granules (57).   The
cells did not appear fully mature by ultrastructure, however, nor
did they synthesize detectable amounts of heparin (57).

Professor Kitamura has reported   the results of a very   interesting study incorporating a different strategy: cultured mast
cells were injected into mast cell-deficient (WB x C57BL/6)F_1 -
W/W^v mice (which exhibit a more profound mast cell deficiency
than do the (B10pd x DBA/2)F_1 - W^f/W^f mice used by Crapper et al.
(65)), and the histochemical characteristics and ultrastructure
of the adoptively transferred mast cell population were   examined
at   different   intervals   up   to   10   wk   later   (66).   Similar
experiments were   performed using   partially purified peritoneal
mast cells from (WB   x C57BL/6)F_1 - +/+ mice. When injected in
vivo, either the cultured mast^1 cells or the freshly isolated
peritoneal mast cells gave rise to mast cells in several
different anatomical sites.   In the peritoneal cavity, skin,
spleen and glandular stomach muscularis propria, these mast cells
exhibited histochemical   features of CTMC.   By   contrast,
adoptively transferred mast cells identified in the mucosa of the
glandular stomach resembled MMC. These results suggest that   the
phenotype of the injected mast cells or their progeny can be
strikingly influenced by microenvironmental factors.

It will be important to repeat these experiments with cloned
mast cells (to evaluate the possibility that the uncloned populations contained mast cells of two or more distinct committed
lineages), and to confirm, by direct biochemical analysis, that
the adoptively transferred cultured cells can give rise to mast
cells capable of synthesizing heparin in vivo. But the evidence
now in hand suggests that at least some populations of cultured
mast cells can generate either MMC or CTMC in vivo, depending on
the microenvironment.   And Professor Kitamura's studies have
revealed what may appear to be an even more startling finding: in
the appropriate microenvironment, peritoneal mast cells (generally regarded as prototypical CTMC) can give rise to mast cells

with at least some of the features of MMC. This observation raises a number of questions which require additional study. Does peritoneal mast cell "transdifferentiation" reflect a change in the major proteoglycan synthesized by the cells or their progeny (i.e. heparin-chondroitin sulfates), or is it due to some other mechanism affecting the cells' histochemistry? To what extent do the other features of the "transdifferentiated" cells resemble those of native MMC? Whatever the answers to these questions turn out to be, Professor Kitamura's experiments illustrate a powerful and potentially useful approach to the issue of mast cell heterogeneity.

Why should we be concerned about the occurrence of "heterogeneity" in murine mast cell populations, whatever mechanisms are involved? Simply because "heterogeneity" of mast cells may be the rule rather than the exception in mammalian species. Even though the greatest amount of data regarding mast cell heterogeneity has been derived from work in mice and rats, there have been recent reports of "heterogeneity" of human mast cell characteristics as well. Human pulmonary mast cells vary in cytoplasmic granule ultrastructure (67,68), size (69) and responses to stimuli of degranulation (69), and human intestinal mast cells exhibit variation in histochemistry (70-72). It should be noted, however, that the separation of mast cell populations into distinct "mucosal" and "connective tissue" subclasses may not be as clear in humans as it is in mice and rats. For example, Dean Befus and his associates found that the majority of mast cells in the submucosa and muscularis propria, as well as in the mucosa of the human large intestine, exhibited histochemical features resembling those of murine "mucosal" mast cells (72); and when Charity Fox and her associates isolated mast cells from human colonic lamina propria, they found that the colonic mast cells resembled purified human lung mast cells in ultrastructure, histamine content, responsiveness to anti-IgE and pattern of arachidonic acid metabolite production (73). The cells did vary in size and ultrastructure; however, these findings were thought possibly to reflect differences in the degree of mast cell maturation (73). It remains to be determined whether the variation in mast cell ultrastructure observed by Fox et al was in any way related to the variation in histochemical staining detected by Befus et al.

Identifying the basis for mast cell heterogeneity in murine rodents, and determining whether similar mechanisms operate in man may be of more than academic interest. If a phenotypically distinct mast cell population analogous to the murine MMC does exist in man, and if this "subset" has an important role in certain human diseases, immunotoxins which can ablate this population based on recognition of surface structures not shared by mature human CTMC may be of significant therapeutic value. However, if the human MMC represents one developmental stage along a single vector of mast cell maturation/differentiation which also produces mast cells of the "connective tissue-type", the clinical consequences of employing such immunotoxins might be very different.

In view of the many important questions about mast cell "heterogeneity" which await definitive resolution, caution is advisable in endorsing any one system of nomenclature for classifying mast cell populations into "subsets". The advantages and disadvantages of the various schemes already proposed have been ably discussed elsewhere (21,23) and will not be reconsidered here. Suffice it to say that a proliferation of competing systems of terminology can be confusing, particularly for scientists not familiar with the field. One approach (certainly not the only one) would be simply to refer to mast cells as "mast cells", taking care to specify their origin (anatomic site and/or type of culture system) and to characterize as many of their features as possible. The construction of a more elaborate system of nomenclature can then be deferred until the natural history of the mast cell, and the mechanisms which account for variation in mast cell phenotype, are better understood.

REFERENCES

1. **Galli, S.J., A.M. Dvorak, and H.F. Dvorak.** 1984. Basophils and mast cells: morhologic insights into their biology, secretory patterns, and function. Prog. Allergy 34:1.

2. **Schwartz, L.B., and K.F. Austen.** 1984. Structure and function of the chemical mediators of mast cells. Prog. Allergy 34:271.

3. **Sonoda, T., Y. Kanayama, H. Hara, C. Hayashi, M. Tadokoro, T. Yonezawa, and Y. Kitamura.** 1984. Proliferation of peritoneal mast cells in the skin of W/W^v mice that genetically lack mast cells. J. Exp. Med. 160:138.

4. **Dvorak, A.M., M.C. Mihm Jr., and H.F. Dvorak.** 1976. Morphology of delayed-type hypersensitivity reactions in man. II. Ultrastructural alterations affecting the microvasculature and the tissue mast cells. Lab. Invest. 34:179.

5. New World Directory of the American Language. Second College Edition. 1980. Edited by Guralnik, D.B.. Simon and Schuster, New York, p. 658.

6. Stedman's Medical Dictionary. Twenty-third edition. 1976. Williams and Wilkins Company, Baltimore.

7. **Maximow, A.** 1960. Uber die Zellformen des lockeren Bindegewebes. Arch. Mikrosk. Anat. EntwMech. 67:680.

8. **Ruitenberg, E.J., and A. Elgersma.** 1976. Absence of intestinal mast cell response in congenitally athymic mice during Trichinella spiralis infection. Nature (London) 264:258.

9. **Burnet, F.M.** 1977. The probable relationship of some or all mast cells to the T-cell system. Cell. Immunol. 30:358.

10. **Tas, J., and R.G. Bernsden.** 1977. Does heparin occur in mucosal mast cells of the rat small intestine? J. Histochem. Cytochem. 25:1058.

11. **Kaliner, M.A.** 1980. Is a mast cell a mast cell a mast cell? J. Allergy Clin. Immunol. 66:1.

12. **Mayrhofer, G.** 1980. Thymus–dependent and thymus–independent subpopulations of intestinal intraepithelial lymphocytes. A granular subpopulation of probable bone marrow origin and relationship to mucosal mast cells. Blood 55:532.

13. **Askenase, P.W.** 1980. Immunopathology of parasitic disease: involvement of basophils and mast cells. Springer Semin. Immunopathol. 2:417.

14. **Enerback, L.** 1981. The gut mucosal mast cell. Monogr. Allergy 17:222.

15. **Metcalfe, D.D., M. Kaliner, and M.A. Donlon.** 1981. The mast cell. CRC Crit. Rev. Immunol. 2:23.

16. **Beaven, M.A., A.H. Soll, and K.J. Lewin.** 1982. Histamine synthesis by intact mast cells from canine fundic mucosa and liver. Gastroenterology 82:254

17. **Befus, A.D., F.L. Pearce, J. Gauldie, P. Horsewood, and J. Bienenstock.** 1982. Mucosal mast cells. I. Isolation and functional characteristics of rat intestinal mast cells. J. Immunol. 128:2475.

18. **Haig, D.M., T.A. McKee, E.E.E. Jarrett, R. Woodbury, and H.R.P. Miller.** 1982. Generation of mucosal mast cells is stimulated in vitro by factors derived from T cells of helminth infected rats. Nature (London) 300:188.

19. **Jarrett, E.E.E., and H.R.P. Miller.** 1982. Production and activities of IgE in helminth infections. Prog. Allergy 31:178.

20. **Crowle, P.K., and D.E. Phillips.** 1983. Characteristics of mast cells in Chediak–Higashi mice: light and electron microscopic studies of connective tissue and mucosal mast cells. Exp. Cell Biol. 51:130.

21. **Befus, D., T. Lee, J. Denburg, and J. Bienenstock.** 1984. "Mucosal" mast cells. Immunol. Today 5:218.

22. **Guy–Grand, D., M. Dy, G. Luffau, and P. Vassali.** 1984. Gut mucosal mast cells. J. Exp. Med. 160:12.

23. **Jarrett, E.E.E., and D.M. Haig.** 1984. Mucosal mast cells in vivo and in vitro. Immunol. Today 5:115.

24. **Pearce, F.L., H. Ali, K.E. Barrett, A.D. Befus, J. Bienenstock, J. Brostoff, M. Ennis, K.C. Flint, B. Hudspith, N.M. Johnson, K.B.P. Leung, and P.T. Peachell.** 1985. Functional characteristics of mucosal and connective tissue mast cells of man, the rat and other animals. Int. Arch. Allergy Appl. Immunol. 77:274.

25. **Shevach, E.M.** 1984. Macrophages and other accessory cells. In: Fundamental Immunology. Edited by Paul, W.E. Raven Press, New York, p. 71.

26. **Cooper, M.D., J. Kearney, and I. Scher.** 1984. B lymphocytes. In: Fundamental Immunology. Edited by Paul, W.E. Raven Press, New York, p. 43.

27. **Kitamura, Y., T. Sonoda, and M. Yokoyama.** 1983. Differentiation of tissue mast cells. In: Hematopoietic Stem Cells,

Alfred Benzon Symposium 18. Edited by Killman, Sy-Aa., E.P. Cronkite, and C.N. Muller-Berat. Munksgaard, Copenhagen, p. 350.

28. **Zucker-Franklin, D., G. Grusky, N. Hirayama, and E. Schnipper.** 1981. The presence of mast cell precursors in rat peripheral blood. Blood 58:544.

29. **Combs, J.W.** 1966. Maturation of rat mast cells. An electron microscopic study. J. Cell. Biol. 31:563.

30. **Combs, J.W.** 1971. An electron microscopic study of mouse mast cells arising in vivo and in vitro. J. Cell. Biol. 48:676.

31. **Combs, J.W., D. Lagunoff, and E.P. Benditt.** 1965. Differentiation and proliferation of embryonic mast cells of the rat. J. Cell. Biol. 25:577.

32. **Pretlow, T.G., and I.M. Cassady.** 1970. Separation of mast cells in successive stages of differentiation using programmed gradient sedimentation. Am. J. Pathol. 61:323.

33. **Rohlich, P., P. Anderson, and B. Uvnas.** 1971. Electron microscopic observations on compound 48/80-induced degranulation in rat mast cells. Evidence for sequential exocytosis of storage granules. J. Cell. Biol. 51:465.

34. **Lagunoff, D.** 1972. Contributions of electron microscopy to the study of mast cells. J. Invest. Dermatol. 58:296.

35. **Dvorak, A.M., S.J. Galli, E.S. Schulman, L.M. Lichtenstein, and H.F. Dvorak.** 1983. Basophil and mast cell degranulation: ultrastructural analysis of mechanisms of mediator release. Fed. Proc. 42:2510.

36. **Ihle, J.N., J. Keller, S. Oroszlan, L.E. Henderson, T.D. Copeland, F. Fitch, M.B. Prystowsky, E. Goldwasser, J.W. Schrader, E. Palaszynski, M. Dy, and B. Lebel.** 1983. Biological properties of homogeneous interleukin 3. I. Demonstration of WEHI-3 growth-factor activity, mast cell growth-factor activity, P cell-stimulating factor activity, colony-stimulating factor activity, and histamine-producing cell-stimulating factor activity. J. Immunol. 131:282.

37. **Fung, M.C., A.J. Hapel, S. Ymer, D.R. Cohen, R.M. Johnson, H.D. Campbell, and I.G. Young.** 1984. Molecular cloning of cDNA for murine interleukin-3. Nature 307:233.

38. **Nabel, G., S.J. Galli, A.M. Dvorak, H.F. Dvorak, and H. Cantor.** 1981. Inducer T lymphocytes synthesize a factor that stimulates proliferation of cloned mast cells. Nature 291:332.

39. **Yokota, T., F. Lee, D. Rennick, C. Hall, N. Arai, T. Mosman, G. Nabel, H. Cantor, and K. Arai.** 1984. Isolation and characterization of a mouse cDNA clone that expresses mast-cell growth-factor activity in monkey cells. Proc. Natl. Acad. Sci. USA 81:1070.

40. **Rennick, D., F.D. Lee, T. Yokota, K.-I. Arai, H. Cantor, and G. Nabel.** 1985. A cloned MCGF cDNA encodes a multilineage hematopoietic growth-factor: multiple activities of interleukin 3. J. Immunol. 134:910.

41. **Yung, Y.-P., R. Eger, G. Tertian, and M.A.S. Moore.** 1981. Long-term in vitro culture of murine mast cells. II. Purification of a mast cell growth-factor and its dissociation from TCGF. J. Immunol. 127:794.

42. **Yung, Y.-P., and M.A.S. Moore.** 1982. Long-term in vitro culture of murine mast cells. III. Discrimination of mast cell growth-factor and granulocyte CSF. J. Immunol. 129:1256.

43. **Yung, Y.-P., and M.A.S. Moore.** 1983. Mast cell growth-factor. Lymphokine Res. 2:127.

44. **Schrader, J.W.** 1981. The in vitro production and cloning of the P cell, a bone marrow-derived null cell that expresses H-2 and Ia-antigens, has mast cell-like granules, and is regulated by a factor released by activated T cells. J. Immunol. 126:452.

45. **Clark-Lewis, I., and J.W. Schrader.** 1981. P cell stimulating factor. Biochemical characterization of a new T cell-derived factor. J. Immunol. 127:1941.

46. **Padawer, J.** 1979. The mast cell and immediate hypersensitivity. In: Immediate Hypersensitivity, Vol. 7. Edited by Bach, M.K. Marcel Dekker, New York, p. 301.

47. **Dvorak, A.M., S.J. Klebanoff, W.R. Henderson, R.A. Monahan, K. Pyne, and S.J. Galli.** 1985. Vesicular uptake of eosinophil peroxidase by guinea pig basophils and by cloned mouse mast cells and granule-containing lymphoid cells. Am. J. Pathol. 118:425.

48. **Dvorak, A.M., T. Ishizaka, and S.J. Galli.** 1985. Ultrastructure of human basophils developing in vitro. Evidence for the acquisition of peroxidase by basophils, and for different effects of human and murine growth factors on human basophil and eosinophil maturation. Lab. Invest. 53:57.

49. **Tertian, G., Y.-P. Yung, D. Guy-Grand, and M.A.S. Moore.** 1981. Long-term in vitro culture of murine mast cells. I. Description of a growth-factor dependent culture technique. J. Immunol. 127:788.

50. **Nagao, K., K. Yokoro, and S.A. Aaronson.** 1981. Continuous lines of basophil/mast cells derived from normal mouse bone marrow. Science 212:333.

51. **Razin, E., C. Cordon-Cardo, and R.A. Good.** 1981. Growth of a pure population of mouse mast cells in vitro with conditioned medium derived from concanavalin A-stimulated splenocytes. Proc. Natl. Acad. Sci. USA 28:2559.

52. **Hasthorpe, S.** 1980. A hemopoietic cell line dependent upon a factor in pokeweed mitogen-stimulated spleen cell conditioned medium. J. Cell. Physiol. 105:379.

53. **Ginsburg, H.** 1963. The in vitro differentition and culture of normal mast cells from mouse thymus. Ann. N.Y. Acad. Sci. 103:20.

54. **Ginsburg, H., and D. Lagunoff.** 1967. The in vitro differentiation of mast cells. Cultures of cells from immunized

mouse lymph nodes and thoracic duct lymph on fibroblast monolayers. J. Cell Biol. 35:685.

55. Ishizaka, T., H. Okudaira, L.E. Mauser, and K. Ishizaka. 1976. Development of rat mast cells in vitro. I. Differentiation of mast cells from thymus cells. J. Immunol. 116:747.

56. Ishizaka, T., T. Adachi, T.-H. Chang, and K. Ishizaka. 1977. Development of mast cells in vitro. II. Biologic function of cultured mast cells. J. Immunol. 118:211.

57. Galli, S.J., A.M. Dvorak, J.A. Marcum, T. Ishizaka, G. Nabel, H. Der Simonian, K. Pyne, J.M. Goldin, R.D. Rosenberg, H. Cantor, and H.D. Dvorak. 1982. Mast cell clones: a model for the analysis of cellular maturation. J. Cell. Biol. 95:435.

58. Schrader, J.W., S.J. Lewis, I. Clark-Lewis, and J.G. Culvenor. 1981. The persisting (P) cell: histamine content, regulation by a T cell-derived factor, origin from a bone marrow precursor, and relationship to mast cells. Proc. Natl. Acad. Sci. USA 78:323.

59. Sredni, B., M.M. Friedman, C.E. Bland, and D.D. Metcalfe. 1983. Ultrastructural, biochemical and functional characteristics of histamine-containing cells cloned from mouse bone marrow: tentative identification as mucosal mast cells. J. Immunol. 131:915.

60. Razin, E., R.L. Stevens, F. Akiyama, K. Schmidt, and K. Austen. 1982. Culture from mouse bone marrow of a subclass of mast cells possessing a distinct chondroitin sulfate proteoglycan with glycosaminoglycans rich in N-acetylgalactosamine-4,6-disulfate. J. Biol. Chem. 257:7229.

61. Razin, E., J.N. Ihle, D. Seldin, J.M. Mencia-Huerta, H.R. Katz, P.A. LeBlanc, A. Hein, J.P. Caulfield, K.F. Austen, and R.L. Stevens. 1984. Interleukin 3: a differentiation and growth factor for the mouse mast cell that contains chondroitin sulfate E proteoglycan. J. Immunol. 132:1479.

62. Katz, H.R., P.A. LeBlanc, and S.W. Russell. 1983. Two classes of mouse mast cells delineated by monoclonal antibodies. Proc. Natl. Acad. Sci. USA 80:5916.

63. Razin, E., J.M. Mencia-Huerta, R.A. Lewis, E.J. Corey, and K.F. Austen. 1982. Generation of leukotriene C4 from a subclass of mast cells differentiated in vitro from mouse bone marrow. Proc. Natl. Acad. Sci. USA 79:4665.

64. Schrader, J.W. 1983. Bone marrow differentiation in vitro. CRC Crit. Rev. Immunol. 4:197.

65. Crapper, R.M., W.R. Thomas, J.W. Schrader. 1984. In vivo transfer of persisting (P) cells: Further evidence for their identity with T-dependent mast cells. J. Immunol. 133:2174.

66. Kitamura, Y., T. Nakano, T. Sonoda, Y. Kanayama, T. Yamamura, and H. Asai. 1985. Probable transdifferentiation between connective-tissue and mucosal mast cells. Proceedings of the First International Symposium on Mast Cell

Heterogeneity. Raven Press. (In press). Also see: Nakano, T., T. Sonoda, C. Hayashi, A. Yamatodani, Y. Kanayama, T. Yamamura, H. Asai, T. Yonezawa, Y. Kitamura, and S.J. Galli. 1985. Fate of bone marrow–derived cultured mast cells after intracutaneous, intraperitoneal and intravenous transfer into genetically mast cell deficient W/W^V mice: evidence that cultured mast cells can give rise to both "connective tissue-type" and "mucosal" mast cells. J. Exp. Med. (In press).

67. Kawanami, O., V.J. Ferrans, J.D. Fulman, and R.G. Crystal. 1979. Ultrastructure of pulmonary mast cells in patients with fibrotic lung disorders. Lab. Invest. 40:717.

68. Dvorak, A.M., I. Hammel, E.S. Schulman, S.P. Peters, D.W. MacGlashan Jr., R.P. Schleimer, H.H. Newball, K. Pyne, H.F. Dvorak, L.M. Lichtenstein, and S.J. Galli. 1984. Differences in the behavior of cytoplasmic granules and lipid bodies during human lung mast cell degranulation. J. Cell. Biol. 99:1678.

69. Schulman, E.S., A. Kagey–Sobotka, D.W. MacGlashan Jr., N. Franklin–Adkinson Jr., S.P. Peters, R.P. Schleimer, and L.M. Lichtenstein. 1983. Heterogeneity of human mast cells. J. Immunol. 131:1936.

70. Strobel, S., H.R.P. Miller, and A. Ferguson. 1981. Human intestinal mucosal mast cells: evaluation of fixation and staining techniques. J. Clin. Pathol. 34:851.

71. Ruitenberg, E.J., L. Gustowska, A. Elgersma, and H.M. Ruitenberg. 1982. Effects of fixation on the light micro-scopical visualization of mast cells in the mucosa and connective tissue of the human duodenum. Int. Arch. Allergy Appl. Immunol. 67:233.

72. Befus, D., R. Goodacre, N. Dyck, and J. Bienenstock. 1985. Mast cell heterogeneity in man. I. Histologic studies of the intestine. Int. Arch. Allergy Appl. Immunol. 76:232.

73. Fox, C.C., A.M. Dvorak, S.P. Peters, A. Kagey–Sobotka, and L.M. Lichtenstein. 1985. Isolation and characterization of human intestinal mucosal mast cells. J. Immunol. 135:483.

Mast Cell Differentiation and Heterogeneity,
edited by A. D. Befus et al.
Raven Press, New York © 1986.

Biochemical Characteristics Distinguish Subclasses of Mammalian Mast Cells

R. L. Stevens, H. R. Katz, D. C. Seldin, and K. F. Austen

Department of Medicine, Harvard Medical School; and Department of Rheumatology and Immunology, Brigham and Women's Hospital, Boston, Massachusetts 02115

Mast cells are IgE receptor-bearing cells that possess prominent cytoplasmic granules which contain histamine and stain metachromatically with cationic dyes. There are at least two distinct subclasses of mast cells in rats and mice (and presumptively in humans) as assessed by morphologic, immunologic, biochemical and functional criteria. In order to understand the growth, differentiation, classification and function of subclasses of mammalian mast cells, biochemical techniques in addition to morphologic and immunologic techniques have been emphasized in this laboratory.

In addition to histamine, all populations of mammalian mast cells that have been examined so far have proteoglycans, serine

This work was supported in part by Grants AI-22563, HL-17382, AM-35907, AM-20580, AM-33328, AM-35985 and RR-05569 from the National Institutes of Health.

Abbreviations: HP-MC, heparin-containing mast cell; ChS-MC, chondroitin sulfate-containing mast cell; Xyl, xylose; Gal, galactose; GalNac, N-acetylgalactosamine; GlcUA, glucuronic acid; GlcNAc,N-acetylglucosamine; IdUA, iduronic acid; DFP, diisopropylfluorophosphate; RMCP-I, rat mast cell protease I, or chymase; LTA_4, LTB_4, LTC_4, LTD_4, LTE_4, leukotrienes A_4, B_4, C_4, D_4 and E_4, respectively; PGD_2, prostaglandin D_2; RMCP-II, rat mast cell protease II; IL-2 and IL-3, interleukin-2 and -3; RBL-1, rat basophilic leukemia cell line.

iduronic acid (IdUA) containing heparin disaccharide, neutral proteases, and acid hydrolases in their secretory granules. When activated immunologically, the mast cells not only release these granule constituents but also generate and release metabolites of arachidonic acid. At least three populations of cells that appear to share these characteristics have been described in several species: heparin-containing mast cells (HP-MC), typified by connective tissue mast cells found in the skin or serosal cavity; chondroitin sulfate-containing mast cells (ChS-MC), represented by mucosal mast cells located in the lamina propria and epithelium of the small intestine, and peripheral blood basophils. Mast cells and basophils have traditionally been distinguished by location (mast cells in tissues and basophils in the blood) and by morphologic and ultrastructural criteria (mast cells are mononuclear cells and contain numerous homogeneous sized granules, while human basophils are polymorphonuclear leukocytes containing less numerous heterogeneously-sized granules). In conditions such as cutaneous basophil hypersensitivity, however, basophils can leave the circulation and enter the tissues. Because non-transformed rat, mouse and human basophils have not been adequately characterized biochemically, it is not yet possible to determine the ontogenic relationship between basophils and ChS-MC. Thus, in this review, circulating basophils will not be discussed.

HP-MC obtained in vivo. Morphologically, the HP-MC varies in size from 10 to 20 μm, contains extensive microvilli, an oval nucleus, and a large number of relatively uniform-sized granules (Table I) (1). In the rat serosal mast cell, these granules are predominantly amorphous, while in humans they are crystalline in appearance. Rat, mouse, and human HP-MC can be detected histochemically by the characteristic metachromasia of their secretory granules when stained by toluidine blue, a phenomenon that is attributed to the presence of heparin proteoglycan (2). A proteoglycan consists of a peptide core substituted with covalently-linked glycosaminoglycan side chains, which are unbranched, highly acidic carbohydrate polymers made up of characteristic repeating sulfated disaccharides (3). During the biosynthesis of mast cell proteoglycans, at least six glycosyltransferases located in the Golgi apparatus are involved in the chain elongation process to produce the initial carbohydrate sequence of the glycosaminoglycan chain. Xylose (Xyl) is 0-linked to serine at specific serine-glycine sites along the peptide core, and then two galactose (Gal) residues and glucuronic acid (GlcUA) are added. After this monosaccharide sequence of GlcUA→Gal→Gal→Xyl-serine has been obtained, mast cells add alternating N-acetylglucosamine (GlcNAc) and GlcUA monosaccharides to produce glycosaminoglycans of the heparin-heparan sulfate family, or alternating N-acetylgalactosamine (GalNAc) and GlcUA monosaccharides to produce glycosaminoglycans of the chondroitin sulfate family. In the case of the HP-MC, five enzymes participate in the subsequent modification of the chain resulting in the conversion of the

Table I

Summary of Characteristics of Heparin-Containing Mast Cells

	Rat HP-MC	Mouse HP-MC	Human Lung HP-MC
Staining	Safranin	Safranin	Alcian blue
Morphology			
Cell diameter (μm)	20	*	10-18
Granule appearance	Uniform size, all amorphous		Crystalline (95%) and amorphous (5%)
Proteoglycan			
Type (μg/10^6 cells)	Heparin (10-20)	Heparin (+)	Heparin (1-4)
Size	750,000	750,000	60,000-200,000
Amines			
Histamine (μg/10^6 cells)	· 10-30	1-7	0.5-2
Serotonin	+	+	—
Neutral proteases			
(μg/10^6 cells)	RMCP-I (chymase) (2430)	*	Tryptase (12)
	Carboxypeptidase A (25)		Chymotrypsin (0.1-1)
Acid hydrolases			
(units/10^6 cells)	β-Hexosaminidase (1.1)	*	β-Hexosaminidase (3.4)
	β-Glucuronidase (0.16)		β-Glucuronidase (0.03)
	Arylsulfatase (0.09)		Arylsulfatase (0.03)
	β-Galactosidase		
Arachidonate metabolites	PGD$_2$	PGD$_2$ + LTC$_4$	PGD$_2$ + LTC$_4$
Activation agents			
IgE	+	+	+
Compound 48/80	+	+ ·	—

* Characteristics have not been determined

majority of the GlcUA→GalNAc disaccharides to the characteristic iduronic acid (IdUA) containing heparin disaccharide, IdUA-2SO$_4$→GlcNSO$_4$-6SO$_4$. Because of the N-sulfation, the hexosamine-uronic acid bond of heparin is particularly susceptible to hydrolysis by nitrous acid, a technique that has been used routinely to identify heparin-containing proteoglycans (4). More recently, bacterial heparinases that also specifically degrade this glycosaminoglycan have become available (5). The final heparin proteoglycan product in both rat (6,7) and mouse (8) HP-MC has a m.w. of 750,000-1,000,000. In rats it consists of approximately 10 glycosaminoglycan side chains of 60,000-80,000 m.w. with each attached to a peptide core that is predominantly a co-polymer of serine and glycine (6). Heparin proteoglycan, with approximately 4,000 sulfate residues and 2,000 carboxylic acid residues per molecule, is the most acidic macro-molecule in the body. Rat intragranular heparin proteoglycan is resistant to degradation by a number of proteolytic enzymes (7), and thus is distinct from all known cell surface and extracellular matrix proteoglycans. Rat HP-MC contain 10-20 μg of heparin proteoglycan/10^6 cells (7), while human lung and skin HP-MC contain 1-4 μg/10^6 cells of a 60,000 - 200,000 m.w. heparin proteoglycan that possesses short glycosaminoglycan chains (9).

Depending on the species, different types and amounts of cationic proteins are present in the secretory granules of the HP-MC. In rats, the HP-MC contains approximately 25 $\mu g/10^6$ cells of carboxypeptidase A (10) and 24-30 $\mu g/10^6$ cells of a 29,000 m.w. single chain neutral protease that binds diisopropylfluorophosphate (DFP) and has chymotrypsin-like specificity (11). This enzyme has been designated rat mast cell protease I (RMCP-I), or chymase. Human lung HP-MC contain approximately 12 ug/10^6 cells of a trypsin-like enzyme, which has been designated as tryptase (12), and small amounts of a chymotrypsin-like enzyme. The molecular weight of human tryptase is 120,000-140,000 and it exists as a tetramer ($\alpha_2\beta_2$) comprised of active subunits of 35,000 and 37,000 m.w. Purified mouse serosal mast cells have recently been shown to contain substantial amounts of a 30,000 m.w. DFP-binding protein as analyzed by SDS-PAGE (DuBuske, Stevens, and Austen, unpublished observation), but the extent of homology of this putative serine protease to RMCP-I or human tryptase is yet to be determined.

Rat, mouse and human HP-MC also contain acid hydrolases, some of which (leucine aminopeptidase + acid phosphatase in rat HP-MC) localize to a non-secretory lysosomal granule. However, β-hexosaminidase, β-galactosidase, β-glucuronidase, and arylsulfatase A have been localized predominantly to the secretory granule by correlation of their net percent release values with that of histamine upon immunologic activation (13). Indeed, activation of mast cells can be readily assessed by measuring the release of acid hydrolases rather than the exocytosis of proteoglycan, neutral proteases, or histamine. The level of each acid hydrolase in mast cells varies considerably among species. For example, the rat serosal HP-MC has approximately five times more β-glucuronidase than the human lung HP-MC, but only one-third the amount of β-hexosaminidase (14).

The HP-MC is particularly rich in histamine, although the amount varies widely among species. Histamine content per 10^6 cells is 1-7 μg for mouse serosal HP-MC (15), 10-30 μg for rat serosal HP-MC (16), and 0.5-2 μg for human lung HP-MC (17). The HP-MC of the rat and mouse, but not of the human, synthesize serotonin (18).

All populations of mast cells studied have plasma membrane IgE receptors. When receptor-bound IgE is cross-linked by specific antigen or by anti-IgE, the cells exocytose their preformed mediators. Upon immunologic activation, mast cells also synthesize and release arachidonic acid-derived mediators (19). Arachidonic acid is liberated from cell membranes by the action of phospholipases and is then metabolized either via the cyclooxygenase pathway to form prostaglandins and thromboxane, or via the 5-lipoxygenase pathway to generate leukotrienes and related

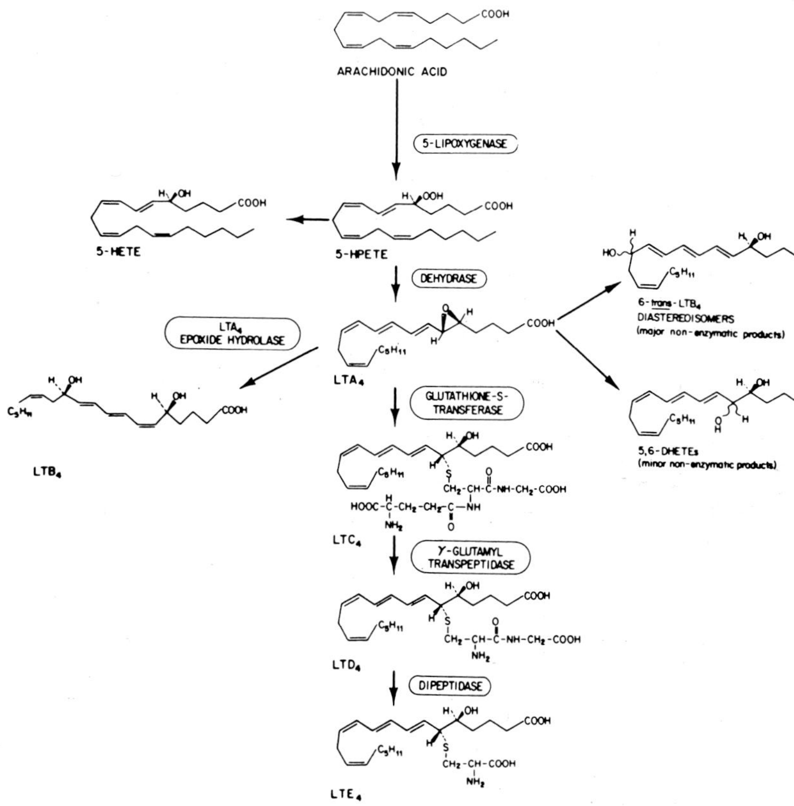

Figure 1.   Biosynthetic pathways of leukotriene generation.
The enzymes of the 5-lipoxygenase pathway are specifically indicated. 5-HETE,   5S-hydroxy-6-trans-8,11,14-cis-eicosatetraenoic
acid; 5-HPETE, 5S-hydroperoxy-6-trans-8,11,14-cis-eicosatetrae-
noic acid; 5,6-di-HETE, 5,6-dihydroxy-eicosatetraenoic   acid;
LTA$_4$, LTB$_4$, LTC$_4$, LTD$_4$, LTE$_4$, leukotrienes A$_4$, B$_4$, C$_4$, D$_4$ and E$_4$,
respectively. (Modified from Lewis and Austen, J. Clin. Invest.
73:889, 1984.)

products (Figure 1). 5-HPETE, the first arachidonate intermediate is processed to leukotriene A$_4$ (LTA$_4$), which can be converted
by an epoxide hydrolase to leukotriene B$_4$ (LTB$_4$). Alternatively,
the glutathione tripeptide can be adducted to LTA$_4$ by the action

of a novel glutathione-S-transferase, forming leukotriene (LTC_4). LTC_4 can be processed to leukotriene (LTD_4) with the removal of a glutamyl residue by γ-glutamyl transpeptidase, and further metabolized to leukotriene E_4 (LTE_4) with the removal of a glycinyl group by a dipeptidase. The three cysteinyl-containing molecules, LTC_4, LTD_4, and LTE_4, form the sulfidopeptide leukotrienes, and collectively are responsible for the vasoactive and bronchoconstricting activity formerly termed "slow reacting substance of anaphylaxis" released in the course of an immediate hypersensitivity reaction. Although many cells can generate leukotrienes or prostaglandins, the products that are synthesized depend on the nature of the stimulus and on the enzymatic capabilities of the particular cell type. Rat serosal HP-MC activated immunologically or by calcium ionophore preferentially metabolize arachidonic acid via the cyclo-oxygenase pathway (20), releasing an average of 13 and 50 $ng/10^6$ cells of prostaglandin D_2 (PGD_2), respectively. Immunologic activation of dispersed human lung cells results in the production of both PGD_2 (20) and sulfidopeptide leukotrienes (17,21), but more than one mast cell subclass may have been present in these preparations.

ChS-MC obtained in vivo. By histochemical and ultrastructural criteria Enerback (22) identified a population of mast cells in the gastrointestinal mucosa of rats that was distinct from the HP-MC (Table II). An increase in the number of intestinal mucosal mast cells is seen in mice infected with Nippostrongylus brasiliensis or Trichinella spiralis, in rats infected with N. brasiliensis, and in humans infected with T. spiralis. Subsequent studies have revealed that the proliferation of this subclass of mast cells in helminth infections is dependent on an intact thymus; nude (23) or thymectomized (24) rats and mice do not exhibit intestinal mucosal mast cell hyperplasia in response to infection with N. brasiliensis. However, the numbers of HP-MC in the skin of athymic mice are similar to those found in normal animals. In rats the mucosal mast cells are smaller than the HP-MC of the serosal cavity, and they possess fewer and more heterogeneously-sized granules.

By comparing the absorbance spectra of granule metachromasia after toluidine blue staining, Tas and Berndsen (2) concluded that rat intestinal mucosal mast cell granules contain little or no heparin, but rather a less-sulfated molecule. Preliminary studies (Stevens, Seldin, Austen, Lee, Befus, and Bienenstock, unpublished observations) of mucosal mast cells from rats infected with N. brasiliensis has revealed that they contain a protease-resistant proteoglycan with a m.w. of approximately 100,000 bearing highly sulfated chondroitin sulfate glycosaminoglycans, and therefore are designated as ChS-MC. This proteoglycan is distinct from all other known mammalian proteoglycans, because approximately 50% of its disaccharides are of the chondroitin sulfate di-B structure, $IdUA-2SO_4-- GalNAc-4SO_4$. Although matrix proteoglycans possessing chondroitin sulfate glycosaminoglycans are synthesized by connective tissue cells

Table II

Summary of characteristics of non-heparin-containing metachromatic cells

	Rat			Mouse	
	Intestinal MMC[a]	Culture-derived	RBL-1	Intestinal MMC	Culture-derived
Staining	Alcian blue	Alcian blue	Alcian blue	Alcian blue	Alcian blue
T-cell factor dependence	+	+	Tumor	+	+
Proteoglycan					
Type	ChS(di-B)	ChS(di-B)	ChS(di-B)[a]	Non-heparin	ChS(E)[a]
Size (m.w.)	100,000	150,000	100,000	*	200,000
Core (amino acid composition)	*	*	SGAE[a]	*	SGE[a]
Protease resistance	+	+	+	*	+
Amines					
Histamine (μg/10[6] cells)	0.1-1	1-2	0.2-0.3	*	0.4-0.6
Serotonin	+	*	+	*	+
Proteases					
Type	RMCP-II	RMCP-II	RMCP-II	*	Serine
Size (molecular weight)	25,000	*	26,000	*	27-31,000
Arachidonate metabolites	*	*	$LTC_4 > PGD_2$	*	$LTC_4 >> PGD_2$
Activation agents					
IgE	+	+	+	*	+
Calcium ionophore	+	+	+	*	+
Compound 48/80	-	*	-	*	-

[a] Abbreviations: ChS(di-B), chondroitin sulfate di-B; ChS(E), chondroitin sulfate E; SGAE, serine, glycine, alanine, glutamic acid; SGE, serine, glycine, alanine.

* Characteristics have not been determined.

+, positive

-, negative

they are made up of monosulfated rather than disulfated disaccharides. Thus, the disulfated chondroitin disaccharides have become a useful phenotypic marker for mast cells. Chondroitin sulfate glycosaminoglycans are degraded by chondroitinase ABC, chondroitinase AC, or hyaluronidase followed by chondroitinase to unsaturated and saturated disaccharides (8,25). In the past, these disaccharides were resolved by laborious techniques such as 14-day paper chromatography, high voltage electrophoresis and thin layer chromatography. In addition, their negative charge often resulted in poor recovery from each technique, thereby hindering their quantification. A high performance liquid chromatography technique using an amino-cyano-substituted straight phase column to rapidly resolve and quantify all known chondroitin sulfate disaccharides has been developed in our laboratory (26). In this procedure, unsulfated, monosulfated, disulfated, and trisulfated disaccharides can be separated and quantified in a single chromatography run. The limit of sensitivity by integrated optical density is 0.1 μg and can be increased 100-fold with the use of radiolabelled proteoglycan.

Rat intestinal mucosal ChS-MC, like HP-MC, contain substantial amounts of a serine protease with chymotrypsin-like specificity, which has been designated rat mast cell protease II (RMCP-II) (27). Although antibodies to RMCP-I and RMCP-II have been raised in rabbits that do not cross-react, these rat proteases have substantial homology in their amino acid sequences (28).

Rat mucosal ChS-MC make little or no serotonin and contain less histamine per cell than rat HP-MC, as measured either from tissues or from isolated cells (29). Rat mucosal ChS-MC contain an estimated 1.3 μg histamine/10^6 cells, while rat serosal HP-MC contain approximately 10-fold more. These findings may reflect the fact that ChS-MC are smaller than HP-MC.

Subpopulations of mast cells do not respond identically to secretagogues, and the ability of pharmacologic agents to inhibit secretion also varies with the mast cell population. Based on histologic and ultrastructural evidence, the in vivo administration of compound 48/80 causes degranulation of rat skin HP-MC, but not mucosal ChS-MC (30). This finding has been confirmed in a comparison of isolated rat serosal HP-MC and rat intestinal mucosal ChS-MC using histamine release as a measure of mast cell activation. In a direct comparison of isolated rat HP-MC and ChS-MC, disodium cromoglycate and theophylline inhibited antigen-induced histamine release from HP-MC but not from ChS-MC, while doxantrazole inhibited release from both populations (31).

ChS-MC obtained in vitro. Metachromatic, histamine-containing cells were first grown from mouse thymocytes cultured on monolayers of mouse embryonic cells (32). Subsequent co-culture of lymph node cells on mouse embryonic monolayers also resulted in mast cell growth, the fastest rate of growth occurring when immune lymph node cells were cultured in the presence of the immunizing antigen. Using lymphocyte conditioned media rather than co-cultures of embryonic monolayers, a number of groups produced metachromatic, histamine-containing cells from mouse hematopoietic and lymphoid tissues (15,33-35). It appears that most, if not all, of these latter cultured cell populations are dependent upon a single growth and differentiation factor (36), termed interleukin 3 (IL-3), produced by lectin-activated T lymphocytes and certain transformed cell lines such as the WEHI-3 myelomonocytic line or the LBRM-33 T-lymphoma line. These culture-derived T-cell-dependent metachromatic cells bear receptors for IgE and secrete preformed granule mediators upon immunologic activation (37) or in response to calcium ionophore A23187 (38), but not in response to compound 48/80 (39). Although these cells have Fc receptors for IgG, they do not phagocytose opsonized or non-opsonized particles. As assessed by antibodies to more than 15 plasma membrane determinants, the phenotype of the mouse culture-derived ChS-MC is distinct from that of any other cell. They express H-2 antigens and the widely distributed Ly-5 antigen. When cultured in the presence of immune type II interferon, they express class II histocompatibility (Ia) antigens (36,40). Mouse culture-derived ChS-MC lack characteristic lymphocyte markers such as Thy-1, Ly-1, Ly-2, and surface immunoglobulin, and do not express the macrophage markers recognized by monoclonal antibodies MAC-1, MAC-2, and MAC-3. However, these cells bind two rat monoclonal antibodies, B23.1 and B54.2, which recognize epitopes on mouse macrophages but not on other cells of the hematopoietic system (41), and do not bind another antibody of the same series, B1.1, which recognizes the mouse serosal HP-MC. The phenotype $B1.1^-/B23.1^+/B54.2^+$ is the same for mast cells derived from progenitors in mouse bone marrow, spleen, lymph node or peripheral blood, as well as for bone marrow-derived cells grown in vitro in either concanavalin A splenocyte-conditioned medium, WEHI-3-conditioned medium or purified IL-3. In contrast, the phenotype of mouse serosal HP-MC is $B1.1^+/B23.1^-/B54.2^+$ (41). We have determined that the B1.1 antibody recognizes the neutral glycosphingolipid globopentaosylceramide (Forssman glycolipid) on the surface of mouse serosal HP-MC, and that mouse T cell factor-dependent, bone marrow-derived mast cells do not contain measurable amounts of this glycosphingolipid (42). However, mouse culture-derived ChS-MC do synthesize the direct precursor, globotetraosylceramide (globoside), and therefore differ from mouse HP-MC in either lacking or having an inactive form of the glycosyltransferase needed to synthesize the more complex globopentaosylceramide.

Mouse cultured ChS-MC do not contain heparin proteoglycan but rather a species of chondroitin sulfate proteoglycan with 30% to 50% of its chondroitin sulfate disaccharides consisting of GlcUA→GalNAc-4,6-disulfate (chondroitin sulfate E) (8). Chondroitin sulfate E proteoglycan, which had not been previously found in mammalian cells, is synthesized by bone marrow-derived mast cells whether they are grown in concanavalin A splenocyte-conditioned medium, WEHI-3-conditioned medium, or purified IL-3. Chondroitin sulfate E proteoglycan is also synthesized by cloned mouse mast cells derived from either fetal liver or lymph node cells, and is released in soluble form by the culture-derived ChS-MC when sensitized with IgE and activated with antigen (37). Recently, this proteoglycan was purified to apparent homogeneity from approximately 6×10^9 cultured ChS-MC by density gradient centrifugation, ion exchange chromatography, and gel filtration chromatography (43). The 200,000 m.w. proteoglycan consists of approximately seven 25,000 m.w. chondroitin sulfate side chains attached to a 10,000 m.w. serine and glycine-rich peptide core, which is homologous to the peptide core of rat heparin proteoglycan. As assessed by gas liquid chromatography, approximately 2% of the dry weight of the chondroitin sulfate E proteoglycan of the cultured ChS-MC consists of neutral sugars, indicating the presence of an as yet unidentified oligosaccharide constituent. Like rat heparin proteoglycan and chondroitin sulfate di-B proteoglycan, mouse chondroitin sulfate E proteoglycan is resistant to proteolytic degradation.

Analysis of the neutral proteases of the cultured mouse ChS-MC has revealed the presence of four DFP-binding proteins comprising approximately 2 μg/10^6 cells (44). These proteins of 27,000, 29,000, 30,000 and 31,000 m.w. are exocytosed by the cell in parallel with histamine upon immunologic activation, and one or more of the exocytosed proteases is capable of digesting human plasma fibronectin at neutral pH. It is not yet known if the four DFP-binding proteases are distinct enzymes or the same enzyme that has undergone varying degrees of post-translational modification. However, it is clear that cultured mouse ChS-MC are similar to in vivo rat HP-MC and ChS-MC in possessing neutral serine proteases in the 25,000 to 30,000 m.w. range.

The mouse cultured ChS-MC was the first pure non-transformed population of mast cells in which it was conclusively demonstrated that a subclass of mast cells could generate and release substantial amounts of leukotrienes when activated by either calcium ionophore or IgE and antigen (37). When these cells are sensizited in vitro with monoclonal IgE, washed, and challenged with specific antigen, they generate a mean of 52 ng/10^6 cells of LTC_4 and 7.5 ng/10^6 cells of LTB_4, but only 0.5 ng of PGD_2/10^6 cells (37,38), unlike rat HP-MC (20). Mouse bone marrow-derived ChS-MC also generate 3-4 ng/10^6 cells of the lipid, acetyl-glyceryl-ether-phosphorylcholine (platelet-activating factor) upon IgE-Fc-dependent activation in vitro (45). These unique properties of the cultured mouse ChS-MC, in contrast to the in vivo mouse HP-MC, are not a consequence of its rapid rate of cell division in culture. Upon inhibition of proliferation by treatment with either sodium butyrate (44) or rat serum (46), the cultured ChS-MC increases its histamine content 5-fold and continues to synthesize chondroitin sulfate E proteoglycan, the four DFP-binding proteases, and both LTC_4 and LTB_4 when activated immunologically.

Both rat bone marrow and mesenteric lymph node cells also contain progenitors of metachromatic, histamine-containing cells with IgE receptors, which have been termed mast cells (47,48). Replication of these mast cells is dependent on a soluble factor(s) from stimulated T lymphocytes (47). Both hematopoietic and lymphoid tissue from rats infected with N. brasiliensis exhibit increased mastopoiesis in vitro compared with tissue from non-infected rats. Several lines of evidence suggest that cultured rat T-cell-dependent mast cells are the in vitro correlate of the rat intestinal mucosal ChS-MC. The two cell types are ultrastructurally similar, with less granulation and more variably shaped nuclei in the rat ChS-MC. Both culture-derived mast cells and intestinal mucosal mast cells stain with alcian blue and not safranin, suggesting that neither cell type synthesizes appreciable amounts of heparin proteoglycan. Indeed, recent preliminary studies have revealed that these rat culture-derived ChS-MC synthesize a protease resistant 150,000–200,000 m.w. chondroitin sulfate proteoglycan, in which 11–13% and 4–6% of the total disaccharides are of the di-B and E type, respectively (Stevens, Austen, Haig, and Jarrett, unpublished observations). The histamine content of the rat cultured ChS-MC is lower than that of the rat serosal HP-MC and, like the rat intestinal mucosal ChS-MC, its proliferation is absolutely dependent upon a thymic or T-cell influence. Perhaps most importantly, as assessed immunochemically, the cultured cells contain the protease RMCP-II (47), which is present in the rat intestinal mucosal ChS-MC.

The growth of metachromatic cells from progenitors in human tissue has also been described; but the distinction between ChS-MC, HP-MC, and basophils cannot be clearly drawn in every case (49). Human fetal liver cells cultured in 10% fetal bovine serum are transiently enriched for cells with IgE receptors and 0.05 – 0.45 μg histamine/10^6 cells (50). Adherent mononuclear cells from peripheral blood cultured in T-cell-conditioned medium and horse serum also give rise to metachromatic cells that bind IgE and contain low levels of histamine (0.05 μg/10^6 cells) (51). Using conditioned medium from lectin-stimulated human peripheral blood T-cells, granulated cells have been grown from human umbilical cord blood (52). These cells have high affinity IgE receptors, synthesize a chondroitin sulfate proteoglycan (Stevens, Seldin, Austen, and Ishizaka, unpublished observation) contain more substantial amounts of histamine (0.48 – 1.6 μg/10^6 cells), and, after sensitization with human IgE, release histamine when exposed to anti-IgE. The growth factor for these cells has been shown to be distinct from human interleukin 2 (IL-2). Metachromatic cells have also been obtained in vitro from normal human bone marrow precursor cells using conditioned medium from lectin-stimulated cells (53). These cells release histamine upon exposure to either calcium ionophore or IgE/anti-IgE, but not to compound 48/80. The factor(s) responsible for differentiation and growth of these cells is distinct from IL-2 and has a m.w. of 25,000–40,000.

Tumor mast cell lines. Because of the difficulty in obtaining large numbers of mast cells from connective tissue, the rat basophilic leukemia (RBL-1) and the mouse mastocytoma (P815) were used for investigations on the mechanism of activation of histamine-containing cells and for characterization of preformed and newly generated mediators. A leukemia composed of basophilic cells was generated in rats with the chemical carcinogen β-chlorethylamine and was subsequently established as the RBL-1 cell line (54). The secretory granules of the RBL-1 cell stain with alcian blue but not with safranin, like the ChS-MC of the rat (55). Cultured RBL-1 cells contain approximately 0.5 μg/10^6 cells of a protease-resistant proteoglycan with a m.w. of 90,000–

150,000, which consists of glycosaminoglycans composed of chondroitin-4-sulfate ($GlcUA \rightarrow GalNAc-4SO_4$) and chondroitin sulfate di-B ($IdUA-2SO_4 \rightarrow GalNAc-4SO_4$) disaccharides (55). Thus, the secretory granule proteoglycan of RBL-1 cells is distinct from the more highly sulfated heparin proteoglycan of rat HP-MC, but is similar to rat intestinal mucosal and culture-derived ChS-MC proteoglycans. RBL-1 cells contain $0.1 - 1.0$ μg histamine/10^6 cells and when activated by calcium ionophore, generate and release both sulfidopeptide leukotrienes and PGD_2. A DFP binding protein of 25,000 m.w., compatible with RMCP-II, was prominent both in extracts of whole RBL-1 cells and in the supernatants of calcium ionophore-activated cells. Characterization of the neutral proteases of the RBL-1 cells by double immunodiffusion analysis has revealed the presence of immunoreactive RMCP-II and the absence of RMCP-I; by radial immunodiffusion the concentration of RMCP-II was 67 ng/10^6 cells. The similarities in staining properties, ultrastructural morphology, proteoglycan structure, immunoreactive neutral protease and generation of leukotrienes suggest that the cultured RBL-1 cell is a transformed mucosal ChS-MC that has escaped T-cell factor dependence.

Mouse mastocytoma lines such as P815 have been used to determine the structure of LTC_4 and its biosynthesis (57), as well as to elucidate the mechanism and order of events in the biosynthesis of heparin glycosminoglycans (3). Because these cells contain very little histamine (<10 ng/10^6 cells) and produce both heparin and chondroitin sulfate proteoglycans, it is not yet possible to determine the relationship of these transformed cells to either the mouse HP-MC or the ChS-MC.

Human cutaneous mastocytoma cells obtained by biopsy contain a heparin proteoglycan (58) analogous to that found in normal human lung HP-MC. Patients with systemic mastocytosis excrete into the urine not only increased amounts of histamine and its metabolites but also large quantities of PGD_2 metabolites (59). These tumor cells have not yet been maintained in culture.

Speculative functions of mast cell mediators. Mast cell proteoglycans are the most negatively charged macromolecules in the body. Thus, intracellularly these proteoglycans may act as an ion exchange resin to concentrate basic molecules such as cationic proteases and histamine in the secretory granules. The relatively small size reported for human heparin proteoglycan may be a requirement for packaging of neutral protease inside the granule, resulting in the crystalline rather than amorphous granule substructure (1). Extracellular matrix proteoglycans are known to regulate the osmotic pressure of connective tissue. It seems likely that the densely packaged secretory granule proteoglycans are underhydrated, and upon activation the osmotic properties of the proteoglycan may be a driving force for granule fusion and exocytosis of mediators deep within the cell. Secretory granule proteoglycans themselves may function as extracellular mediators with the specific function of each being dependent on the structure of the molecule. Chondroitin sulfate

E suppresses amplification of the alternative pathway of complement in vitro by inhibiting the formation of the activated properdin-stabilized amplification C3 convertase C3b,Bb,aP (60), whereas heparin acts by competing with factor B for binding to C3b (61). Both activate the Hageman factor-dependent contact system of plasma in vitro, which results in coagulation and kinin formation (62). Heparin also functions as an anticoagulant in vitro and in vivo (63).

The tight ionic association between RMCP-I, carboxypeptidase A, and heparin proteoglycan in rat HP-MC results in partial inhibition of protease activity against large m.w. substrates (64). Because of the size of this proteoglycan-two enzyme complex, it appears to remain near the activated cell (65). A postulated role for mast cell proteases must consider substrate size, tissue distribution of the exocytosed enzyme-proteoglycan complex, and a mechanism of solubilization of the complex in vivo. Purified RMCP-I can, at physiologic concentrations, stimulate exocytosis of mediators from rat HP-MC in vitro (64) and in vivo. This finding suggests the RMCP-I itself can amplify an inflammatory event. The release of RMCP-II into the intestinal lumen of N. brasiliensis-infected rats after systemic injection of worm antigen is associated with increased mucosal permeability (66), which may be a reflection of the ability of this enzyme to degrade tissue components. Indeed, RMCP-I, RMCP-II, and at least one of the DFP-binding proteases in the cultured mouse ChS-MC have been shown to degrade fibronectin in vitro (44,67). Both RMCP-I and RMCP-II degrade Type IV collagen in vitro (68). Since both of these matrix proteins participate in cell-cell and cell-matrix adherence, their degradation might allow for increased diapedesis of various inflammatory cells and/or their mediators into the intestinal lumen during parasitic infections, thereby augmenting the local host response. In some strains of rats, expulsion of N. brasiliensis is accompanied by depletion of mucosal ChS-MC granule contents as determined by histochemical and ultrastructural criteria, suggesting a role for RMCP-II and other mediators in parasite expulsion. However, because worm expulsion in most rat strains occurs just before the peak of mucosal ChS-MC hyperplasia that characterizes the inflammatory response to the parasite, mast cell mediators may have an important role in tissue repair processes. Studies on the cleavage of low m.w. synthetic substrates by RMCP-I and RMCP-II have revealed that these enzymes have a preferred substrate specificity (69). RMCP-I and carboxypeptidase A probably function in vivo in concert. Since RMCP-I is an endopeptidase with specificity for aromatic amino acids and carboxypeptidase A is an exopeptidase with the same specificity, carboxypeptidase A would hydrolyze the aromatic residue of the native substrate exposed by RMCP-I.

On a weight basis, the acid hydrolases are a minor component of the secretory granule protein. Because these enzymes are active only at acid pH, and because many are exoenzymes which must act in proper sequence to hydrolyze molecules, it is unlikely that they participate substantially in degradation of the extracellular matrix around mast cells. Acid hydrolases contain mannose phosphate residues and bind to mannose phosphate receptors on the surface of cells (70). Studies on patients with mucopolysaccharidoses, who are genetically deficient in one or more of these acid hydrolases, have shown that their fibroblasts in culture can endocytose normal acid hydrolases, thereby increasing the intracellular content of these enzymes and reversing the defect (71). Likewise, exocytosed acid hydrolases from mast cells may serve to increase the content of these degradative enzymes in the lysosomes of adjacent cells. Alternatively, since some acid hydrolases are in a non-secretory compartment, these enzymes may participate in the internal metabolism of mast cell mediators by mast cells themselves. Rat HP-MC are long lived in connective tissue and continuously synthesize mediators; thus, the cells must also catabolize their mediators so as not to exceed storage capacity.

Normal mast cells in vivo contain substantial amounts of histamine in their secretory granules. Upon exocytosis, this low molecular weight amine dissociates from heparin and chondroitin sulfate proteoglycans, allowing it to bind to histamine cell surface receptors on other cells. Histamine induces a multitude of effects (72), including contraction of bronchial and gastro-intestinal smooth muscle; stimulation of gastric acid secretion; inhibition of secretion of mediators from neutrophils, basophils, and other mast cells; induction of airway permeability by altering epithelial cell-cell contacts; stimulation of chemotaxis of neutrophils and eosinophils; and induction of expression of $C3b$ receptors on human eosinophils.

The lipid mediators have a variety of effects, particularly upon vascular beds and smooth muscle. PGD_2 induces vasodilation and increases vascular permeability (73). LTB_4 induces chemo-taxis of leukocytes (74), most notably neutrophils, and augments adherence of leukocytes to endothelial cells. The sulfidopeptide leukotrienes, LTC_4, LTD_4, and LTE_4 are potent smooth muscle con-strictors, causing bronchospasm (19) and arteriolar constriction (75). They enhance venular permeability and increase the rate of bronchial mucus secretion. Platelet-activating factor, as its name implies, induces platelet activation and aggregation.

Mast cell mediators may act in a concerted manner (Figure 2). For example, in an intestinal helminth infection, where T-cell activation and IL-3 production would stimulate ChS-MC prolifera-tion, mast cell mediators might have profound effects on the course of the infection. Histamine and LTC_4 would induce permeability of the intestinal vasculature, while LTB_4 might stimulate chemotaxis of cells possessing cytotoxicity against the parasites, and later of extracellular matrix-producing cells that participate in repair of the damaged intestine. The exocytosed proteoglycan might prevent complement activation and excessive host cell damage. Leukotriene generation could induce mucus

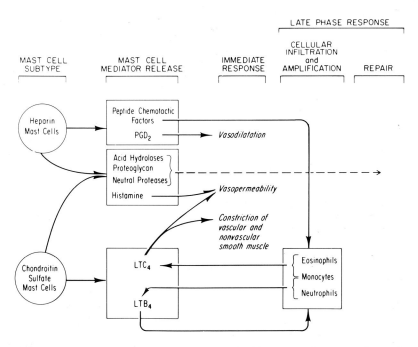

Figure 2. Schematic role for mediators in IgE-dependent reactions.

production, resulting in more effective expulsion of the parasite and of debris in the damaged intestine. Neutral proteases may participate in degradation and repair of the connective tissue. Histamine has been observed to stimulate fibroblast growth in culture (76) and fibroblast monolayers maintain the phenotypic characteristics of HP-MC obtained in vivo and added for co-culture (77) suggesting an interaction relevant to the maintenance of connective tissue.

REFERENCES

1. **Caulfied, J.P., R.A. Lewis, A. Hein, and K.F. Austen.** 1980. Secretion in dissociated human pulmonary mast cells: evidence for solubilization of granule contents before discharge. J. Cell. Biol. 85:299.
2. **Tas, J., and R.G. Berndsen.** 1977. Does heparin occur in mucosal mast cells of the rat small intestine? J.

Histochem. Cytochem. 25:1058.
3. **Roden, L.** 1980. Structure and metabolism of connective tissue proteoglycans. In: The Biochemistry of Glycoproteins and Proteoglycans. Edited by W.J. Lennarz. Plenum Press, New York, p. 267.
4. **Cifonelli, J.A.** 1968. Reaction of heparitin sulfate with nitrous acid. Carbohydr. Res. 8:233.
5. **Linker, A., and P. Hovingh.** 1972. Heparinase and heparitinase from Flavobacteria. Meth. Enzymol. 28:902.
6. **Robinson, H.C., A.A. Horner, M. Hook, S. Ogren, and U. Lindahl.** 1978. A proteoglycan form of heparin and its degradation to single-chain molecules. J. Biol. Chem. 253:6687.
7. **Yurt, R.W., R.W. Leid Jr., K.F. Austen, and J.E. Silbert.** 1977. Native heparin from rat peritoneal mast cells. J. Biol. Chem. 252:518.
8. **Razin, E., R.L. Stevens, F. Akiyama, K. Schmid, and K.F. Austen.** 1982. Culture from mouse bone marrow of a subclass of mast cells possessing a distinct chondroitin sulfate proteoglycan with glycosaminoglycans rich in N-acetylgalactosamine-4,6-disulfate. J. Biol. Chem. 257:7229.
9. **Metcalfe, D.D., R.A. Lewis, J.E. Silbert, R.D. Rosenberg, S.I. Wasserman, and K.F. Austen.** 1979. Isolation and characterization of heparin from human lung. J. Clin. Invest. 64:1537.
10. **Everett, M.-T., and N. Neurath.** 1980. Rat peritoneal mast cell carboxypeptidase: localization, purification and enzymatic properties. FEBS Lett. 110:292.
11. **Lagunoff, D., and P. Pritzl.** 1976. Characterization of rat mast cell granule proteins. Arch. Biochem. Biophys. 173:554.
12. **Schwartz, L.B., R.A. Lewis, and K.F. Austen.** 1981. Tryptase from human pulmonary mast cells. Purification and characterization. J. Biol. Chem. 256:11939.
13. **Schwartz, L.B., R.A. Lewis, D. Seldin, and K.F. Austen.** 1981. Acid hydrolases and tryptase from secretory granules of dispersed human lung mast cells. J. Immunol. 126:1290.
14. **Schwartz, L.B., and K.F. Austen.** 1980. Enzymes of the mast cell granule. J. Invest. Dermatol. 74:349.
15. **Galli, S.J., A.M. Dvorak, J.A. Marcum, T. Ishizaka, G. Nabel, H. Der Simonian, K. Pyne, J.M. Goldin, R.D. Rosenberg, H. Cantor, and H.F. Dvorak.** 1982. Mast cell clones: a model for the analysis of cellular maturation. J. Cell. Biol. 95:435.
16. **Benditt, E.P., M. Arase, and M.E. Roeper.** 1956. Histamine and heparin in isolated rat mast cells. J. Histochem. Cytochem. 4:419.
17. **Paterson, N.A.M., S.I. Wasserman, J.W. Said, and K.F. Austen.** 1976. Release of chemical mediators from partially purified human lung mast cells. J. Immunol. 117:1356.
18. **Benditt, E.P., R.L. Wong, M. Arase, and E. Roeper.** 1955.

5-Hydroxytryptamine in mast cells. Proc. Soc. Exp. Biol. Med. 90:303.

19. **Lewis, R.A., and K.F. Austen.** 1984. The biologically active leukotrienes: biosynthesis, metabolism, receptors, functions and pharmacology. Perspectives. J. Clin. Invest. 73:889.

20. **Lewis, R.A., N.A. Soter, P.T. Diamond, K.F. Austen, J.A. Oates, and L.J. Roberts II.** 1982. Prostaglandin D$_2$ generation after activation of rat and human mast cells with anti-IgE. J. Immunol. 129:1627.

21. **Peters, S.P., D.W. MacGlashan, E.S. Schulman, R.P. Schleimer, E.C. Hayes, J. Rokach, N.F. Adkinson, and L.M. Lichtenstein.** 1984. Arachidonic acid metabolism in purified human lung mast cells. J. Immunol. 132:1972.

22. **Enerback, L.** 1966. Mast cells in rat gastrointestinal mucosa. II. Dye binding and metachromatic properties. Acta Pathol. Microbiol. Scand. 66:303.

23. **Mayrhofer, G., and H. Bazin.** 1981. Nature of the thymus dependency of mucosal mast cells. III. Mucosal mast cells in nude mice and nude rats, in B rats and in a child with the Di George syndrome. Int. Arch. Allergy Appl. Immunol. 64:320.

24. **Mayrhofer, G.** 1979. The nature of the thymus dependency of mucosal mast cells. II. The effect of thymectomy and of depleting recirculating lymphocytes on the resonse to Nippostrongylus brasiliensis. Cell. Immunol. 47:312.

25. **Saito, H., T. Yamagata, and S. Suzuki.** 1968. Enzymatic methods for the determination of small quantities of isomeric chondroitin sulfates. J. Biol. Chem. 243:1536.

26. **Seldin, D.C., N. Seno, K.F. Austen, and R.L. Stevens.** 1984. Analysis of polysulfated chondroitin disaccharides by high performance liquid chromatography. Anal. Biochem. 141:291.

27. **Woodbury, R.G., and H. Neurath.** 1978. Purification of an atypical mast cell protease and its levels in developing rats. Biochemistry 17:4298.

28. **Woodbury, R.G., N. Katunuma, K. Kobayashi, K. Titani, and H. Neurath.** 1978. Covalent structure of a group-specific protease from rat small intestine. Biochemistry 17:811.

29. **Befus, A.D., F.L. Pearce, J. Gauldie, P. Horsewood, and J. Bienenstock.** 1982. Mucosal mast cells. I. Isolation and functional characteristics of rat intestinal mast cells. J. Immunol. 128:2475.

30. **Enerback, L.** 1966. Mast cells in rat gastrointestinal mucosa. III. Reactivity towards compound 48/80. Acta Pathol. Microbiol. Scand. 66:313.

31. **Pearce, F.L., A.D. Befus, J. Gauldie, and J. Bienenstock.** 1982. Mucosal mast cells. II. Effects of anti-allergic compounds on histamine secretion by isolated intestinal mast cells. J. Immunol. 128:2481.

32. **Ginsburg, H.** 1963. The in vitro differentiation and culture of normal mast cells from the mouse thymus. Ann. N.Y. Acad. Sci. 103:20.

33. **Razin, E., C. Cordon–Cardo, and R.A. Good.** 1981. Growth of a pure population of mouse mast cells in vitro with conditioned medium derived from concanavalin A–stimulated splenocytes. Proc. Natl. Acad. Sci. 78:2559.

34. **Schrader, J.W.** 1981. The in vitro production and cloning of the P cell, a bone marrow–derived null cell that expresses H-2 and Ia-antigens, has mast cell–like granules, and is regulated by a factor released by activated T cells. J. Immunol. 126:452.

35. **Tertian, G., Y.–P. Yung, D. Guy–Grand, and M.A.S. Moore.** 1981. Long-term in vitro culture of murine mast cells. I. Description of a growth factor-dependent culture technique. J. Immunol. 127:788.

36. **Razin, E., J.N. Ihle, D. Seldin, J.–M. Mencia–Huerta, H.R. Katz, P.A. LeBlanc, A. Hein, J.P. Caulfield, K.F. Austen, and R.L. Stevens.** 1984. Interleukin 3: a differentiation and growth factor for the mouse mast cell that contains chondroitin sulfate E proteoglycan. J. Immunol. 132:1479.

37. **Razin, E., J.–M. Mencia–Huerta, R.L. Stevens., R.A. Lewis, F.–T. Liu, E.J. Corey, and K.F. Austen.** 1983. IgE–mediated release of leukotriene C_4, chondroitin sulfate E proteoglycan, β–hexosaminidase, and histamine from cultured bone marrow–derived mast cells. J. Exp. Med. 157:189.

38. **Mencia–Huerta, J.–M., E. Razin, E.W. Ringel, E.J. Corey, D. Hoover, K.F. Austen, and R.A. Lewis.** 1983. Immunologic and ionophore–induced generation of leukotriene B_4 from mouse bone marrow–derived mast cells. J. Immunol. 130:1885.

39. **Sredni, B., M.M. Friedman, C.E. Bland, and D.D. Metcalfe.** 1983. Ultrastructural, biochemical and functional characteristics of histamine–containing cells clones from mouse bone marrow: tentative identification as mucosal mast cells. J. Immunol. 131:915.

40. **Wong, G.H.W., I. Clark–Lewis, J.L. McKimm–Breschkin, and J.W. Schrader.** 1982. Interferon–γ–like molecule induces Ia antigens on cultured mast cell progenitors. Proc. Natl. Acad. Sci. 79:6989.

41. **Katz, H.R., P.A. LeBlanc, and S.W. Russell.** 1983. Two classes of mouse mast cells delineated by monoclonal antibodies. Proc. Natl. Acad. Sci. 80:5916.

42. **Katz, H.R., G.A. Schwarting, P.A. LeBlanc, K.F. Austen, and R.L. Stevens.** 1985. Identification of the neutral glycosphingolipids of murine mast cells: expression of Forssman glycolipid by the serosal but not the bone marrow–derived subclass. J. Immunol. 134:2617.

43. **Stevens, R.L., K. Otsu, and K.F. Austen.** 1985. Purification and analysis of the core glycopeptide of the protease-resistant intracellular chondroitin sulfate E proteoglycan from the interleukin 3–dependent mouse mast cells. J. Biol. Chem. (in press).

44. **DuBuske, L., K.F. Austen, J. Czop, and R.L. Stevens.** 1984. Granule-associated serine neutral proteases of the mouse

bone marrow–derived mast cell that degrade fibronectin: their increase after sodium butyrate treatment of the cells. J. Immunol. 133:1535.

45. **Mencia–Huerta, J.–M., R.A. Lewis, E. Razin, and K.F. Austen.** 1983. Antigen–initiated release of platelet–activating factor (PAF–acether) from mouse bone marrow–derived mast cells sensitized with monoclonal IgE. J. Immunol. 131:2958.

46. **Stevens, R.L., W.F. Bloes, D.C. Seldin, E. Razin, H.R. Katz, and K.F. Austen.** 1984. Inhibition of proliferation of mouse T–cell–dependent bone marrow–derived mast cells by rat serum does not change their unique phenotype. J. Immunol. 133:2674.

47. **Haig, D.M., T.A. McKee, E.E.E. Jarrett, R. Woodbury, and H.R.P. Miller.** 1982. Generation of mucosal mast cells is stimulated in vitro by factors from T cells of helminth–infected rats. Nature 300:188.

48. **Denburg, J.A., A.D. Befus, and J. Bienenstock.** 1980. Growth and differentiation in vitro of mast cells from mesenteric lymph nodes of Nippostrongylus brasiliensis–infected rats. Immunology 41:195.

49. **Seldin, D.C., and K.F. Austen.** 1985. Mast cell heterogeneity: the T cell factor–dependent mast cell in vitro and in vivo. In: Proc. Urticarias Symposium. Edited by Greaves, M.W., A.K. Black, R.J. Pye, and R.H. Champion. Cambridge, England, Churchill–Livingstone, p.12.

50. **Razin, E., A.B. Rifkind, C. Cordon–Cardo, and R.A. Good.** 1981. Selective growth of a population of human basophil cells in vitro. Proc. Natl. Acad. Sci. 78:5793.

51. **Czarnetzki, B.M., G. Kruger, and W. Sterry.** 1983. In vitro generation of mast cell–like cells from human peripheral mononuclear phagocytes. Int. Arch. Allergy Appl. Immunol. 71:161.

52. **Ogawa, M., T. Nakahata, A.G. Leary, A.R. Sterk, K. Ishizaka, and T. Ishizaka.** 1983. Suspension culture of human mast cells/basophils from umbilical cord blood mononuclear cells. Proc. Natl. Acad. Sci. 80:4494.

53. **Tadokoro, K., B.M. Stadler, and A.L. De Weck.** 1983. Factor-dependent in vitro growth of human normal bone marrow–derived basophil–like cells. J. Exp. Med. 158:857.

54. **Eccleston, E., B.J. Leonard, J.S. Lowe, and H.J. Welford.** 1973. Basophilic leukaemia in the albino rat and a demonstration of the basopoietin. Nature New Biol. 244:73.

55. **Seldin, D.C., S. Adelman, K.F. Austen, R.L. Stevens, A. Hein, J.P. Caulfield, and R.G. Woodbury.** 1985. Homology of the rat basophilic leukemia cell and the rat mucosal mast cell. Proc. Natl. Acad. Sci., 82:3871.

56. **Seldin, D.C., K.F. Austen, and R.L. Stevens.** 1985. Purification and characterization of protease-resistant secretory granule proteoglycans containing chondroitin sulfate Di–B and heparin–like glycosaminoglycans from rat basophilic leukemia cells. J. Biol. Chem. (in press).

57. **Murphy, R.C., S. Hammarstrom, and B. Samuelsson.** 1979. Leukotriene C: a slow reacting substance from murine mastocytoma cells. Proc. Natl. Acad. Sci. 76:4275.

58. **Metcalfe, D.D., N.A. Soter, S.I. Wasserman, and K.F. Austen.** 1980. Identification of sulfate mucopolysaccharides including heparin in the lesional skin of a patient with mastocytosis. J. Invest. Dermatol. 74:210.

59. **Roberts, L.J. II, B.J. Sweetman, R.A. Lewis, K.F. Austen, and J.A. Oates.** 1980. Increased production of prostaglandin D_2 in patients with systemic mastocytosis. N. Engl. J. Med. 303:1400.

60. **Wilson, J.G., D.T. Fearon, R.L. Stevens, N. Seno, and K.F. Austen.** 1984. Inhibition of the function of activated properdin by squid chondroitin sulfate E glycosaminoglycan and murine bone marrow–derived mast cell chondroitin sulfate E proteoglycan. J. Immunol. 132:3058.

61. **Weiler, J.M., R.W. Yurt, D.T. Fearon, and K.F. Austen.** 1978. Modulation of the formation of the amplification convertase of complement C3b,Bb, by native and commercial heparin. J. Exp. Med. 147:409.

62. **Hojima, Y., C.G. Cochrane, R.C. Wiggins, K.F. Austen, and R.L. Stevens.** 1984. In vitro activation of the contact (Hageman factor) system of plasma by heparin and chondroitin sulfate E. Blood 63:1453.

63. **Lam, L.H., J.E. Silbert, and R.D. Rosenberg.** 1976. The separation of active and inactive forms of heparin. Biochem. Biophys. Res. Commun. 69:570.

64. **Schick, B., K.F. Austen, and L.B. Schwartz.** 1984. Activation of rat serosal mast cells by chymase, an endogenous secretory granule protease. J. Immunol. 132:2571.

65. **Schwartz, L.B., C. Riedel, J.P. Caulfield, S.I. Wasserman, and K.F. Austen.** 1981. Cell association of complexes of chymase, heparin proteoglycan, and protein after degranulation by rat mast cells. J. Immunol. 126:2071.

66. **King, S.J., and H.R.P. Miller.** 1984. Anaphylactic release of mucosal mast cell protease and its relationship to gut permeability in Nippostrongylus–primed rats. Immunology 51:653.

67. **Vartio, T., H. Seppa, and A. Vaheri.** 1981. Susceptibility of soluble and matrix fibronectins to degradation by tissue proteinases, mast cell chymase, and cathepsin G. J. Biol. Chem. 256:471.

68. **Sage, H., R.G. Woodbury, and P. Bornstein.** 1979. Structural studies on human type IV collagen. J. Biol. Chem. 254:9893.

69. **Yoshida, N., M.T. Everitt, H. Neurath, R.G. Woodbury, and J.C. Powers.** 1980. Substrate specificity of two chymotrypsin–like proteases from rat mast cells. Biochemistry 19:5799.

70. **Sly, W.S., H.D. Fischer, A. Gonzalez–Noriega, J.H. Grubb, and M. Natowicz.** 1981. Role of 6–phosphomannosyl–enzyme receptor in intracellular transport and adsorptive

pinocytosis of lysosomal enzymes. Methods Cell. Biol. 23:191.

71. Brot, F.E., J.H. Glaser, K.J. Roozen, W.S. Sly, and P.D. Stahl. 1974. In vitro correction of deficient human fibroblasts by β-glucuronidase from different human sources. Biochem. Biophys. Res. Commun. 57:1.

72. Beaven, M.A. 1978. In: Histamine: its role in physiological and pathological processes. New York, Karger.

73. Wasserman, M.A., D.W. DuCharme, R.L. Griffin, G.L. DeGraaf, and F.G. Robinson. 1977. Bronchopulmonary and cardiovascular effects of prostaglandin D_2 in the dog. Prostaglandins 13:255.

74. Ford–Hutchinson, A.W., M.A. Bray, M.V. Doig, M.E. Shipley, and M.J.H. Smith. 1980. Leukotriene B, a potent chemokinetic and aggregating substance released from polymorphonuclear leukocytes. Nature (Lond.) 286:264.

75. Pfeffer, M.A., J.M. Pfeffer, R.A. Lewis, E. Braunwald, E.J. Corey, and K.F. Austen. 1983. Systemic hemodynamic effects of leukotrienes C_4 and D_4 in the rat. Am. J. Physiol. 244:H628.

76. Russell, J.D., S.B. Russell, and K.M. Trupkin. 1977. The effect of cultured fibroblasts isolated from normal and keloid tissue. J. Cell. Physiol. 93:389.

77. Levi–Schaffer, F., K.F. Austen, J.P. Caulfield, A. Hein, W.F. Bloes, and R.L. Stevens. 1985. Fibroblasts maintain the phenotype and viability of the rat heparin–containing mast cell in vitro. J. Immunol. (in press).

Mast Cell Differentiation and Heterogeneity,
edited by A. D. Befus et al.
Raven Press, New York © 1986.

Histologic and Functional Properties of Mast Cells in Rats and Humans

D. Befus, T. Lee, T. Goto, R. Goodacre, F. Shanahan, and J. Bienenstock

Department of Pathology, McMaster University, Hamilton, Ontario L8N 3Z5, Canada

The widely accepted view that mast cells in different sites are functionally heterogeneous has often been based upon studies which incorporated comparisons of cells from different species as well. For example, many recent studies of cultured murine mast cells derived from bone marrow, fetal liver, or intestine incorporate comparisons with peritoneal mast cells, not derived from the mouse, but derived from the rat (1). This is despite good evidence that rat peritoneal mast cells are not functionally identical to mouse peritoneal mast cells (2). In this chapter we will describe the properties of mast cells derived from the gastrointestinal and respiratory tracts of a single species, namely the rat (3–5). Procedures to induce mast cell hyperplasia in these sites have been established and methods for the isolation of the mast cells developed. More recently, we have studied the histochemical properties of mast cells derived from the intestine (6), lung and nose (7) of man, and have developed procedures for the isolation and functional characterization of human intestinal mast cells (8).

Supported by the Medical Research Council of Canada, Tobacco Research Council, Canadian Foundation for Ileitis and Colitis, Fisons Pharmaceuticals and the Rockefeller Foundation. J. Butera provided skilled secretarial support.

Dr. Befus's and Dr. Lee's present address is: Department of Microbiology and Infectious Disease, Health Sciences Centre, University of Calgary, Calgary, Alberta T2N 4N1, Canada.

RAT INTESTINAL MAST CELLS

Earliest studies of mast cell heterogeneity in the rat were based upon differences in the sensitivity of mast cells from different tissues to formaldehyde or other fixatives (9). Mast cells in the peritoneal cavity could be easily stained with a variety of procedures following the use of a variety of fixatives. However, mast cells from the intestinal lamina propria could not be stained by a variety of procedures and could only be shown when basic lead acetate or Carnoy's fixative was employed, and low pH procedures with toluidine blue or alcian blue were used for staining. Recent evidence suggests that mast cells grown in vitro from rat bone marrow (10) or mesenteric lymph node (11) have similar histochemical properties and fixation sensitivities to mast cells in the intestinal lamina propria.

Other more recent studies of mast cell heterogeneity in the rat have investigated the effects of a variety of secretagogues on mediator secretion by mast cells derived from the peritoneal cavity or intestinal lamina propria (12). Both mast cell types bear IgE receptors and secrete histamine in response to specific sensitizing antigen or anti-IgE antibody (3). A range of secretagogues including compound 48/80, bee venom peptide 401, the neurointestinal peptides, vasoactive intestinal peptide, somatostatin, bradykinin and neurotensin (13), as well as the opiates dynorphin, beta-endorphin and alpha-neoendorphin (14); are potent stimulators of mediator release by peritoneal cells but not by isolated intestinal mast cells. However, both mast cell types respond to various ionophores (3), and interestingly, both secrete histamine in response to stimulation with the neurointestinal peptide, substance P (13). A range of potential secretagogues which are active on the peritoneal mast cell, including neutrophil cationic protein, anaphylatoxin, and dextran have not been studied for their effects on isolated intestinal mast cells.

There are also some interesting results on modulation of mediator secretion from rat mast cells of different sources. For example, adenosine potentiates mediator secretion by both peritoneal and intestinal mucosal mast cells (unpublished data), whereas phosphatidylserine enhances secretion by peritoneal but not isolated intestinal mast cells (3). The anti-allergic compounds disodium cromoglycate, AH9679 and theophylline inhibit mediator secretion by the peritoneal but not the intestinal mast cell (4). However, two anti-allergic compounds, namely doxantrazole (4) and quercetin (15), are potent inhibitors of histamine secretion by both mast cell types.

The majority of these studies on secretagogues and anti-allergic compounds have employed mixed populations of intestinal cells containing mast cell purities of <30%. This obviously presents some difficulties as the responses of the mast cells could be modulated by other cell types contaminating the preparations. Thus, we developed methodology to enrich and purify isolated rat intestinal mucosal mast cells (16). This procedure incorporates two gradients; one a discontinuous Percoll gradient which separates cells on the basis of density and produces mast cell populations of between 40% and 70% purity; the other a continuous BSA gradient with separation conducted at unit gravity for 4 hr. This gradient separates on the basis of cell size and produces fractions containing >95% pure mast cells. In these high purity fractions one recovers ¯84% of the recoverable intestinal mast cells. These purified mast cells are representative of the mast cells present in vivo on the basis of cell size, histamine content and rat mast cell protease content. Moreover, in these parameters they are distinct from purified peritoneal mast cells (Table I).

TABLE I. Characteristics of purified rat intestinal mast cells

Character	Intestinal Mast Cell Source			Peritoneal Mast Cells (Purified)
	Initial Isolation	Percoll Enriched	Staput Purified	
% Mast cells	35	66	96	98
Diameter (μM)	14	13	13	19
Histamine (pg/cell)	1.6	1.6	1.5	15
Protease (RMCP-II pg/cell)	26	21	23	0
Secretagogues and drug responsiveness	|---------------similar-----------|			distinct

Interestingly, isolated intestinal mast cells have significantly fewer (40,000) IgE receptors than peritoneal mast cells (200,000 to 300,000) and given differences in cell size, and assuming spherical shape, this establishes that the density of IgE receptors on intestinal mast cells is approximately $1200/\mu m^2$, whereas on purified peritoneal mast cells the receptor density is $2800/\mu m^2$. Moreover, studies of antigen responsiveness, mediator release and response to substance P and 48/80, show that the purified intestinal mast cell is functionally comparable to the original isolated intestinal mast cells (Table I). Now that highly purified and representative mucosal mast cells are available, detailed biochemical analyses can be conducted which ultimately will contribute greatly to our understanding of mast cell heterogeneity.

MAST CELLS IN THE RAT RESPIRATORY TRACT

To determine the distribution and abundance of histochemically distinct mast cell subtypes in the respiratory tract of the normal rat we prepared sections of trachea, bronchus, lung parenchyma, and mediastinal lymph node following fixation in Mota's basic lead acetate or 10% neutral buffered formalin. Tissues were embedded in methacrylate, sectioned at 2 μm thickness and stained with toluidine blue, pH 6.9 (5). The abundance of mast cells in all sites, with the exception of the lamina propria underlying the cartilaginous portion of the trachea, were similar, regardless of the nature of the fixative employed (Table II). This indicates that only a single, histochemical subtype of mast cell, analogous to the so-called connective tissue type of mast cell derived from the peritoneal cavity and elsewhere, is present in the normal rat respiratory tract. In the lamina propria underlying the cartilaginous rings, the numbers of mast cells differed depending on the nature of the fixative employed (Table II). Mast cell numbers were greater following Mota's basic lead acetate fixation, which implies that two histochemically distinct mast cell populations exist in this site, one analogous to the so-called connective tissue mast cell and the other histochemically similar to the mast cell which is present in the intestinal lamina propria, especially following parasitic infection.

TABLE II. Distribution and abundance of mast cells in normal rats*

Tissue	Fixation	
	Formalin	Mota's
Trachea		
mucosa:cartilaginous part	940 $\pm$ 160[**]	1580 $\pm$ 160[**]
mucosa:membranous part	5300 $\pm$ 440	5060 $\pm$ 390
serosal layer:membranous part	5710 $\pm$ 210	6320 $\pm$ 240
Bronchus:mucosa	1750 $\pm$ 90	2110 $\pm$ 140
Lung parenchyma	470 $\pm$ 30	460 $\pm$ 30
Mediastinal lymph node	2140 $\pm$ 360	2300 $\pm$ 470

* Mast cells/mm^3; mean $\pm$ SEM; n = 14–19

** $p < 0.05$

In an effort to isolate and functionally characterize mast cells from the rat respiratory tract, we treated the rats with intratracheal bleomycin sulphate, which is known to induce pulmonary fibrosis (17). Associated with this fibrotic response is a marked hyperplasia of mast cells throughout the pulmonary parenchyma (5). Using methods comparable to those we developed to isolate mast cells from the gastrointestinal tract, we isolated mast cells from the pulmonary parenchyma of rats with the bleomycin-induced fibrosis. At approximately 30 days following bleomycin instillation, when mast cell hyperplasia is at its maximum, we isolated populations containing about 2% mast cells. The functional properties of these isolated pulmonary mast cells, which have the histochemical characteristics of the mast cells derived from the rat peritoneal cavity, have been studied (Table III). Mast cells derived from the fibrotic lung have a histamine content intermediate between that of peritoneal and intestinal mast cells. Hyperplasia of pulmonary mast cells is independent of an intact thymus as it occurs in bleomycin-treated congenitally athymic nude rats. Mast cells from the fibrotic lung are similar to those from the peritoneal cavity in their responsiveness to various secretagogues and phosphatidyl-serine. In summary, mast cells derived from the fibrotic respiratory tract are histochemically and functionally similar to those of the peritoneal cavity, and distinct from those derived from the rat intestinal mucosa.

TABLE III. Mast cell heterogeneity in the rat

Character	Peritoneal Cavity	Mast Cell Source Fibrotic Lung	Parasitized Intestine
Formalin sensitive	−	−	+
Histamine (pg/cell)	15	5	1–2
Thymus dependency	−	−	+
Phosphatidylserine responsiveness	+	+	−
48/80, VIP, Somatostatin, Bradykinin	++	+	−
A23187, Substance P	++	+	+

To determine whether parasitic infection of the respiratory tract would induce hyperplasia of a mast cell with the histochemical properties of the intestinal mucosal mast cell, we infected rats four times with the nematode Nippostrongylus brasiliensis. These multiple infections induced marked mast cell hyperplasia in the bronchus and lung parenchyma (50-70 fold), but not in the mediastinal lymph nodes and the distribution and abundance of mast cells varied markedly from that in bleomycin-treated animals (unpublished data). Interestingly, the mast cells from parasitized rats had histochemical properties analogous to those derived from the peritoneal cavity. However, upon careful morphological examination many of these mast cells in the parasitized lung (10 days after a fourth infection) had large intracytoplasmic globules analogous to those in the globule leukocyte/intestinal intraepithelial mast cells (18). To our knowledge this is the first report of a globule leukocyte-like mast cell with the histochemical properties of peritoneal mast cells. Whether these mast cells derived from the parasitized respiratory tract are functionally similar to the peritoneal mast cell or the intestinal mucosal mast cell remains to be studied.

MAST CELL POPULATIONS IN MAN

To determine whether histochemically distinct populations of mast cells are present in various tissues in man we have compared the distribution and abundance of mast cells in the intestine and respiratory tract following fixation with Mota's basic lead acetate or 10% neutral buffered formalin. Two histochemically distinct subpopulations are present in both the gastrointestinal and respiratory tracts and their distribution and abundance varies from one anatomical site to another (6). We have conducted similar studies on metachromatic cells in the human nose and have similar evidence of histochemically distinct mast cell subtypes in that site (7).

Using collagenase digestion procedures comparable to those used for the rat intestine, we have isolated cells from the human gastrointestinal mucosa with a yield of approximately 8% mast cells and 1 to 2 pg of histamine per mast cell (8). As in the histochemical assessment in situ, mast cell subpopulations in our isolated intestinal cell preparations are of two types: approximately 65% of the isolated mast cells have the histochemical properties analogous to the intestinal mucosal mast cell of the rat, whereas approximately 35% have the histochemical properties of the mast cell subtype histochemically analogous to that in the rat peritoneal cavity.

Functional analyses of the mast cell populations derived from the human intestine (populations containing mixtures of the two histochemically distinct subtypes) show that they secrete histamine in response to various dilutions of anti-IgE antibody, and have a different dose-response curve than human peripheral blood basophils, being less responsive at greater dilutions of anti-IgE antibody (8). Interestingly, when the anti-allergic compound disodium cromoglycate is used to inhibit histamine secretion induced by anti-IgE antibody, no statistically significant inhibition occurs. However, approximately 20-30% inhibition was present and whether this represents the results of experimental variation or reflects that one mast cell subtype is responsive to this anti-allergic agent and another is unresponsive, remains to be established using isolated, histochemically distinct populations. In total, we have assayed a variety of secretagogues and anti-allergic compounds (Table IV). The results indicate that these mixed mast cell populations from the human gastrointestinal tract have an interesting functional profile that remains to be further characterized.

TABLE IV.   Characteristics of isolated human intestinal
            (lamina propria) mast cells

Characteristic	Definition
Formalin sensitivity	Resistant (35%)/Sensitive 65%
48/80	Unresponsive
Anti-IgE	Responsive
A23187	Responsive
Cromoglycate (10^{-4}M)	20-30% inhibition
Theophylline	>75% inhibition

EPILOGUE

We have shown that there are histochemically distinct mast cell subpopulations in the intestine and respiratory tract of rat, man and also monkey (not discussed). Their distribution and abundance varies with the site and species. In the rat these mast cell subtypes are functionally distinct, but as yet this has not been established in man or monkey, and whether the histochemical distinctions recognized in these latter species predict functional differences, as in the rat, remains to be established. It is essential that histochemically distinct mast cell subpopulations from sites such as the human gastrointestinal tract be separated and purified, and their functional qualities assessed before these important questions of mast cell heterogeneity in man can be answered.

REFERENCES

1. Sredni, B., M.M. Friedman, C.E. Bland, and D.D. Metcalfe. 1983. Ultrastructural, biochemical, and functional characteristics of histamine-containing cells cloned from mouse bone marrow: tentative identification as mucosal mast cells. J. Immunol. 131:915.
2. Barrett, K.E., and F.L. Pearce. 1983. A comparison of histamine secretion from isolated peritoneal mast cells of the mouse and rat. Int. Archs. Allergy Appl. Immunol. 72:234.
3. Befus, A.D., F.L. Pearce, J. Gauldie, P. Horsewood, and J. Bienenstock. 1982. Mucosal mast cells. I. Isolation and functional characteristics of rat intestinal mast cells. J. Immunol. 128:2475.
4. Pearce, F.L., A.D. Befus, J. Gauldie, and J. Bienenstock. 1982. Mucosal mast cells. II. Effects of anti-allergic compounds on histamine secretion by isolated intestinal mast cells. J. Immunol. 128:2481.
5. Goto, T., D. Befus, R. Low, and J. Bienenstock. 1984. Mast cell heterogeneity and hyperplasia in rats with bleomycin-induced pulmonary fibrosis. Am. Rev. Respir. Dis. 130:797.
6. Befus, A.D., R. Goodacre, N. Dyck, and J. Bienenstock. 1985. Mast cell heterogeneity in man. I. Histologic studies of the intestine. Int. Archs. Allergy Appl. Immunol. 76:232.
7. Otsuka, H., J. Denburg, J. Dolovich, D. Hitch, P. Lapp, R.S. Rajan, J. Bienenstock, and D. Befus. 1986. Heterogeneity of metachromatic cells in human nose: significance of mucosal mast cells. J. Allergy Clin. Immunol. (In press).
8. Befus, A.D., R. Goodacre, N. Dyck, and J. Bienenstock. 1984. Isolation and characterization of human intestinal mast cells. Fed. Proc. 43:1973.
9. Enerback, L. 1981. The gut mucosal mast cell. Monogr. Allergy 17:222.
10. Haig, D.M., T.A. McKee, E.E.E. Jarrett, R. Woodbury, and H.R.P. Miller. 1982. Generation of mucosal mast cells is stimulated in vitro by factors derived from T cells of helminth-infected rats. Nature 300:188.
11. Shanahan, F., T.D.G. Lee, J.A. Denburg, J. Bienenstock, and D. Befus. 1986. Functional characterization of mast cells generated in vivo from the mesenteric lymph node of rats infected with Nippostrongylus brasiliensis. Immunology (In press).
12. Befus, A.D., F.L. Pearce, and J. Bienenstock. 1985. Intestinal mast cells in pathology and host resistance. In: Food Allergy and Intolerance. Edited by J. Brostoff and S.J. Challacombe. Saunders (In press).
13. Shanahan, F., J. Denburg, J. Fox, J. Bienenstock, and A.D. Befus. 1985. Mast cell heterogeneity: effects of neuro-enteric peptides on histamine release. J. Immunol. 135:1331.

14. **Shanahan, F., T.D.G. Lee, J. Bienenstock, and A.D. Befus.** 1984. The influence of endorphins on peritoneal and mucosal mast cell secretion. J. Allergy Clin. Immunol. 74:499.
15. **Pearce, F.L., A.D. Befus, and J. Bienenstock.** 1984. Mucosal mast cells. III. Effect of quercetin and other flavonoids on antigen-induced histamine secretion from rat intestinal mast cells. J. Allergy Clin. Immunol. 73:819.
16. **Lee, T.D.G., F. Shanahan, H.R.P. Miller, J. Bienenstock, and A.D. Befus.** 1985. Intestinal mucosal mast cells: Isolation from rat lamina propria and purification using unit gravity velocity sedimentation. Immunology 55:721.
17. **Thrall, R.S., J.R. McCormick, R.M. Jack, R.A. McReynolds, and P.A. Ward.** 1979. Bleomycin-induced pulmonary fibrosis in the rat. Am. J. Pathol. 95:117.
18. **Huntley, J.F., G. Newlands, and H.R.P. Miller.** 1984. The isolation and characterization of globule leucocytes: their derivation from mucosal mast cells in parasitized sheep. Parasite Immunol. 6:371.

Mast Cell Differentiation and Heterogeneity,
edited by A. D. Befus et al.
Raven Press, New York © 1986.

Functional Differences Between Mast Cells from Various Locations

Frederick L. Pearce

Department of Chemistry, University College London, London WC1H 0AJ, United Kingdom

It is now generally accepted that mast cells from different species, and even from diverse tissues within a given animal, are functionally heterogeneous. In particular, they may vary in their responses to given secretory stimuli and to particular anti-allergic agents (for reviews, see 1-7). Such investigations have been greatly facilitated by the development of methods for the enzymic dispersion of mast cells from a number of target tissues including the heart (8), intestine (9), lung (10-12), mesentery (13) and skin (14) of experimental animals and man. The use of these preparations complements studies with leukemic basophil cell lines, freshly isolated human basophils and the serosal mast cells of rodents. The latter cell types are widely available and have been extensively used to characterize the detailed biochemical events involved in mediator release. In addition, free mast cells may be obtained by human bronchoalveolar lavage (15). The present paper will compare some of the functional properties of these different cell types.

DIFFERENTIAL EFFECTS OF HISTAMINE LIBERATORS

The pathophysiological stimulus for the release of histamine from the mast cell is provided by the combination of specific antigen with reaginic antibody fixed to the cell surface. In addition, secretion may be induced by a variety of pharmacological agonists which act independently of the immunological mechanism. Some of these agents have a broad spectrum of

Acknowledgement: Work from the author's laboratories was supported by grants from Fisons Pharmaceuticals Ltd., the MRC, NATO, SERC and Wellcome Trust.

activity, whereas others exhibit a high degree of specificity in their action. The present article will consider representatitve examples of the latter.

Polyamines. The synthetic polyamine compound 48/80 is one of the best studied and widely used histamine liberators and is often described as a 'classical mast cell degranulating agent'. In fact, the amine is an extremely selective secretagogue. As is well known, the compound is a potent releaser of histamine from rat peritoneal and pleural mast cells. It is, however, rather less active against peritoneal cells of the hamster and notably less effective against those of the mouse (Fig. 1a). Enzymically dispersed mesenteric, and to a lesser extent lung and skin mast cells of the rat also show significant reactivity, whereas heart and intestinal mast cells of this species are totally unresponsive (Fig. 1a). Tissue mast cells of the guinea pig are also completely unreactive, as are human basophils and human lung mast cells obtained by enzymic dissociation of the tissue or by bron-choalveolar lavage (Fig. 1b).

Other polybasic histamine liberators, notably peptide 401 (the mast cell degranulated [MCD] peptide from bee venom) and polymyxin, closely resemble compound 48/80 in their specificity of action (2,3,6). Polylysine also shows similar properties in that it is much more effective against mast cells from the rat than those from the guinea pig or man, but appears overall to have a broader profile of activity. In particular, it is an effective releaser of histamine from peritoneal mast cells of the mouse and from human basophil leukocytes (16,17).

Plasma substitutes. A number of plasma volume expanders and substitutes may act as effective histamine liberators in appropriate experimental situations. Of these, the polysaccharide dextran is the best characterized and provides a particularly good example of a highly selective mast cell activator. Thus, dextran is a potent releaser of histamine from peritoneal and pleural mast cells of the rat and has some effect on isolated connective tissue, but not intestinal mucosal mast cells of this animal (1-4,6,14). However, it is completely inactive against pulmonary or mesenteric mast cells of the guinea pig and man, and against human basophil leukocytes and peritoneal mast cells of the hamster and mouse (2,3,6,11-13,16,18). Correspondingly, parenteral administration of dextran produces an acute anaphylactoid reaction and histamine intoxication in the rat, but is completely without effect in any other species including the dog, guinea pig, hamster, pigeon and rabbit (for references, see 2).

In contrast to the selective effects of dextran in the rat, polyvinylpyrrolidone (PVP) produces histamine release only in the dog. Moreover, the response involves the selective degranulation of cutaneous mast cells. Strikingly, PVP is completely without effect in the ape, cat, guinea pig, hamster, man, mouse, rat and rabbit (for references, see 2).

Phosphatidylserine (PS). While PS alone does not directly release histamine, the lipid markedly enhances secretion of the amine from mouse or rat peritoneal mast cells activated by IgE-directed or related ligands. The potentiation is specific for

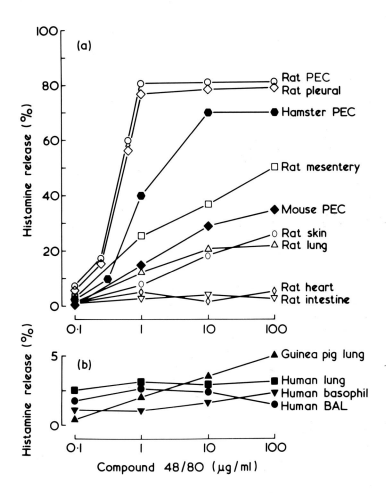

Figure 1. Histamine release induced by compound 48/80 from isolated mast cells from various sources. Results are based on at least four experiments and error bars are omitted for clarity. Mast cells were obtained by enzymic dissociation of the named tissues or by direct lavage. PEC denotes peritoneal exudate cells and BAL indicates those cells obtained by bronchoalveolar lavage. Basophils were isolated by conventional techniques. Modified from (5).

these agonists and the compound has no effect, or even inhibits the release induced by most chemical histamine liberators. The potentiating effect of PS is highly selective for the named mast cells and the compound has a limited or negligible effect on isolated mast cells from the intestine, lung, mesentery and skin of the rat (6,9,13,14), the lung and mesentery of the guinea pig (12,13), the peritoneum of the hamster (18) and the lung of man (11). Human basophils are also refractory to the compound.

Calcium ionophores. Ionophores are organic molecules capable of forming lipid soluble complexes with metal cations. They may then directly transport these ions across hydrophobic barriers, including artificial and biological membranes, independently of any receptor-mediated mechanisms. Moreover, according to the geometry and nature of the co-ordination site, ionophores may exhibit differing degrees of cation specificity. In view of the central role of calcium as a second messenger in stimulus-secretion coupling, considerable interest has focused on ionophores which preferentially transport this cation. The antibiotics A23187, ionomycin and chlortetracycline exhibit such a selectivity and all three agents are potent liberators of histamine from rat serosal mast cells. Peritoneal mast cells of the hamster (18) and mouse (16) are at least equally sensitive to A23187 and ionomycin, but tissue mast cells including those from heart, intestine, lung and mesentery and skin of the rat (8,9,13,14,19), the heart, lung and mesentery of the guinea pig (8,12,13,19) and the lung of man (11) are generally hyporesponsive to these agents, releasing smaller amounts of histamine and at higher concentrations of secretagogue. The ionophore chlortetracycline shows an even higher degree of selectivity in its action. Mast cells from the hamster peritoneum, rat skin and mesentery respond only at very high concentrations, whereas pulmonary and cardiac mast cells from the rat and guinea pig are essentially unreactive (8,14,18,20). In total, these results demonstrate that the differential reactivity of various mast cells is not confined to receptor-mediated ligands but may reflect basic differences in membrane composition or in the secretory mechanism.

DIFFERENTIAL EFFECTS OF ANTI-ALLERGIC AGENTS

Methylxanthines and sympathomimetic amines. Methylxanthines and sympathomimetic amines are widely used in the treatment of human bronchial asthma. They owe their efficacy both to their intrinsic bronchodilator activity and to their ability to inhibit mediator release from certain types of mast cell. Thus theophylline, or its more highly water-soluble derivative aminophylline, inhibits histamine secretion from human basophils and from pulmonary and cutaneous mast cells of the guinea pig, man, monkey and the rat (for references, see 6). In total contrast, theophylline has no effect on mediator release from rat intestinal mucosal mast cells (9) and does not prevent IgE-mediated intestinal anaphylaxis in this species.

Mast cells also differ in their sensitivity towards beta-adrenoceptor agonists. Sympathomimetic amines are thus potent inhibitors of mediator release from pulmonary and cutaneous mast cells of a number of species including man. However, free peritoneal mast cells of the rat, mouse and hamster are resistant to the inhibitory effects of catacholamines, suggesting that they lack coupled, functional beta-receptors (6,21).

Disodium cromoglycate. The introduction of the drug disodium cromoglycate undoubtedly provided a major advance in the treatment and prophylaxis of asthma and other allergic conditions. While the mode of action of the drug may be more complex than initially thought, its clinical utility is generally ascribed at least in part to its ability to inhibit mediator release from tissue mast cells. In fact, the compound exhibits a high degree of specificity in its action. The chromone is thus a potent inhibitor of anaphylactic histamine release from peritoneal or pleural mast cells of the rat, moderately effective against these cells from the hamster but totally inactive against the mouse (Fig. 2a). Surprisingly, the compound is only weakly active against enzymically dispersed human lung mast cells but is noticeably more effective against those cells obtained by human bronchoalveolar lavage (Fig. 2a). This finding may have considerable clinical significance since the latter cells are presumably located within or immediately adjacent to the airways and alveoli. They then represent the subpopulation which comes into immediate, direct contact with inhaled antigens and may thus play a major role in the pathogenesis of human asthma.

In contrast to its activity against serosal mast cells of the rat, cromoglycate is totally ineffective against intestinal mucosal mast cells of this species and is similarly inactive against human basophil leukocytes and tissue mast cells of the guinea pig (Fig. 2b). Further examples of the specificity of the chromone have recently been discussed (6,22,23). Perhaps most strikingly, with the exception of the rat, the compound appears to be ineffective against the cutaneous mast cells of most species tested including the cow, dog, guinea pig, man, monkey, mouse and rabbit.

CLINICAL CONSEQUENCES OF MAST CELL HETEROGENEITY

The functional heterogeneity of mast cells clearly raises fundamental questions concerning the development of appropriate models for identifying potential histamine liberators in man and in characterizing novel anti-allergic compounds. The former effect is obviously pertinent since a number of compounds in common clinical use such as plasma substitutes, radiocontrast media, solubilizing agents for lipophilic drugs and anti-histamines may act as histamine releasers in appropriate experimental situations. Given the variation in response of different preparations of mast cells to such agents, their

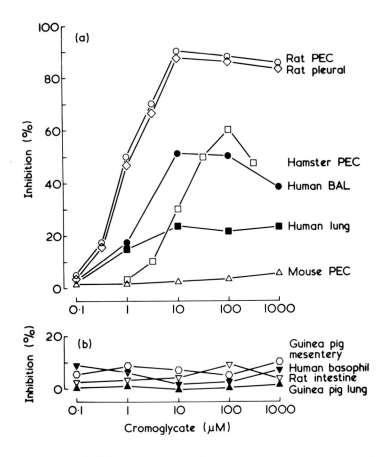

Figure 2. Inhibition by disodium cromoglycate of anaphylactic histamine release from isolated mast cells from various sources. Results are based on at least four experiments and error bars are omitted for clarity. For further details, see legend to Fig. 1. modified from (5).

potential effects and risk in man cannot currently be predicted by in vitro tests but only on the basis of controlled clinical trials.   The selective   action of cromoglycate-like   drugs similarly complicates the development of new   anti-allergic compounds. It is now clear that an agent active against mast cells in one site may be totally ineffective against mast cells in another location. Under these circumstances, it is essential to carry out tests   using the target cells of   interest.

Ultimately, the availability of preparations of isolated human mast cells from defined locations such as the airways, gut and skin should facilitate the screening of novel drugs directed against specific inflammatory and allergic disorders.

REFERENCES

1. **Bienenstock, J., A.D. Befus, J. Denburg, R. Goodacre, F. Pearce, and F. Shanahan.** 1983. Mast cell heterogeneity. Monogr. Allergy 18:124.
2. **Pearce, F.L.** 1982. Functional heterogeneity of mast cells from different species and tissues. Klinische Wochenschr. 60:954.
3. **Pearce, F.L.** 1983. Mast cell heterogeneity. Trends Pharmacol. Sci. 4:165.
4. **Pearce, F.L.** 1985. Intestinal mast cells. In Gut Defences in Clinical Practice. M.S. Losowsky and R.V. Heatley, eds. Churchill-Livingstone, London (In press).
5. **Pearce, F.L., H. Ali, K.E. Barrett, A.D. Befus, J. Bienenstock, J. Brostoff, M. Ennis, K.C. Flint, B. Hudspith, N.McI. Johnson, K.B.P. Leung, and P.T. Peachell.** 1985. Functional characteristics of mucosal and connective tissue mast cells of man, the rat and other animals. Int. Arch. Allergy Appl. Immunol. 77:274-276.
6. **Pearce, F.L., H. Ali, K.E. Barrett, A.D. Befus, J. Bienenstock, J. Brostoff, M. Ennis, K.C. Flint, N.McI. Johnson, K.B.P. Leung, and P.T. Peachell.** 1985. Mast cell heterogeneity. Differential responsivity to histamine liberators and anti-allergic drugs. In Advances in the Biosciences, Vol. 51, Frontiers in Histamine Research. C.R. Ganellin and J.-C. Schwartz, eds. Pergamon Press, Oxford pp.411-421.
7. **Metcalfe, D.D.** 1983. Effector cell heterogeneity in immediate hypersensitivity reactions. Clin. Rev. Allergy 1:311.
8. **Ali, H., and F.L. Pearce.** 1985. Isolation and properties of cardiac and other mast cells from the rat and guinea pig. Agents Actions (In press).
9. **Pearce, F.L., A.D. Befus, J. Gauldie, and J. Bienenstock.** 1982. Mucosal mast cells. II. Effects of anti-allergic compounds on histamine secretion by isolated intestinal mast cells. J. Immunol. 128:2481.
10. **Schulman, E.S., D.W. MacGlashan, S.P. Peters, R.P. Schleimer, H.H. Newball, and L.M. Lichtenstein.** 1982. Human lung mast cells: purification and characterization. J. Immunol. 129:2662.
11. **Ennis, M.** 1982. Histamine release from human pulmonary mast cells. Agents Actions 12:60.
12. **Barrett, K.E., M. Ennis, and F.L. Pearce.** 1983. Mast cells isolated from guinea-pig lung: characterization and studies on histamine secretion. Agents Actions 13:122.
13. **Ennis, M., and F.L. Pearce.** 1980. Differential reactivity of

isolated mast cells from the rat and guinea pig. Eur. J. Pharmacol. 66:339.

14. Barrett, K.E., H. Ali, and F.L. Pearce. 1985. Studies on histamine secretion from enzymically dispersed mast cells from rat skin. J. Invest. Dermatol. 84:22.

15. Flint, K.C., K.B.P. Leung, F.L. Pearce, B.N. Hudspith, J. Brostoff, and N.McI. Johnson. 1985. Human mast cells recovered by bronchoalveolar lavage: their morphology, histamine release and the effects of sodium cromoglycate. Clin. Sci. 68:427.

16. Barrett, K.E., and F.L. Pearce. 1983. A comparison of histamine secretion from isolated peritoneal mast cells of the mouse and rat. Int. Arch. Allergy Appl. Immunol. 72:234.

17. Foreman, J.C., and L.M. Lichtenstein. 1980. Induction of histamine secretion by polycations. Biochim. Biophys. Acta 629:587.

18. Leung, K.B.P., and F.L. Pearce. 1984. A comparison of histamine secretion from peritoneal mast cells of the rat and hamster. Brit. J. Pharmacol. 81:693.

19. Truneh, A., M. Ennis, and F.L. Pearce. 1982. Some characteristics of histamine secretion from mast cells treated with ionomycin. Int. Arch. Allergy Appl. Immunol. 69:86.

20. Pearce, F.L., K.E. Barrett, and J.R. White. 1983. Histamine secretion from mast cells treated with chlortetracycline (aureomycin): a novel calcium ionophore. Agents Actions 13:117.

21. Leung, K.B.P., K.E. Barrett, and F.L. Pearce. 1984. Differential effects of anti-allergic compounds on peritoneal mast cells of the rat, mouse and hamster. Agents Actions 14:461.

22. Church, M.K. 1978. Cromoglycate-like anti-allergic drugs: a review. Drugs of Today 14:281.

23. Garland, L.G., A.F. Green, and H.F. Hodson. 1978. Inhibitors of the release of anaphylactic mediators. In Handbook of Experimental Pharmacology. Vol. 50/II. G.V.R. Born, A. Farah, H. Herken, and G.D. Welch, eds. Springer, Berlin. p.467.

Mast Cell Differentiation and Heterogeneity,
edited by A. D. Befus et al.
Raven Press, New York © 1986.

Neuromodulation of Mast Cell and Basophil Function

Edward J. Goetzl, Tania Chernov-Rogan, *Kiyoshi Furuichi,
Laura M. Goetzl, Johnny Y. Lee, and Frederic Renold

*Howard Hughes Medical Institute Laboratories and Departments of Medicine and
Microbiology, University of California Medical Center, San Francisco, California 94143;
and *National Institute of Arthritis, Diabetes, Digestive and Kidney Diseases,
Bethesda, Maryland 20205*

Many in vivo pulmonary, cutaneous, gastrointestinal, and other target organ reactions to neuropeptides of sensory nerves resemble corresponding components of immediate hypersensitivity responses (1,2). Sensory neuropeptides exert both direct actions on smooth muscles, blood vessels, glands, and leukocytes and indirect actions attributable to the activities of mediators released from mast cells and basophils stimulated by the neuropeptides. The high content of neuropeptides, such as substance P (SP) and somatostatin (SOM), in peripheral endings of C fibers and A–delta fibers of sensory nerves carrying nociceptive signals, provides tissue concentrations sufficient to mediate direct and indirect effects.

Sensory neuropeptide–mediated reactions share with immediate hypersensitivity responses a rapid time–course and the capacity for circumscribed expression in local target tissues. However, the interactions and combined capabilities of neural and immune

---

Dr. Goetzl's present address is: Division of Allergy and Immunology, University of California, San Francisco, California 94143.

This work was supported in part by Grants AI–19784 and HL–31809 from the National Institutes of Health.
Abbreviations: SOM, somatostatin; SP, substance P; VIP, vasoactive intestinal peptide; BMMC, bone marrow–derived mast cells; SMC, serosal mast cells; DNP, dinitrophenyl; RBL, rat leukemic basophils.

systems introduce mechanisms of specificity unique to reactions involving contributions from both systems. The possibility for bidirectional regulation is provided by the stimulatory effect of SP and the inhibitory activity of SOM on some mast cells, while the role of IgE antibodies and antigens is restricted to activation of the mast cells. SP and SOM are distributed selectively in two distinct sets of nerve fibers, may be released separately and in various concentration ratios, and manifest different preferences for basophils and subpopulations of mast cells. At immunologically relevant concentrations, SP and SOM have more potent effects on mouse cultured mast cells than on rat peritoneal mast cells, while only SOM appears capable of influencing the function of basophils. The actions of SP, SOM and other neuropeptides on mast cells and basophils will be described, with an emphasis on the specific susceptibilities of different populations of mast cells to the effects of neuropeptides. The recent finding that mast cells and basophils contain factors immunologically and functionally similar to SOM and SP suggests that such peptides may represent an endogenous as well as a neural mechanism for modulation of immediate hypersensitivity.

Direct Physiological and Pathopharmacological Effects of Sensory Neuropeptides on Non-neural Tissues. SP rapidly elicits a flare and wheal in human skin, with a potency approximately 100–400 times greater than histamine (2,3). The wheal appears to be attributable largely to a direct increase in capillary-venular permeability by the carboxy-terminal octapeptide substituent of SP (2). In contrast, the early cutaneous vasodilation of the flare response to SP is mediated predominantly by the amino-terminal tetrapeptide substituent of SP, which evokes histamine release from rat peritoneal mast cells in vitro and in human skin in vivo (2). Current data suggest that the combined effects of SP and of histamine and other mast cell-derived mediators are required to achieve the immediate hypersensitivity responses elicited by antigen challenge. The observations that capsaicin depletion of SP from human skin (2) and prior administration of aerosols of SP antagonists in rodent airways (4,5) suppress the local actions of histamine on smooth muscles and microvasculature indicate that mast cell-derived histamine may release functionally critical quantities of SP from sensory nerve fibers in target organs.

The flare and wheal reactions of human skin to SOM and vasoactive intestinal peptide (VIP) are due largely to release of SP from cutaneous nerve endings (6), while those elicited by neurotensin, dynorphin, and beta-endorphin are attributable principally to mediators released from mast cells and other cells activated by the neuropeptides (7,8). The effects of SP and other tachykinins on non-vascular smooth muscle and the leukocytic components of inflammation have been reviewed recently (9,10).

Mechanisms of Action of Sensory Neuropeptides on Mast Cells and Basophils. Some of the cellular and biochemical characteristics of the stimulation of rat serosal or connective tissue

mast cells by neuropeptides have been elucidated in vitro. SP
and SOM rapidly release histamine from partially purified preparations of such mast cells by a non-cytotoxic mechanism (2,11,12).
A level of histamine release from the mast cells equal to 50% of
the maximum is attained by concentrations of either peptide of
$2-8 \times 10^{-6}$M. Critical peptide determinants of connective tissue
mast cell-activating activity are the amino-terminal tetrapeptide
for SP (2,11) and the amino acids Ala^1, Gly^2, and $Lys^{4,9}$ for SOM
(12). A dependence of mast cell-activating activity on the
conformation of SOM in solution is suggested by the lower potency
of dihydro-SOM than native SOM, which has one intrachain disulfide bond (12). That the mechanism of activation of rat peritoneal mast cells by sensory neuropeptides has distinctive cellular
features was demonstrated by the finding that SP and SOM elicited
histamine release from mast cells rendered unresponsive to IgE-
dependent and anaphylatoxin-mediated activation by prior exposure
to the respective stimuli.

The observation that human gastrin I elicits significant
release of histamine from human cutaneous mast cells in vitro and
in vivo at concentrations 10,000- to 100,000-fold lower (13) than
those required to attain similar levels of release of histamine
from rat peritoneal mast cells implied that subpopulations of
mast cells might vary in sensitivity to the stimulatory or
inhibitory effects of neuropeptides. To examine the cellular
specificity of SOM, cultured rat leukemic basophils of the 2H3
line (RBL), mouse bone marrow-derived mast cells (BMMC) induced
to differentiate by interleukin 3 (14), and rat peritoneal mast
cells (PMC) purified by centrifugation on Ficoll gradients to
75-82% were sensitized with mouse monoclonal IgE antibodies to
dinitrophenyl ligand (DNP). The cells were washed and preincubated for 10 min at room temperature without or with 10^{-14}M to
10^{-7}M tetradecapeptide SOM 14, and challenged by incubation for
20 min at 37°C with 1 μg/ml of DNP-bovine gammaglobulin (RBL and
BMMC) or rabbit anti-mouse F(ab')$_2$ (PMC). The IgE-dependent
release of histamine and leukotriene C_4 (LTC_4) from BMMC was
suppressed significantly by 3×10^{-9}M to 3×10^{-8}M and 10^{-9}M to
3×10^{-8}M SOM 14, respectively (Fig. 1). The release of the same
mediators from BMMC challenged with 1 μM ionophore A 23187 was
not inhibited by any of the concentrations of SOM 14. RBL and
human basophils were more susceptible than BMMC to the inhibitory
effects of SOM 14, which suppressed significantly the release of
histamine and LTD_4 at respective concentrations of 3×10^{-13}M to
10^{-9}M and 10^{-13}M to 10^{-11}M (15). In contrast, none of the
concentrations of SOM 14 examined had any effect on release of
mediators from PMC challenged by an IgE-dependent mechanism. The
rank order of susceptibility to inhibition of mediator release by
SOM 14 was not a function of differences in the level of release
of mediators by the respective populations of mast cells in the
absence of SOM 14. SOM 28 was less potent than SOM 14 in
inhibiting release of histamine from BMMC and had no effect on
release from PMC.

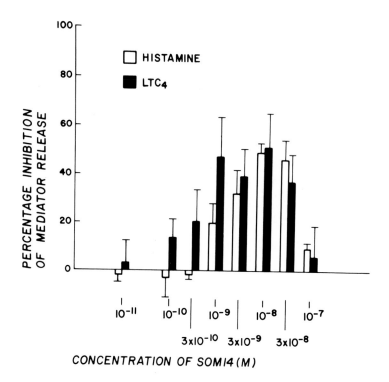

Figure 1. Inhibition by SOM 14 of the release of mediators from mouse bone marrow–derived mast cells. Each bar and bracket depicts the mean ± S.D. of the results of three studies of different preparations of mast cells. Mast cells of over 90% purity developed in cultures of mouse bone marrow cells incubated as described (14) with interleukin 3 (cloned gene product provided generously by Dr. Donna Rennick[7] of Dnax Research, Inc., Palo Alto, CA). Suspensions of 1×10^7 mast cells/ml were sensitized for 2 hr at 4°C with 5 μg/ml of mouse monoclonal IgE anti–DNP, washed and resuspended at 1×10^6/ml in Hanks' solution – 0.2 g% human serum albumin, preincubated without or with SOM 14, and challenged with 1 μg/ml of DNP–bovine gamma globulin. The range of net release of histamine in 20 min was 17–23% and of LTC$_4$ was 42–65 pmoles per 10^6 mast cells. The inhibition of release of histamine was significant at $p < 0.01$ with 3×10^{-9} M to 3×10^{-8} M SOM 14 and of LTC$_4$ at 10^{-9} M to 3×10^{-8} M SOM 14, respectively.

TABLE I

Sensory Neuropeptide Immunoreactivity in Extracts of Leukocytes

Cell Type	Somatostatin	Substance P
	$(pg/10^7$ cells)	
Rat basophilic leukemia cells	272[*]	23
Mouse bone marrow–derived, cultured mast cells	212	64
Human blood monocytes	325	14
U937 cells	365	10
Molt 4 cells	381	13
Human blood T–lymphocytes	4.4	2.5
Human blood B–lymphocytes	6.8	1.2

[*]
Each value is the mean of the results of duplicate determinations of neuropeptide content in extracts of three different preparations of each type of leukocyte. The mean total number of cells in the suspensions extracted were: $RBL_7= 1.4 \times 10^8$, mouse mast cells= 2.5×10^7, monocytes= 3.2×10^7, U 937= 3.6×10^8, Molt 4= 1.5×10^8, T–lymphocytes= 2.0×10^8, and B–lymphocytes= 3.8×10^7.

Endogenous Neuropeptide–like Factors of Basophils and Mast Cells. Suspensions of human and rodent leukocytes were purified as described (14–16), preincubated with 5 mM diisopropylfluorophosphate to block native proteolytic activity, washed, and extracted with ethanol:0.2M acetic acid in water (3:1, v:v). The extracts were dried in vacuo, redissolved in 2 ml of Hanks' solution, and assessed with radioimmunoassays for content of immunoreactive SOM and SP (Table I). Extracts of rat leukemic basophils, mouse cultured mast cells, and human mononuclear phagocytes, but not of human blood lymphocytes had substantial quantities of immunoreactive SOM and lesser amounts of SP. Mast cells had the highest ratio of SP to SOM. Filtration of extracts of RBL on a column of Sephadex G–25 in 0.05M acetic acid in water revealed two principal peaks of SOM–like peptides of m.w. 2000–4000 and 500–1200. The larger SOM–like peptide was resolved from SOM 28 and the smaller from SOM 14 by reverse-phase high performance liquid chromatography (17).

The results of preliminary studies indicate that the RBL–

derived SOM-like peptides inhibit mediator release from RBL at concentrations of 10^{-11} to 10^{-10}M, but the roles of such peptides have not been defined adequately.

REFERENCES

1. **Payan, D.G., J.D. Levine, and E.J. Goetzl.** 1984. Modulation of immunity and hypersensitivity by sensory neuropeptides. J. Immunol. 132:1601.
2. **Foreman, J.C., and C.C. Jordan.** 1983. Histamine release and vascular changes induced by neuropeptides. Agents Actions 13:105.
3. **Hagermark, O., T. Hokfelt, and B. Pernow.** 1978. Flare and itch induced by substance P in human skin. J. Invest. Dermatol. 71:233.
4. **Lundberg, J.M., and A. Saria.** 1983. Capsaicin-induced desensitization of airway mucosa to cigarette smoke, mechanical and chemical irritants. Nature 302:251.
5. **Lundberg, J.M., A. Saria, E. Brodin, S. Rosell, and K. Folkars.** 1983. A substance P antagonist inhibits vagally induced inflammation and bronchial smooth muscle contraction in the guinea pig. Proc. Natl. Acad. Sci. USA 80:1120.
6. **Anand, P., S.R. Bloom, and G.P. McGregor.** 1983. Topical capsaicin pretreatment inhibits axon reflex vasodilatation caused by somatostatin and vasoactive intestinal peptide in human skin. Br. J. Pharmacol. 78:665.
7. **Foreman, J.C., C.C. Jordan, and W. Piotrowski.** 1982. Interaction of neurotensin with the substance P receptor mediating histamine release from rat mast cells and the flare in human skin. Br. J. Pharmacol. 77:531.
8. **Casale, J.B., S. Bowman, and M. Kaliner.** 1984. Induction of human cutaneous mast cell degranulation by opiates and endogenous opioid peptides: Evidence for opiate and non-opiate receptor participation. J. Allergy Clin. Immunol. 73:775.
9. **Bernstein, J.E., and J.R. Hamill.** 1981. Substance P. J. Invest. Dermatol. 77:250.
10. **Pernow, B.** 1983. Substance P. Pharm. Rev. 35:85.
11. **Mazurek, N., I. Pecht, V.I. Teichburg, and S. Blumberg.** 1981. The role of the N-terminal tetrapeptide in the histamine-releasing action of substance P. Neuropharmacol. 20:1025.
12. **Theoharides, T.C., and W.W. Douglas.** 1981. Mast cell histamine secretion in response to somatostatin analogues: Structural considerations. Eur. J. Pharmacol. 73:131.
13. **Tharp, M.D., R. Thirlby, and T.J. Sullivan.** 1984. Gastrin induces histamine release from human cutaneous mast cells. J. Allergy Clin. Immunol. 74:159.
14. **Nabel, G., S.J. Galli, A.M. Dvorak, H.F. Dvorak, and H. Cantor.** 1981. Inducer T-lymphocytes synthesize a factor that stimulates proliferation of cloned mast cells. Nature (Lond.) 291:332.

15. **Goetzl, E.J., and D.G. Payan.** 1984. Inhibition by somatostatin of the release of mediators from human basophils and rat leukemic basophils. J. Immunol. 133:3255.

16. **Payan, D.G., D.R. Brewster, A. Missirian–Bastien, and E.J. Goetzl.** 1984. Substance P recognition by a subset of human T lymphocytes. J. Clin. Invest. 74:1532.

17. **Goetzl E.J., T. Chernov–Rogan, M.P. Cooke, F. Renold, and D.G. Payan.** 1985. Endogenous somatostatin–like peptides of rat basophilic leukemia cells. J. Immunol. 135:2707.

Mast Cell Differentiation and Heterogeneity,
edited by A. D. Befus et al.
Raven Press, New York © 1986.

Mast Cell Heterogeneity: Studies in Non-Human Primates

Kim E. Barrett and Dean D. Metcalfe

Allergic Diseases Section, Laboratory of Clinical Investigation, National Institute of Allergy and Infectious Diseases, National Institutes of Health, Bethesda, Maryland 20205

Mast cells are found in the human body in many tissues. Their numbers are particularly high in those sites which contact the external environment, such as the lungs and airways, the skin, and the gastrointestinal tract. This distribution is consistent with the pathophysiologic role of mast cells in a wide variety of diseases, including asthma, rhinitis and inflammatory diseases of the skin and gastrointestinal tract.

Numerous studies have now firmly established that the morphologic, cytochemical and functional properties of mast cells are dependent on the tissue and species from which they are derived. Work in this area has been recently reviewed by the present authors (1,2) and others (3,4). It has become increasingly apparent that rat peritoneal mast cells, widely used as a model of mast cell activation, do not totally reflect the responsiveness and characteristics of other mast cells even in the rat. The properties of rat peritoneal mast cells therefore would not be expected to necessarily parallel those of the human mast cell populations in health and in disease, and thus the existence of mast cell heterogeneity has fundamental implications for the understanding and treatment of allergic conditions, and the elucidation of the physiological role of mast cells.

Various researchers have sought to isolate mast cells from tissues and to describe their location-specific properties. Such studies have generated much valuable information on how mast cell populations may differ. However, the extent to which mast cell heterogeneity exists in humans and higher animals is unknown. Data on human mast cells are limited, and additionally, information on heterogeneous responses between cells from various tis-

Acknowledgements: We thank Drs. William London and Eva Szucs for their assistance in the studies reported here, and Shirley Starnes for typing the manuscript.

231

sues may be confounded by methodological and inter-individual variations which could produce altered mast cell properties arte-factually. In rodent models, parallel processing of multiple tissues from the same animal has shown that heterogeneity persists even if all mast cells are isolated in a similar fashion (5,6).

We have recently developed a method for the isolation of mast cells from normal monkey intestinal and pulmonary tissue (7). Monkeys are known to resemble humans closely in many immunological systems, and have the advantage that several tissues can be obtained from the same animal. In the absence of multiple tissue samples from individual human subjects, monkey tissues may then provide information on the possible existence of mast cell heterogeneity in humans. The preliminary results of a study which compares the responses of these monkey lung and intestinal mast cells to anti-allergic drugs are presented here.

CYTOCHEMICAL HETEROGENEITY

Mast cells exert effects on surrounding tissues during hypersensitivity reactions by the release and de novo generation of a wide range of chemical mediators. The results of histochemical studies have led to the conclusion that the complement of such mediators which may be obtained from a given mast cell is dependent on the derivation of the cell, although direct measurement of released cell products has been limited owing to the technical difficulties inherent in obtaining tissue mast cell preparations which are free of other cell types.

The question of cytochemical heterogeneity has been addressed with reference to the three classes of mast cell mediators. These classes are; firstly, preformed mediators which are readily eluted from the granule under physiologic conditions; secondly, mediators which remain granule-associated; and thirdly, mediators which are newly generated secondary to the activation process (8). Of these groups, granule-associated mediators have been most closely studied in the context of heterogeneity. In particular, the proteoglycan content appears to be a distinct marker for different granulated cell types (9).

Proteoglycans are a major granule constituent in all mast cells so far studied, and are comprised of a protein core covalently linked to many similar sugar side chains known as glycosaminoglycans. The alternating sugars making up the glycosaminoglycan chains define the class of proteoglycan. These side chains have a characteristic degree of sulphation, conferring on the different classes of proteoglycans various degrees of overall negative charge. This negative charge is thought to account for the striking metachromatic staining properties of mast cells and basophils.

Many mast cells contain the highly sulphated molecule heparin, which readily displays metachromasia with a number of dyes. However, working in rats, Enerback noted that the mast cells located in the mucosal layers of the small intestine required special fixation and staining conditions for their demonstration (10,11). This led him and others to speculate that rat intestinal mucosal mast cells contain a proteoglycan of lower charge than heparin (12,13). The absence of heparin in these cells was confirmed by spectrophotometric studies (14). Studies in our laboratory extend these observations to non-human primates. If staining conditions appropriate to the demonstration of heparin-containing cells are used, only the mast cells located submucosally in sections of monkey small intestine are stained. Additional mast cells are visualized in the mucosal layer if different staining conditions are employed (Table I).

TABLE I.   <u>Mast cell numbers in sections of monkey small intestine observed under different staining conditions</u>[a]

Stain	Location of Cells Stained	Mast Cells per mm^{3b}
. Toluidine blue (pH 4–5)	Submucosal	8,635 ± 605
. Alcian blue (pH <1)	Throughout section	18,794 ± 992
. Cells staining only with alcian blue (pH <1) (2–1)	Mucosal	10,159 ± 1,898

[a] Samples of intestinal tissue were fixed in Carnoy's fixative overnight before sectioning and staining.

[b] Means ± S.E.M. for 7 animals.

Interestingly, the isolation technique employed in our laboratory for monkey intestinal mast cells appears to select for these mucosal cells in the final cell preparation.

We conclude from these observations that the two heterogeneous mast cell populations found in the intestinal tissues of rats are also seen in monkeys. It is then possible that monkey intestinal mucosal mast cells contain a proteoglycan of lower sulphation

than heparin, which could be an "over-sulphated" chondroitin sulphate as has been shown in studies on cultured mouse mast cells
(15,16).

FUNCTIONAL HETEROGENEITY

The aspect of mast cell heterogeneity most relevant to the
management and therapy of allergic disease is the area of
functional heterogeneity. Thus, mast cells from different
species and tissues may vary in their response to activating
stimuli and in their susceptibility to inhibition by
anti-allergic drugs. This has obvious implications for the
pathogenesis of mast cell-related disorders and for the
development of treatment strategies for such diseases.

Functional heterogeneity among rat mast cells in vivo was
described as early as 1966 by Enerback, who noted that connective
tissue mast cells could be degranulated by the classical mast
cell secretagogue compound 48/80, whereas those in the mucosal
layers of the gastrointestinal tract were unaffected (17).
Subsequent studies in vitro have described many heterogeneous
functional responses of mast cells from both experimental animals
(5,6,18-22) and humans (23-25).

Information on the specific functional properties of mast cell
subsets may allow us to infer details of the modulation of the
release process in the populations studied. All mast cells so
far examined respond to an anaphylactic stimulus, i.e. the
cross-linking of cell-surface IgE by specific antigen or anti-IgE
antibodies. However, the level of the response may vary markedly. In our studies with isolated monkey lung and intestinal mast
cells, where the cells are dispersed from the tissues in identical fashion, the lung mast cells are consistently more responsive
to anti-IgE than intestinal cells derived from the same animal
(histamine release 25.3 $\pm$ 3.2% for the lung cells vs 8.9 $\pm$ 3.7%
for the intestinal cells, means $\pm$ S.E.M., n=3).

Functional mast cell heterogeneity also extends to the
susceptibility of the release process to inhibition by anti-
allergic drugs. Hence, mast cells isolated from different
animals, or from different tissues, may vary in their responses
to agents such as theophylline and disodium cromoglycate. The
development of anti-allergic medications is complicated by the
existence of this aspect of heterogeneity, since ideally,
potential therapies should be tested on mast cells derived from
the intended target tissue. Even if this goal is achieved,
problems may be encountered with inter-species variation. For
example, mucosal mast cells from the gastrointestinal tract of
the rat are reportedly unaffected by theophylline (26), whereas
histamine release from monkey mucosal mast cells could be
completely inhibited by the agent (7).

As alluded to above, interpretation of experiments where mast
cells from different tissues are compared following isolation
from separate animals and by different methods is hampered by

possible methodological artefacts. We have therefore isolated mast cells from the lung and small intestine of individual monkeys in parallel and then compared their responses to anti-allergic drugs. As shown in Table II, histamine release from the two cell populations was comparably inhibited by dibutyryl cyclic AMP, isoprenaline and theophylline. Quercetin was also active in inhibiting release, but appeared somewhat more potent against the intestinal mast cells than those from the lung. A preliminary study with disodium cromoglycate revealed variable inhibition of release from the intestinal cells, whereas the lung-derived cells were minimally affected. These results highlight inter-species variations in mast cell function, since rat intestinal mucosal mast cells are reportedly refractory to the action of disodium cromoglycate and theophylline (26), but are responsive to quercetin (27). They provide in addition clear evidence of the existence of functional heterogeneity between the mast cells of different tissues in higher animals as well as rodents. However, they also point to the fact that mast cells within an animal may share several characteristics if care is taken to isolate different populations in the same way.

TABLE II. Comparison of the efficacy of various anti-allergic drugs in inhibiting anaphylactic histamine release from monkey lung and intestinal mast cells[b]

Drug	Approximate ID_{50} against	
	Lung mast cells	Intestinal mast cells
Dibutyryl cyclic AMP	0.5 mM	0.5 mM
Isoprenaline	0.5 μM	0.5 μM
Theophylline	0.2 mM	0.2 mM
Quercetin	7.0 μM	2.0 μM

[a] Cells were isolated from normal monkey lung and small intestine and challenged with anti-human IgE in the presence or absence of anti-allergic drug.

CONCLUSIONS

This article has surveyed some areas of how mast cells may differ in terms of their cytochemical and functional properties. Mast cell heterogeneity may also encompass other areas not covered here, such as morphology, ontogeny, and growth factor requirements. A summary of the various aspects of mast cell heterogeneity is given in Table III.

It should be noted that the studies described here were performed on tissue obtained from normal animals. Care should be taken in the interpretation of studies which utilize tissues from diseases animals or humans, since the disease process may conceivably alter the properties of some of the mast cell populations obtained. For example, we have preliminary evidence that mast cell function in different tissues varies during different stages of parasitic infections in monkeys. During chronic infection, lung mast cells, but not those from the intestine, are able to respond to parasite antigens. During a low level active infection the intestinal cells become responsive. Since tissues from healthy humans are unavailable, and delays may often occur before processing of other human specimens may be initiated (which may also affect mast cell properties), normal non-human primates may provide an ideal model of mast cell heterogeneity and function.

TABLE III. Aspects of mast cell heterogeneity

Morphology	–	Granule size and number
	–	Granule ultrastructure
	–	Membrane structures; processes, folds
	–	Nuclear structure
Mediators	–	Readily elutable, e.g. vasoactive amines
	–	Granule-associated, e.g. proteoglycans
	–	Generated, e.g. arachidonate metabolites
Ontogeny	–	Stem cells
	–	Growth factors
	–	Environmental influences
Function	–	Secretagogues
	–	Anti-allergic drugs

The finding that heterogeneity exists in monkey mast cell populations is suggestive of similar heterogeneity occurring among human mast cells. Thus, we should not expect the development of an anti-allergic drug which will inhibit the activation of all mast cell types; rather, the development of therapies for specific conditions may result from the use of model mast cell systems from defined locations. An understanding of why and how mast cells differ in diverse tissues may illuminate not only mast cell pathology and allergic diseases, but also a possible physiologic role for mast cells.

REFERENCES

1. **Barrett, K.E., and D.D. Metcalfe.** 1984. Mast cell heterogeneity: Evidence and implications. J. Clin. Immunol. 4:253.

2. **Metcalfe, D.D.** 1983. Effector cell heterogeneity in immediate hypersensitivity reactions. Clin. Rev. Allergy 1:311.

3. **Bienenstock, J., A.D. Befus, J. Denburg, R. Goodacre, F.L. Pearce, and F. Shanahan.** 1983. Mast cell heterogeneity. Monogr. Allergy 18:124.

4. **Pearce, F.L.** 1982. Functional heterogeneity of mast cells from different species and tissues. Klinische Wchsr. 60:954.

5. **Ali, H., and F.L. Pearce.** 1985. Isolation and properties of cardiac and other mast cells from the rat and guinea pig. Agents Actions 16:138.

6. **Barrett, K.E., and F.L. Pearce.** 1982. A comparative study of histamine secretion from rat peritoneal and pleural mast cells. Agents Actions 12:186.

7. **Barrett, K.E., and D.D. Metcalfe.** 1985. This histologic and functional characterization of enzymatically–dispersed intestinal mast cells of non–human primates: effects of secretagogues and anti–allergic drugs on histamine secretion. J. Immunol. 135:2020.

8. **Marom, Z., and T.B. Casale.** 1983. Mast cells and their mediators. Ann. Allergy 50:367.

9. **Bland, C.E., K.L. Rosenthal, D.H. Pluznik, G. Dennert, H. Hengartner, J. Bienenstock, and D.D. Metcalfe.** 1984. Glycosaminoglycan profiles in cloned granulated lymphocytes with natural killer function and in cultured mast cells: their potential use as biochemical markers. J. Immunol. 132:1937.

10. **Enerback, L.** 1966. Mast cells in rat gastrointestinal mucosa. I. Effects of fixation. Acta Path. Microbiol. Scand. 66:289.

11. **Enerback, L.** 1966. Mast cells in rat gastrointestinal mucosa. 2. Dye–binding and metachromatic properties. Acta Path. Microbiol. Scand. 66:303.

12. **Enerback, L.** 1981. The gut mucosal mast cell. Monogr. Allergy 17:222.

13. **Mayrhofer, G.** 1980. Fixation and staining of granules in mucosal mast cells and intraepithelial lymphocytes in the rat jejunum, with special reference to the relationship between the acid glycosaminoglycans in the two cell types. Histochem. J. 12:513.

14. **Tas, J., and R.G. Berndsen.** 1977. Does heparin occur in mucosal mast cells of the rat small intestine? J. Histochem. Cytochem. 25:1058.

15. **Sredni, B., M.M. Friedman, C.E. Bland, and D.D. Metcalfe.** 1983. Ultrastructural, biochemical and functional

characteristics of histamine-containing cells cloned from mouse bone marrow: tentative identification as mucosal mast cells. J. Immunol. 131:915.

16. **Razin, E., R.L. Stevens, F. Akiyama, K. Schmid, and K.F. Austen.** 1982. Culture from mouse bone marrow of a subclass of mast cells possessing a distinct chondroitin sulfate proteoglycan with glycosaminoglycans rich in N-acetyl-galactosamine-4,6-disulfate. J. Biol. Chem. 257:7229.

17. **Enerback, L.** 1966. Mast cells in rat gastrointestinal mucosa. III. Reactivity towards compound 48/80. Acta Path. Microbiol. Scand. 66:313.

18. **Befus, A.D., F.L. Pearce, J. Gauldie, P. Horsewood, and J. Bienenstock.** 1982. Mucosal mast cells. I. Isolation and functional characterization of rat intestinal mast cells. J. Immunol. 128:2475.

19. **Barrett, K.E., H. Ali, and F.L. Pearce.** 1985. Studies on histamine secretion from enzymically dispersed mast cells from rat skin. J. Invest. Dermatol. 84:22.

20. **Barrett, K.E., M. Ennis, and F.L. Pearce.** 1983. Mast cells isolated from guinea pig lung: characterization and studies on histamine secretion. Agents Actions 13:122.

21. **Heymanns, J., H. Behrendt, and W. Schmutzler.** 1982. Comparative studies of mast cells from normal (non-immunized) and actively sensitized dogs. Agents Actions 12:192.

22. **Pearce, F.L., and M. Ennis.** 1980. Isolation and some properties of mast cells from the mesentery of the rat and guinea pig. Agents Actions 10:124.

23. **Ennis, M.** 1982. Histamine release from human pulmonary mast cells. Agents Actions 12:60.

24. **Church, M.K., G.J.-K. Pao, and S.T. Holgate.** 1982. Characterization of histamine secretion from mechanically dispersed human lung mast cells: effects of anti-IgE, calcium ionophore A23187, compound 48/80, and basic polypeptides. J. Immunol. 129:2116.

25. **Church, M.K., R.A.K. Mageed, and S.T. Holgate.** 1983. Human tonsillar mast cells. Characteristics of histamine secretion and methods of dispersion. Int. Arch. Allergy Appl. Immunol. 72:188.

26. **Pearce, F.L., A.D. Befus, J. Gauldie, and J. Bienenstock.** 1982. Mucosal mast cells. II. Effects of anti-allergic compounds on histamine secretion by isolated intestinal mast cells. J. Immunol. 128:2481.

27. **Pearce, F.L., A.D. Befus, and J. Bienenstock.** 1984. Mucosal mast cells. III. Effect of quercetin and other flavonoids on antigen-induced histamine secretion from rat intestinal mast cells. J. Allergy Clin. Immunol. 73:819.

Mast Cell Differentiation and Heterogeneity,
edited by A. D. Befus et al.
Raven Press, New York © 1986.

Intestinal Mucosal Mast Cells in Normal and Parasitized Rats

H. R. P. Miller, S. J. King, S. Gibson, J. F. Huntley,
G. F. J. Newlands, and *R. G. Woodbury

*Department of Pathology and Immunology, Moredun Institute, Edinburgh, Scotland; and
Department of Biochemistry, University of Washington, Seattle, Washington 98195

Taliaferro and Sarles (1) were first to provide a detailed ↲ account of mast cell changes in the intestine, lung, and skin of rats during primary infection with the nematode parasite Nippostrongylus brasiliensis. They noted the massive accumulation in the intestinal lamina propria of what they called connective tissue basophils which arose by differentiation and division from a stromal cell population. They also described the appearance of a cell-type containing eosinophilic granules, the globule leukocyte (GL) within the enteric epithelium. The populations of connective tissue basophils and GL increased even more rapidly after a second infection, with cell division occurring in both cell types (1).

It was not until the 1960s that Taliaferro and Sarles' observations (1) were confirmed and extended. The principal reason for the recrudescence of interest was the publication of the now classic experiments by Enerback (reviewed in 2) which showed that mast cells in the rat intestine were peculiarly sensitive to fixation, and differed in their content of amine and in their responsiveness to the histamine liberator 48/80 when compared with mast cells elsewhere in the body. Application of the histochemical technique described by Enerback revealed that the population of enteric mast cells expanded exponentially during primary infection with N. brasiliensis (3) and at the same time their granules became depleted of proteoglycan (4). Ultrastructural studies confirmed both that the mast cells arose in situ from a population of undifferentiated blast cells comparable

Acknowledgements. We thank Mrs. A. Baird for secretarial help. This work was, in part, supported by a grant from the Wellcome Trust.

to the stromal cells originally described by Taliaferro and Sarles (1), and that the granules of these maturing cells had a depleted appearance (5,6).

Further studies of the histochemistry and ultrastructure of parasitized intestine led to the conclusion that GL were in fact intraepithelial mast cells because, like mast cells, they contained proteoglycan, monoamines, and basic protein (4,7). Ultrastructurally GL were indistinguishable from mast cells except that their granules were even more depleted and, histochemically, they were less able to metabolize and store L–DOPA than the mast cells (4–6).

In 1977 Nawa and Miller (8) described an accelerated intestinal mast cell response in infected rats adoptively immunized with immune thoracic duct lymphocytes and 2 years later three independent reports confirmed the crucial role that immune lymphocyte (9–11) and specifically T lymphocyte (9,10), played in mast cell hyperplasia in the parasitized gut. The term "mucosal mast cell" (MMC) rather than intestinal mast cell was coined by Mayrhofer (12) and received the stamp of approval from Enerback (2). It would appear, therefore, that the rat connective tissue basophil, recognized as being different from the skin mast cell by Taliaferro and Sarles (1) is, in fact, the distinctive intestinal mucosal mast cell (IMMC). The latter contains non-heparin glycosaminoglycan (GAG) (13) and a variant serine proteinase (RMCP II), which is distinct from the proteinase (RMCP I) found in connective tissue mast cells (14).

The in vitro culture of murine mast cells, a technique developed by Ginsburg and colleagues (15), and extended by Jarrett and Haig (16) has confirmed that in the rat intestinal mucosal mast cells are bone marrow-derived and that T-cell factors, probably equivalent to murine IL-3, are responsible for their differentiation (13). Finally, Befus and colleagues have shown that rat intestinal mucosal mast cells, when compared with rat peritoneal mast cells, are relatively unresponsive to many secretagogues and to anti-allergic drugs (17,18).

Having explained the context in which our interest in mucosal mast cells has developed, we would now like to describe some of our more recent results obtained exclusively from in vivo studies of normal and parasitized rats, with our main interest being the role that these cells might have in the pathogenesis of enteric parasitism.

Mucosal Mast Cell Characteristics

Fixation and histochemistry. Because the fixatives currently used for demonstration of IMMC are acidic and do not provide optimal preservation of serine esterase activity, nor of the antigenicity of mast cell proteinase (19 and unpublished

observations), we have recently employed short-term (6 hr)
fixation of tissues in paraformaldehyde (19). Using this
fixation it was possible to further compare the histochemical
properties of IMMC and GL. The results summarized in Table I add
further support to our original hypothesis that GL are intraepithelial mast cells (4,7).

TABLE I. A comparison of the properties of mucosal mast cells in
parasitized rat jejunum

	Mucosal Mast Cell	Globule Leukocyte
Fixation		
Paraformaldehyde (6 hr)	++	++
Formalin (>24 hr)	−	−
Granule Histochemistry		
Glycosaminoglycan	++	++ to +
Monoamine	+	+
RMCP II[a]	++	++
RMCP I[b]	−	−−
Serine Esterase	++	++
Aryl Sulphatase[c]	+	+
β Hexosaminidase[c]	+	+
Sensitivity to corticosteroids	Yes	Yes

All granule histochemistry and immunocytochemistry was carried
out on paraformaldehyde fixed (6 hr) tissues apart from aryl sul-
fatases which were detected ultrastructurally in glutaraldehyde-
fixed tissues.
a. RMCP II was detected by immunoperoxidase cytochemistry in IMMC
and GL using affinity purified rabbit anti-RMCP II F(ab')$_2$.
It was not present in connective tissue mast cells.
b. RMCP I was detected in connective tissue mast cells but not in
IMMC or GL using affinity purified rabbit anti-RMCP I F(ab')$_2$
and an immunoperoxidase conjugate. Both primary antibodies
were cross-absorbed; anti-RMCP I on RMCP II-sepharose and
anti-RMCP II on RMCP I-sepharose, and were monospecific for
their respective enzymes by Western Blot analysis (Gibson and
Miller, unpublished).
c. Preliminary data

Rat mast cell proteinase II.   The exclusive localization of RMCP II in IMMC and GL (Table I) was indirectly confirmed by analysis of the distribution of this enzyme in normal rat tissue (Figure 1).   Samples were homogenized and assayed as described previously (20) and it is clear that the major concentration of RMCP II is in the intestinal tract of normal rats (Fig. 1). Trace amounts of this enzyme were present in the respiratory tract and in thymus (Fig. 1).   Attempts to extract RMCP II from peritoneal mast cells, skin, muscle, liver, kidney, bone marrow and spleen either with low (0.15M KCl) or high (1.5M KCl) concentration of salts were without success.

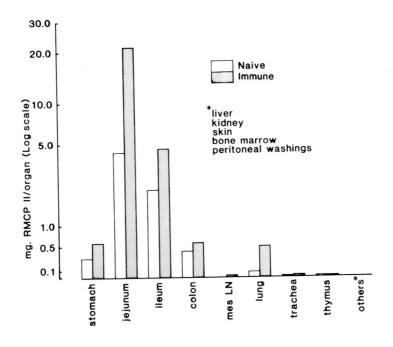

Figure 1.   Distribution of RMCP II in different organs and tissues of naive rats and of rats previously infected with N. brasiliensis (immune).

Mucosal mast cells in corticosterid treated rats. Anaphylactic sensitivity is abrogated in rats treated with corticosteroids (21,22) and it had also previously been shown that mast
cell hyperplasia in N. brasiliensis infection was blocked by corticosteroids (23). Therefore, we studied the effect of corticosteroids on IMMC in normal and Nippostrongylus-immune rats (24).
Treatment of rats with as little as 1 mg methylprednisolone
acetate (Depomedrone Upjohn, Kalamazoo) per kg body weight 48 and
24 hr before sacrifice caused a highly significant reduction in
the numbers of IMMC and GL (Table 1, Figure 2) and of the concentration of RMCP II in the jejunum (Figure 2) (24). We had also
shown that pretreatment of sensitized rats with methylprednisolo-
lone abrogates the anaphylactic release of RMCP II (24).

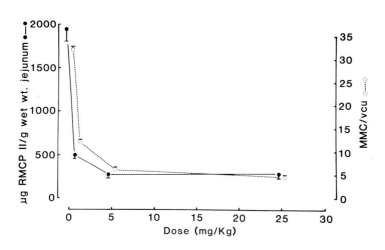

Figure 2. Quantification of RMCP II and of IMMC in the
jejunum of rats immune to N. brasiliensis and pretreated with
different doses of methylprednisolone (data from 24) (vertical
bars represent ± SE of mean).

To summarize, the characteristics of IMMC have been further
defined by using short term paraformaldehyde fixation which
permits enzyme histochemistry of the granules and immunolocalization of granule constituents, as well as demonstration of
glycosaminoglycans with toluidine blue. Use of these techniques

has shown that GL have all of the histochemical properties of IMMC, thereby further confirming the probable relationship between these two cell types (4,7). Importantly, the serine proteinase RMCP II is most abundant in the small intestine, and although found in small amounts in lung and thymus, is virtually absent from other major organs and tissues. Finally, IMMC are highly sensitive to corticosteroids, although it is not known whether this is a direct effect of the drugs on the cells themselves or whether it is mediated indirectely via other cells or their products.

Mucosal Mast Cells and Parasitism

Hyperplasia of IMMC is known to occur during infection with the nematodes N. brasiliensis and Trichinella spiralis and with the enteric protozoan parasite Eimeria nieschulzi (3,25,26). Measurements of the serum levels of RMCP II in each of the three infections revealed a rather similar sequence of events (Figure 3). Although not shown here, both species of nematodes caused

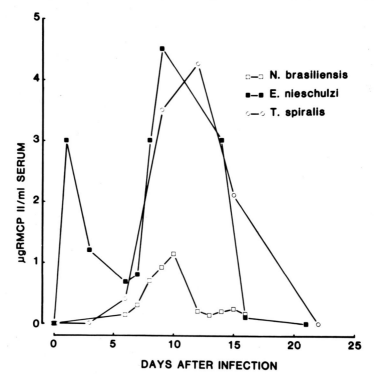

Figure 3. Concentration of RMCP II in the sera of rats infected with N. brasiliensis ( ), T. spiralis (O–O) or E. nieschulzi ( ). (data from 29 and 30).

the systemic release of RMCP II during the first hours of
infection (27,28) and, for E. neischulzi, there was an initial
systemic release of RMCP II which was maximal 24 hr after
infection (Figure 3) (29). The response for all three parasites
was, therefore, biphasic, the second peak being maximal 9–11 days
after challenge (Figure 3) with as much as 3–4 μg RMCP II/ml
serum in rats harbouring T. spiralis or E. nieschuzi, and 1 μg/ml
for N. brasiliensis (Figure 3).

There was an increase in the population of IMMC during all
three infections and an increase in the mucosal concentrations of
RMCP II (28–30). For both N. brasiliensis and E. nieschulzi the
rise in mast cells preceded the accumulation of RMCP II (29,30)
and we have suggested that the maturing cells could be releasing
rather than storing the enzyme (30). This was supported by the
fact that in an earlier study (3) we had only been able to detect
RMCP II in the more mature cells late in infection. However, by
using paraformaldehyde fixation and affinity purified F(ab')2-
peroxidase conjugates, we have now been able to show that
comparable numbers of cells are detected with toluidine blue, by
serine esterase histochemistry and by anti-RMCP II–peroxidase
conjugate (Figure 4). Clearly, therefore, if they contain
granules, they also contain GAG, serine esterase and RMCP II.
The fact that the immature cells contain fewer granules (3,4,5)
may explain the delayed increase in the mucosal concentration of
RMCP II early in infection (30).

Rats which have experienced previous infection with N. brasiliensis, T. spiralis, or E. neischulzi have, for several weeks, an
expanded IMMC population in the jejunal mucosa (3,25,26). When
rechallenged with the homologous parasite there is, in immune
rats, rapid systemic release of RMCP II. The peak of the
response occurs 1 hr after challenge with N. brasiliensis, but
enzyme is still present at 4 hr (27). Again, with both N.
brasiliensis and T. spiralis, the release of RMCP II is more
rapid and of a greater magnitude in immune than in naive rats
(27,28). For E. nieschulzi, the responses were of the same
magnitude in both naive and immune rats (29). This latter rather
surprising result may reflect the inability of the sporozoites to
penetrate into the epithelium of immune rats since substantially
fewer parasites were detected in the jejunal epithelial cells of
immune rats when compared with controls (29).

In summary, two episodes of systemic release of RMCP II are
observed following enteric infection with parasites. The first
of these is during the establishment phase of infection and, in
rats already immune to Nippostrongylus or Trichinella is an
anamnestic response, being of a greater magnitude and occurring
more rapidly than in naive controls. The second episode occurs
at a time when the host is developing resistance to infection,
but for both Nippostrongylus and Eimeria reaches a peak before
jejunal IMMC hyperplasia is itself maximal whereas, for
Trichinella, maximal systemic release of RMCP II coincides with
maximal IMMC hyperplasia.

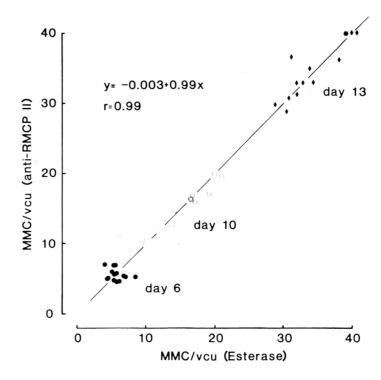

Figure 4.   Simple regression analysis of the jejunal IMMC counts during primary N. brasiliensis infection which shows the highly significant correlation (r=0.99) between the numbers of cells detected by immunocytochemical staining for RMCP II and the numbers containing serine esterase (naphthol AS-D-Chloroacetate/ Fast Garnet GBC Salt) in 5 μm paraffin sections. All tissues fixed with 4% paraformaldehyde for 6 hr.

Anaphylactic Release of RMCP II

A major consequence of helminth infection is the development of high titres of parasite-specific, and non-specific IgE (32). The functional role of IgE will be discussed later in this review but it has long been evident that the host, sensitized by helminth infections, becomes highly susceptible to anaphylactic shock if challenged with homologous soluble antigens (32). Because the major shock organ in the rat is the small intestine, evidence for the participation of IMMC in this reaction was sought.

Several studies of the anaphylactic release of RMCP II have been undertaken, and an example of the reaction is shown in Figure 5. The appearance of RMCP II in serum is both time- and dose-dependent (20), although we have found that considerable variation exists between different batches of rats. For example, in several experiments systemic release peaked 30-60 min after challenge (e.g. Figure 5 and Ref. 2), whereas in other studies the reaction has been extremely rapid and severe with maximal systemic release of RMCP II occurring within 5-10 min of challenge.

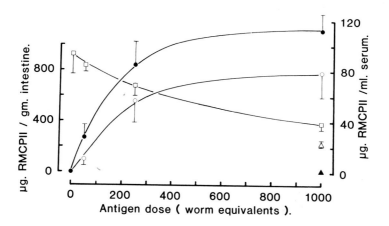

Systemic release of RMCPII during intestinal anaphylaxis.

Figure 5. Dose response curves illustrating the appearance, in serum, of RMCP II 1 hr (O-O) and 4 hr (O-O) after challenge of immune rats with varying doses of soluble worm antigen. The extent of depletion of this enzyme from the jejunal mucosa is shown at 4 hr (   ). No RMCP II was detected in sera of normal rats challenged with the highest dose of antigen (Δ). Also shown is the concentration of RMCP II in the intestines of non-immune rats given 1000 worm equivalents of soluble antigen (Δ). Vertical bars depict ± SEM.

The features which distinguished anaphylactic release of RMCP II from enteric infection with parasites were: (a) the magnitude of the response, with up to 1 mg RMCP II/ml of serum detected in some rats (20), (b) the substantial depletion of both RMCP II and IMMC from the mucosa (Figure 5), (c) massive mucus secretion and epithelial shedding (27), and (d) greatly increased mucosal permeability (20). Interestingly, secretion of RMCP II into the gut lumen occurred very rapidly (20), whereas translocation of Evan's blue from blood to gut lumen occurred relatively slowly, thereby dissociating epithelial and vascular permeability (20). A hypothetical view of the underlying mechanisms are shown in Figure 6.

Intestinal anaphylaxis in the rat : sequence of events ?

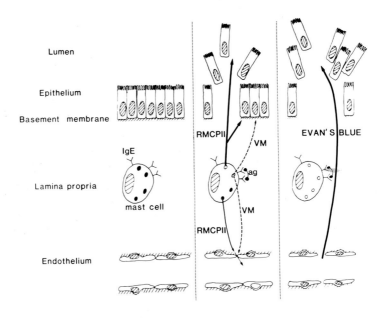

Figure 6.   Schematic view of mast cell–mediated changes in rats experiencing gut anaphylaxis.   Panel on the left shows the resting state in the intestine of an immune animal sensitized by nematode infection.   Immediately following intravenous challenge with worm antigen (o–o) RMCP II is released and acts on the basal lamina of the epithelium and endothelium causing shedding of the epithelial cells (central panel).   Other vasoactive mediators (VM) may promote capillary or venular permeability (central panel) and Evan's blue, which was injected intravenously with the antigen, escapes into the gut lumen (right panel). Using this technique it was possible to compare RMCP II levels in blood concentrations of enzyme and Evan's blue in the intestinal lumen. (For details see 20).

Site-specific Recruitment of Mast Cells During Parasitic Infection

Analysis of the distribution of RMCP II in N. brasiliensis-infected rats has shown that there is a 5-fold (from 4 mg to 22 mg) increase in the amount of RMCP II in the proximal half of the small intestine (Figure 1). Increases also occurred in the gastric, ileal and colonic mucosae, although the concentration of enzyme in these organs was substantially lower than in the jejunum (Figure 1). The gut is, therefore, the major source of RMCP II in both normal and immune rats (Figure 1) even though increased levels of enzyme occurred in immune lung and MLN (Figure 1). The contribution to the body's pool of RMCP II by trachea and thymus is minuscule and none was detected in a wide range of other tissues and organs (Figure 1).

Experiments currently in progress with the cestode Mesocestoides corti have shown that the mass of fibrotic tissue arising from the pancreas as a consequence of infection is very rich in mast cells and RMCP II. This infection is also associated with hepatic and intestinal accumulation of RMCP II and mast cells (33). In this regard Mesocestoides infection resembles Taenia taeniaeformis (34) in promoting recruitment of IMMC even though infection is extraintestinal. In contrast with enteric infections, we have been unable to detect RMCP II in significant amounts in the sera of M. corti infested rats.

In summary, recruitment of RMCP II-containing mast cells is maximal at the site of parasite infection, although it would appear that tissues which are not directly parasitized but which may encounter parasite antigens are also sites of mast cell recruitment.

General Overview of MMC Function in Parasitized Rats

So great is the profusion of mast cells in the intestines of parasitized rats that it has always been tempting to conclude that they play a part in the protective response against infection. The debate has, in the past, centred on histological assessment of IMMC infiltration in relation to expulsion or survival of the parasites (35). Unfortunately, such techniques take no account of the secretory activities of the cells, nor is it possible to determine the role of IMMC by measuring histamine levels at the site of infection, because no allowance is made for turnover of this amine or of potential non-mast cell sites of histamine storage.

With regard to histology, the adoption of paraformaldehyde fixation opens new avenues for histochemical and immunocytochemical characterization of IMMC. Thus, it is now possible to identify the enzyme profiles of newly-developing, as well as mature IMMC in both the lamina propria and epithelium. Their granules contained proteoglycan, monoamine (4), serine esterase and RMCP II. Immunolocalization of RMCP II, a biochemical fingerprint unique to IMMC, confirms the mast cell identity of GL.

That RMCP II is predominantly within IMMC was also suggested by the significant correlation between enzyme levels and IMMC counts in normal and immune gut, and also in intestines depleted of both IMMC and RMCP II either by anaphylactic shock (36) or by corticosteroids (see Figure 2). Furthermore, the release of RMCP II into blood within minutes or hours of intestinal infestation by nematodes or protozoa would again indicate the enteric origin of this enzyme.

Growth of IMMC in vitro is accompanied by release of RMCP II into the culture supernatant (37). In our in vivo studies of parasite infection, the systemic secretion of enzyme was temporally associated with the appearance in the mucosa of immature mast cells. These, as we have already pointed out (5,6), have the subcellular components necessary for an active synthetic and secretory function. Our findings are, therefore, compatible with the in vitro demonstration of the secretory capacities of maturing IMMC. By using the cell isolation techniques developed by Befus, Lee and Bienenstock (17,18,38), it should now prove feasible to determine whether immature cells are actively synthesizing and secreting RMCP II, or merely releasing stored enzyme from their sparse granules.

Quantification of systemic levels of RMCP II is, we believe, a highly sensitive method of monitoring the functional activity of IMMC. Furthermore, in shocked rats there is concomitant release of GAG (39). The concentrations of GAG in the plasma of such rats correlate well with the serum levels of RMCP II (39), and it is possible that both are released from IMMC. Studies are currently in progress to characterize the plasma GAG and to compare it with GAG from peritoneal mast cells and from IMMC derived by in vitro culture.

Whilst there is little conceptual difficulty in proposing a role for parasite-specific IgE in the generation of anaphylactic lesions in the gut, it is not certain that IgE is involved in systemic secretion of RMCP II during primary nematode or protozoal infection. Little or no parasite-specific IgE is detectable at this early stage of N. brasiliensis infection (32), nor is there evidence to suggest that Eimeria-specific IgE is generated in the rat (40). An alternative mechanism, described in the mouse, is a T-cell-derived antigen-specific factor which causes mast cell secretion without frank degranulation (41) and it may be that such a system also exists in the rat.

The observation that IMMC in immune rats internalize IgE, whereas those in normal rats do not contain this immunoglobulin (34,42) is intriguing and, so far, unexplained. It does, however, point to additional differences between mast cell types. Furthermore, the studies by Mayrhofer and colleagues on IgE internalization suggest that mast cells analogous to IMMC are recruited to bronchial mucosa during N. brasiliensis infection (42). Our data on the distribution of RMCP II in the lung tend to support this view and, in addition, show that there are increased levels of this enzyme in mesenteric lymph node. It is interesting in this regard that hepatic infection with the cestode T. taeniaeformis is associated with increased recruitment of IgE-containing IMMC in the jejunum (34). Our current collaborative studies with Chernin and McLaren (31) have shown that M. corti infection raises hepatic, pancreatic and intestinal concentrations of RMCP II and it is likely that IMMC are recruited to these sites too. The possible mechanisms of recruitment are discussed elsewhere (43).

Finally, it is pertinent to ask what role IMMC and their mediators might play in the protective response against enteric parasites. One view was that IMMC-derived mediators caused increased mucosal permeability, thereby facilitating the translocation of plasma-derived macromolecules into the gut lumen (44). These mediators may also stimulate mucus release and thereby promote mucus trapping of the parasites (45) and smooth muscle hyper-reactivity (46), or they may have a direct effect on the parasite themselves. Apart from histamine and RMCP II, the inflammatory mediators generated or released by IMMC have not been characterized and it is too early to speculate at this stage whether leukotrienes or prostaglandins are generated by activated IMMC.

We are currently comparing the catalytic activities of RMCP I and II on native proteins. Our results have confirmed previous observations that type IV collagen of basement membrane is catabolized by RMCP II (47) and, for this reason, we believe that RMCP II may be involved in the generation of mucosal epithelial permeability (see Figure 6). Indeed, IMMC-mediated sloughing of enteric mucosal epithelium may prove to be a useful method of eliminating those parasites which live on the surface of, or penetrate into, the epithelium. Similarly, the secretion of this enzyme by immature IMMC, or their precursors, may be one reason why so many IMMC migrate across the basal lamina and reside within the epithelium of parasitized intestine. By moving to this location they are placed in a pole position for the generation of a rapid, but anatomically superficial reaction against noxious stimuli.

REFERENCES

1. Taliaferro, W.H., and M.P. Sarles. 1939. The cellular reactions in the skin, lungs, and intestine of normal and immune rats after infection with Nippostrongylus muris. J. Infect. Dis. 64:157.

2. Enerback, L. 1981. The gut mucosal mast cell. Monogr. Allergy 17:222.

3. Miller, H.R.P., and W.F.H. Jarrett. 1971. Immune reactions in membranes. I. Intestinal mast cell response during helminth expulsion in the rat. Immunology 20:277.

4. Miller, H.R.P., and R. Walshaw. 1972. Immune reactions in

membranes. IV. Histochemistry of intestinal mast cells during helminth expulsion in the rat. Am. J. Pathol. 69:195.

5. **Miller, H.R.P.** 1971. Immune reactions in mucous membranes. II. The differentiation of intestinal mast cells during helminth expulsion in the rat. Lab. Invest. 24:339.

6. **Miller, H.R.P.** 1971. Immune reactions in mucous membranes. III. The discharge of intestinal mast cells during helminth expulsion in the rat. Lab. Invest. 24:348.

7. **Murray, M., H.R.P. Miller, and W.F.H. Jarrett.** 1968. The globule leukocyte and its derivation from the subepithelial mast cell. Lab. Invest. 19:222.

8. **Nawa, Y., and H.R.P. Miller.** 1977. In: Workshop No. 46, Progress in Immunology III, p.840.

9. **Nawa, Y., and H.R.P. Miller.** 1979. Adoptive transfer of the intestinal mast cell response in rats infected with Nippostrongylus brasiliensis. Cell. Immunol. 42:225.

10. **Mayrhofer, G.** 1979. The nature of the thymus dependency of mucosal mast cells. II. The effect of thymectomy and of depleting recirculating lymphocytes on the response to Nippostrongylus brasiliensis. Cell. Immunol. 47:32.

11. **Befus, A.D., and J. Bienenstock.** 1979. Immunologically mediated intestinal mastocytosis in Nippostrongylus brasiliensis infected rats. Immunology 38:95.

12. **Mayrhofer, G.** 1979. The nature of the thymus dependency of mucosal mast cells. I. An adoptive secondary response to challenge with Nippostrongylus brasiliensis. Cell. Immunol. 47:304.

13. **Jarrett, E.E.E., and D.M. Haig.** 1984. Mucosal mast cells in vivo and in vitro. Immunology Today 5:115.

14. **Woodbury, R.G., G.M. Gruzenski, and D. Lagunoff.** 1978. Immunofluorescent localization of a serine protease in rat small intestine. Proc. Natl. Acad. Sci. USA 73:2785.

15. **Ginsburg, H., and L. Sachs.** 1963. Formation of pure suspensions of mast cells in tissue culture by differentiation of lymphoid cells from mouse thymus. J. Natl. Cancer Inst. 31:1.

16. **Haig, D.M., T.A. McKee, E.E.E. Jarrett, R.G. Woodbury, and H.R.P. Miller.** 1982. Generation of mucosal mast cells is stimulated in vitro by factors derived from T cells of helminth-infected rats. Nature 300:188.

17. **Befus, A.D., F.L. Pearce, J. Gauldie, P. Horsewood, and J. Bienenstock.** 1982. Mucosal mast cells. I. Isolation and functional characteristic of rat intestinal mast cells. J. Immunol. 128:2475.

18. **Pearce, F.L., A.D. Befus, J. Gauldie, and J. Bienenstock.** 1982. Mucosal mast cells. II. Effects of anti-allergic compounds on histamine secretion by isolated intestinal mast cells. J. Immunol. 128:2481.

19. **Newlands, G.F.J., J.F. Huntley, and H.R.P. Miller.** 1984. Concomitant detection of mucosal mast cells and eosinophils

in the intestines of normal and Nippostrongylus brasiliensis-immune rats. A re-evaluation of histochemical and immunocytochemical techniques. Histochemistry 81:585.

20. **King, S.J., and H.R.P. Miller.** 1984. Anaphylactic of mucosal mast cell protease and its relationship to gut permeability in Nippostrongylus-primed rats. Immunology 51:653.

21. **Briggs, N.T., and D.L. Degiusti.** 1966. Generalized allergic reaction in Trichinella-infected mice: the temporal association of host immunity and sensitivity to exogenous antigens. Am. J. Trop. Med. Hyg. 15:919.

22. **Church, M.K., and P. Miller.** 1978. Time courses of the anti-anaphylactic and anti-inflammatory effects of dexamethesone in the rat and mouse. Br. J. Pharmac. 62:481.

23. **Jarrett, W.F.H., E.E.E. Jarrett, H.R.P. Miller, and G.M. Urquhart.** 1967. Quantitative studies on the mechanism of self-cure in Nippostrongylus brasiliensis infections. In: The Reaction of the Host to Parasitism. Edited by E.J.L. Soulsby. Elwert Margrub, Lahn, p.191.

24. **King, S.J., H.R.P Miller, G.F.J. Newlands, and R.G. Woodbury.** 1985. Anaphylactic release of mucosal mast cell protease and its suppression by corticosteroids. Proc. Nat. Acad. Sci. (USA) 82:1214.

25. **Alizadeh, H., and D. Wakelin.** 1982. Comparison of rapid expulsion of Trichinella spiralis in mice and rats. Int. J. Parasit. 12:65.

26. **Rose, M.E., B.M. Ogilvie, and J.W.A. Bradley.** 1980. Intestinal mast cell responses in rats and chickens to coccidiosis, with some properties of chicken mast cells. Int. Arch. Allergy Appl. Immunol. 63:21.

27. **Miller, H.R.P., R.G. Woodbury, J.F. Huntley, and G.F.J. Newlands.** 1983. Systemic release of mucosal mast cell protease in primed rats challenged with Nippostrongylus brasiliensis. Immunology 49:471.

28. **Moqbel, R., S.J. King, D. Wakelin, H.R.P. Miller, and A.B. Kay.** 1985. Generation of leukotrienes in rat gastrointestinal tract during the rapid expulsion of the nematodes N. brasiliensis and T. spiralis. (Manuscript in preparation).

29. **Huntley, J.F., G.F.J. Newlands, H.R.P. Miller, M. McLauchlan, M.E. Rose, and P. Hesketh.** 1985. Systemic release of mucosal mast cell protease during infection with the intestinal protozoal parasite, Eimeria nieschulzi. Studies in normal and nude rats. In press.

30. **Woodbury, R.G., H.R.P. Miller, J.F. Huntley, G.F.J. Newlands, A.C. Palliser, and D. Wakelin.** 1984. Mucosal mast cells are functionally active during the spontaneous expulsion of primary intestinal nematode infections in the rat. Nature 312:450.

31. **Woodbury, R.G., and H.R.P. Miller.** 1982. Quantitative analysis of mucosal mast cell protease in the intestines of Nippostrongylus-infected rats. Immunology 46:487.

32. **Jarrett, E.E.E., and H.R.P. Miller.** 1982. Production and

activities of IgE in helminth infection. Prog. Allergy 31:178.

33. **Chernin, J., D.J. McLaren, G.F.J. Newlands, and H.R.P. Miller.** 1985. Recruitment of mucosal mast cells to non-mucosal sites in rats infected with Mesocestoides corti. (Manuscript in preparation).

34. **Lindsay, M.C., D.B. Blaies, and J.F. Williams.** 1983. Taenia taeniaeformis: immunoglobulin E-containing cells in intestinal and lymphatic tissues of infected rats. Int. J. Parasit. 13:91.

35. **Askenase, P.W.** 1980. Immunopathology of parasitic disease: involvement of basophils and mast cells. Springer Semin. Immunopathol. 2:417.

36. **King, S.J., H.R.P. Miller, R.G. Woodbury, and G.F.J. Newlands.** 1985. Gut mucosal mast cells in Nippostrongylus-primed rats are the major source of secreted rat mast cell proteinase II following systemic anaphylaxis. Submitted for publication.

37. **Haig, D.M., C. McMenamin, C. Gunneberg, R.G. Woodbury, and E.E.E. Jarrett.** 1983. Stimulation of mucosal mast cell growth in normal and nude rat bone marrow cultures. Proc. Natl. Acad. Sci. USA 80:4499.

38. **Lee, T.D.G., F. Shanahan, H.R.P. Miller, J. Bienenstock, and A.D. Befus.** 1985. Intestinal mucosal mast cells: Isolation from rat lamina propria and purification using unit gravity velocity sedimentation. Immunology 55:721.

39. **King, S.J., K. Reilly, J. Dawes, and H.R.P. Miller.** 1985. The presence in blood of both glycosaminoglycan and mucosal mast cell protease following systemic anaphylaxis in the rat. Int. Arch. Allergy Appl. Immunol. 76:286.

40. **Rose, M.E.** 1982. Host immune responses In: The Biology of the Coccidia. Edited by P.L. Long. Baltimore, University Park Press, p.330.

41. **Askenase, P.W., R.W. Rosenstein, and W. Ptak.** 1983. T cells produce an antigen-binding factor with in vivo activity analogous to IgE antibody. J. Exp. Med. 157:862.

42. **Mayrhofer, G., H. Bazin, and J.L. Gowans.** 1976. Nature of cells binding anti-IgE in rats immunized with Nippostrongylus brasiliensis: IgE synthesis in regional nodes and concentration in mucosal mast cells. Eur. J. Immunol. 6:537.

43. **Haig, D.N., C. McMenamin, and E.E.E. Jarrett.** 1985. Mast cell development in the rat. In: Symposium on Mast Cell Differentiation and Heterogeneity. Edited by D. Befus, J. Denburg, and J. Bienenstock. This volume.

44. **Murray, M.** 1972. Immediate hypersensitivity effector mechanisms. II. In vivo reactions. In: Immunity to Animal Parasites. Edited by E. Sadun. New York, Academic Press, p. 155.

45. **Miller, H.R.P., J.F. Huntley, and G.R. Wallace.** 1981. Immune exclusion and mucus trapping during the rapid expulsion of

Nippostrongylus brasiliensis from primed rats.   Immunology 44:419.

46. **Castro, G.A.** 1980. Regulation of pathogenesis in disease caused by gastrointestinal parasites.   In: The Host Invader Interplay. Edited by Van den Bossche. Amsterdam, Elsevier, p. 457.

47. **Sage, H., R.G. Woodbury, and P. Bornstein.** 1979. Structural studies on human type IV collagen. J. Biol. Chem. 254:9893.

Mast Cell Differentiation and Heterogeneity,
edited by A. D. Befus et al.
Raven Press, New York © 1986.

Rat Mast Cell/Basophil Heterogeneity Related to Cell Maturation

Elizabeth WoldeMussie and Michael A. Beaven

Laboratory of Chemical Pharmacology, National Heart, Lung and Blood Institute, Bethesda, Maryland 20205

Although it is now well recognized that mast cells from different species and from different tissues within the same species may have quite diverse properties (for reviews see refs. 1–3), it is now evident that maturational changes may give rise to heterogeneous populations of mast cells within a single tissue. Early studies of rat connective tissue, as well as peritoneal mast cells, pointed to sequential changes in cytochemical and morphological characteristics (4,5) and to progressive increases in intracellular content of mediators, heparin, beta–glucuronidase and chymase (6,7) as cells matured. There is also evidence from recent work that maturation of rat peritoneal (8) and human (9) mast cells is accompanied by increased mediator content and responsiveness to degranulating stimuli.

Combs and co–workers defined four recognizable stages of growth (Stages 1 to 4) (4). Stage 1 cells contained small numbers of cytoplasmic granules, which stained with alcian blue, and large nuclei which showed signs of post–mitotic activity. As cells progressed through the remaining stages, the numbers of granules and proportion of granules that stained red with safranin increased. By Stage 3 the majority of, and by Stage 4 all granules stained red. The red stain indicated sulfation of granule constituents. These changes correlated with other signs of cell maturation, namely, uptake of radiolabelled thymidine (Stage 1 cells) and the incorporation of [^{35}S] sulfate into proteoglycans (Stage 2 and 3 cells).

The focus of our current work has been the mechanism of mast cell/basophil degranulation but it has become apparent to us that heterogeneity arising from cell maturation has an important bearing on our work. This has led us to evaluate the relation–

ship between secretory response and cell maturation in two widely studied models, rat peritoneal mast cells and cultured rat leukemic basophil (2H3) cells (10–12). As discussed below, both cell types exhibit an apparent heterogeneity in regard to histamine content, histamine turnover and ability to respond to secretory stimulants. We believe, however, that once the age-dependent heterogeneity is properly understood, the phenomenon could be used in elucidating mechanisms of mast cell secretion.

Our awareness of the phenomenon resulted from studies with Andrew Soll (13,14) in which histamine-containing cells from various tissues were separated by countercurrent elutriation. This process separates cells by size into sequential fractions of progressively larger cells. We noted that the distribution of mast cells did not always correlate with that of histamine and histamine synthetic activity (8). We review here these, as well as our ongoing studies.

METHODS

Collection and separation of cell fractions. Cells were collected from the peritoneal cavity of male Sprague Dawley rats (6–8 weeks of age) by lavage and from confluent monolayer cultures of 2H3 cells by treatment with trypsin (8,11). The cells were suspended (10 to 120×10^6 cells/20 ml) in Hanks' medium supplemented with 10 mM N-2-hydroxyethylpiperazine-N'-2-ethanesulfonic acid and 0.1% bovine serum albumin (supplemented Hanks' medium). They were then separated by elutriation into 12, 100 ml fractions. The protocols, which differed slightly for rat and 2H3 cells, have been described in detail (8,11). Cells in each fraction were counted and their size determined by use of a Particle Data Counter. Cell viability was assessed by ethidium bromide/fluorescein diacetate dyes. In studies with rat cells, two sets of air-dried slides were prepared from each fraction. One set of slides was stained with Wright's stain, the other was fixed in Newcomer's fixative and stained sequentially with alcian blue and safranin (4). Recently, we have substituted Newcomer's with Carnoy's fixative to improve the quality of slides. Mast cells were categorized as Stage 1, 2, 3 or 4 cells as described by Combs and co-workers (4).

To harvest cells, the cell suspensions were centrifuged (200 x g for 10 min). The cell pellets were washed once in the supplemented Hanks' medium and were then redispersed in the same medium (0.15 to 0.4 $\times 10^6$ cells/100 μl) for assay of histamine synthetic activity either by release of $^{14}CO_2$ from L-[carboxyl-^{14}C]-histidine or by formation of radiolabelled histamine from L-[2 ring-^{14}C]-histidine. For studies of uptake and metabolism of the ring labelled histidine in 2H3 cells, the cells were resuspended in culture medium before plating in 24 well cluster plates (0.2 x 10^6 cells/well). After incubating them (37°C in 5%

CO_2/95% air) for 2 hr, the adherent cells were washed twice with supplemented Hanks' medium before addition of the same medium (200 μl) and labelled histidine. Studies of histamine release were conducted with washed-cell suspensions of mast cells (10^6 cells/ml) or monolayer cultures of 2H3 cells in cluster plates (0.2 x 10^6 cells/well) in supplemented Hanks' medium (see refs. 8, 11 for specific details).

Assay procedures. Histamine was assayed in elutriated fractions of cells or, for histamine-release studies, in both cell extracts and medium, by radioenzymatic assay procedures (8). Histamine synthetic activity was determined in intact cell suspensions (or sonified cell extracts) by the procedures discussed above. Lactate dehydrogenase, DNA and histidine were assayed by standard procedures (8,11).

RESULTS AND COMMENTS

Studies with rat peritoneal mast cells. When rat peritoneal cells were fractionated by elutriation, monocytes were distributed mainly in fractions 3 to 5. Stage 1 mast cells were most prevalent in fractions 3 to 5 and constituted 70-80% of the mast cells in these fractions. The numbers of Stage 1 cells declined markedly thereafter (i.e. fractions 6 through 12). Stage 2 and 3 cells showed peaks of distribution in fractions 7 and 9 respectively. Stage 4 cells were predominant (45-55%) in fractions 10 to 12. In the latter fractions greater than 80% of the cells were Stage 3 and 4 mast cells. Fractions of small mast cells contained approximately 1 pg histamine/cell, whereas fractions of large cells contained >12 pg histamine/cell. Fractions of small cells also possessed high histamine synthetic activity. With increasing cell size histamine synthetic activity diminished (Table I) (8).

Interestingly, fractions of small cells sequestered newly formed histamine poorly. When cells were incubated in supplemented Hanks' medium at 37°C, histamine levels in the medium steadily increased over the course of 3 hr, while the intracellular histamine content remained unchanged. Histamine was released, however, in greater proportions from immature cells (fractions 5 and 7) than from fractions of mature cells (fractions 8 to 11). As rates of release were markedly reduced (by 66-90%) in the presence of the histidine decarboxylase inhibitor α-fluoromethylhistidine (10 μM), we assumed that rates of release were a measure of histamine synthesis and turnover. The calculated fractional turnover rates ranged from 0.12 $\pm$ 0.02/hr for fraction 5 to 0.03/hr or lower for fractions 8 to 11. The turnover rates correlated with the abundance of Stage 1 cells in each fraction (unpublished data).

Small mast cells thus exhibit high rates of histamine turnover, high synthetic activity but low histamine content. In

TABLE I. Changes in histidine-uptake, histamine synthesis and histamine content
in relation to cell cycle and stage of cell development

Parameter measured	2H3 Cells[1]			Mast Cells[2]	
	Elutriated cells		Confluent cultures	Immature (fraction 5)	Mature (fraction 9)
	Early G_1 (fraction[3])	G_1/S interface (fraction 6 or 7)			
Cell vol (Femtoliter)	730	1,170	1,100	–	790
Histidine content (mM)	–	–	1.8 ± 0.5	–	–
Histamine content					
(mM)	0.08 ± 0.01	2.4 ± 0.2	3.5 ± 0.6	–	159 ± 12
(pg/cell)	0.07 ± 0.01	0.31 ± 0.03	0.42 ± 0.07	0.5	12.6 ± 1.1
(nmoles/10^6 cells)	0.06 ± 0.01	2.8 ± 0.3	3.8 ± 0.6	4.5	114 ± 1.0
Histidine uptake					
K_m (µM)	23	17	24 ± 4.0	–	–
V_{max} (nmole/10^6 cells/min)	0.8	2.9	5.1 ± 1.2	–	–
Histamine synthetic activity (pmoles/10^6 cells/min)[3]	0.15 ± 0.02	1.3 ± 0.3	2.0 ± 0.5	4.2 ± 1.3	1.6 ± 0.4
Histamine turnover rate[4] (pmoles/10^6 cells/min)	–	–	3.7 ± 0.25	9.0	–
(hr)	–	–	12 ± 1.0	10 ± 1.0	>30

Values are mean of 2 experiments or mean $\pm$ S.E. of more than 3 experiments.

1. Data from ref 11.
2. Data from refs 8, 13 and unpublished data.
3. As determined by release of $^{14}CO_2$ from L-[^{14}C-carboxyl] labelled histidine added to intact cell suspensions.
4. Calculated from rates of spontaneous histamine release in cell cultures or cell suspensions.

fractions of larger cells, the progressive increases in numbers
of granules and degree of sulfation of granule proteoglycans (as
indicated by safranin staining) was associated with an improved
ability to sequester newly formed histamine. In addition to the
slow rates of histamine turnover noted above in fractions of
large cells, a stoichiometric relationship was observed between
$^{14}CO_2$ release from L-[carboxyl-^{14}C]-histidine and intracellular
accumulation of radiolabelled histamine (from [ring-^{14}C] labelled
histidine).   There was no detectable loss of newly formed
histamine from fractions of mature cells by 60 min (unpublished
data).
     A gradation in responsiveness of mast cells to compound 48/80
has been noted: fractions of small cells were resistant, those
of intermediate sized cells were partially responsive and those
of mature cells were fully responsive to the histamine-releasing
action of compound 48/80 (8).   A similar profile of responsiveness was observed with A23187 (0.5 µM). This could be indicative
of inadequate intracellular responses to the calcium signal in
small mast cells (unpublished data).
     Studies with rat leukemic basophil (2H3) cultures.   These
cells, as do rat peritoneal mast cells, take up and decarboxylate
histidine to form histamine. Uptake is mediated by a temperature-
dependent system with a high affinity (apparent K_m, 24 µM) for
histidine. Transfer of newly formed histamine from cytosol to
secretory granules is slow: 66% of the newly formed histamine is
incorporated into the granular pool by 60 min (11), whereas more
than 90% is incorporated into this pool by 5 min in mast cells
(15).   As with immature peritoneal mast cells, substantial
amounts of histamine but not LDH are released from confluent 2H3
cell cultures into the medium.   The calculated fractional

turnover rate was similar to that for small mast cells (Table I). There were characteristic changes in histamine content, histidine uptake and histidine decarboxylation during the 2H3 cell cycle. Elutriation studies indicated low rates of histidine uptake and decarboxylation in the smallest 2H3 cells, a progress-ive increase in the ability to take up and decarboxylate histi-dine As the cells approached 'S' phase of growth and a marked decline in these functions as cells approached cell division. There were similar changes in histamine content. The kinetic constants suggested that the changes in histidine uptake were due to variation in numbers of sites of histidine transport (11).

The data for 2H3 cells and rat mast cells are compared in Table I. These data indicated that during cell division all components associated with histamine synthesis -- histidine uptake and decarboxylation -- were reduced. Recent studies have also shown that the secretory response of 2H3 cells is blocked when cells are arrested in mitosis. It is rapidly restored once cells resume growth through the G_1 phase of the cycle (12). Cells arrested in mitosis still generate a calcium signal, which is associated with rapid breakdown of plasma membrane phospho-inositides (16), in response to antigenic stimulation. Apparently the systems involved in transduction of membrane signals remain intact. Intracellular secretory mechanisms beyong the calcium signal appear to be suppressed (12).

CONCLUSIONS

Maturation of mast cells leads to important functional as well as morphological changes. A population of mast cells in one tissue or even a single clone of cultured cells may not represent a homogenous population of cells for the purpose of studying mast cell function. It is apparent from our studies that systems involved in histamine synthesis and storage are expressed at different stages of maturation. The ability to sequester newly formed histamine appears to be dependent on the abundance of granules and a sufficient supply of sulfated proteoglycans within them. In this respect, similarities exist between the immature mast cell and transformed 2H3 cells. Both have relatively few granules and a low complement of sulfated proteoglycans and both exhibit high rates of histamine turnover. A possible clinical implication is that in lesions with developing colonies of young or transformed mast cells, histamine turnover as well as degranulation might contribute to the inflammatory reactions associated with such lesions.

The secretory response of cells develops at a late stage of maturation and, in cell cultures, is suppressed during mitosis. The sequence of biochemical events that lead to mast cell degranulation is still unclear, but advantage could be taken of the present findings. Is the lack of responsiveness, for example, of small mast cells due to an inability to transduce membrane signals or the cell's inability to respond to intracellular signal(s)? Could useful information be gained if we knew the stages of development at which cells become competent in generating a calcium signal? As the studies with 2H3 cells have shown, the generation of the calcium signal is uncoupled from secretion at one stage of the cell cycle.

REFERENCES

1. Barrett, K.E., and D.D. Metcalfe. 1984. Mast cell heterogeneity: Evidence and implications. J. Clin. Immunol. 4:253.
2. Bienenstock, J., A.D. Befus, J. Denburg, R. Goodacre, F.L. Pearce, and F. Shanahan. 1983. Mast cell heterogeneity. Monogr. Allergy 18:124.
3. Pearce, F.L. 1982. Functional heterogeneity of mast cells from different species and tissues. Klin. Wochenschr. 60:954.
4. Combs, J.W., D. Lagunoff, and E.P. Benditt. 1965. Differentiation and proliferation of embryonic mast cells of the rat. J. Cell. Biol. 25:577.
5. Wilhelm, D.L., L.C.J. Yong, and S.G. Watkins. 1978. The mast cell: distribution and maturation in the rat. Agents Actions 8:146.
6. Enerback, L., and L. Mellblom. 1978. Growth related changes in the content of heparin and 5-hydroxytryptamine of mast cells. Cell Tissue Res. 187:367.
7. Kruger, P.G., and D. Lagunoff. 1981. Effect of age on mast cell granules. Int. Arch. Allergy Appl. Immunol. 65:291.
8. Beaven, M.A., D.L. Aiken, E. WoldeMussie, and A.H. Soll. 1983. Changes in histamine synthetic activity, histamine content and responsiveness to compound 48/80 with maturation of rat peritoneal mast cells. J. Pharm. Exptl. Therap. 224:620.
9. Schulman, E.S., A. Kagey-Sobotka, D.W. MacGlashan, N.F. Adkinson, S.P. Peters, R.P. Schleimer, and L.M. Lichtenstein. 1983. Heterogeneity of human mast cells. J. Immunol. 131:1936.
10. Siraganian, R.P., A. McGivney, E.L. Barsumian, F.T. Crews, F. Hirata, and J. Axelrod. 1982. Variants of the rat basophilic leukemia cell line for the study of histamine release. Fed. Proc. 41:30.
11. WoldeMussie, E., D.L. Aiken, and M.A. Beaven. 1985. Changes in histidine uptake and histamine synthesis during the growth cycle of rat basophilic leukemia (2H3) cells. J. Pharm. Exptl. Therap. 232:20.
12. Hesketh, T.R., M.A. Beaven, J. Rogers, B. Burke, and G.B. Warren. 1984. Stimulated release of histamine by rat basophilic leukemia 2H3 cells is inhibited during mitosis. J. Cell Biol. 98:2250.
13. Soll, A.H., K.J. Lewin, and M.A. Beaven. 1981. Isolation of histamine containing cells from rat gastric mucosa:

Mast Cell Differentiation and Heterogeneity,
edited by A. D. Befus et al.
Raven Press, New York © 1986.

Bronchoalveolar Mast Cells from Macaques Infected with *Ascaris Suum*

R. P. Eady, B. Greenwood, S. T. Harper, J. Mann, T. S. C. Orr, and E. Wells

Fisons Plc - Pharmaceutical Division, Loughborough, Leicestershire LE11 0RH, England

The immunological sensitivity of wild caught non-human primates to Ascaris antigen has been utilized as a model of asthma (1). As only a low percentage of monkeys are naturally reactive to bronchial challenge with purified antigen, laboratory bred macaques were infected with the nematode Ascaris suum to produce a colony of animals sensitive to aerosolized antigen (2,3).

Following oral administration of A. suum ova, animals developed a blood eosinophilia and specific antibodies to Ascaris antigen (2). IgE antibodies were detected by a radiometric assay but both a heat labile (56°C) and a heat stable antibody were demonstrated by the passive cutaneous anaphylaxis test (2). Bronchial provocation with purified extracts of Ascaris evoked a bronchoconstriction in infected animals which correlated closely with a rise in arterial plasma histamine levels (3).

Several reports (4–7) have described cells with the characteristics of mast cells occurring in the bronchoalveolar lavage of monkeys naturally sensitive to A. suum. This report summarizes studies on the bronchoalveolar mast cells obtained from Ascaris infected macaques.

GENERAL METHODS

Embryonated A. suum ova were prepared by the in vitro culture of decorticated ova in 0.1N H_2SO_4. Macaca arctoides monkeys were infected by the oral instillation of a calculated volume of egg suspension (660 embryonated ova per kg body weight). The present studies were carried out on animals which had received four to five infections over a period of 2 years. Bronchoalveolar lavage was performed on sedated animals essentially as described by Patterson et al. (5).

RESULTS

Identification of cells. Lung lavage cells from macaques reinfected with A. suum comprised mainly macrophages, epithelial cells and mast cells. The number of eosinophils varied between animals depending on the stage of the infection. All lavages from infected macaques contained a large percentage (10–35%) of basophiloid-like cells as defined by Kimura's stain (8). In 21 lung lavages, from a pool of 10 reinfected animals, the mean distribution of cells was 8.5% eosinophils, 3.4% neutrophils, 8.8% small mononuclear cells, 35.7% macrophages, 20.8% epithelial cells and 22.8% mast cells.

In an attempt to determine the type of mast cell in the bronchoalveolar lavage of the macaque, a histological study was carried out comparing lavage mast cells with mast cells in tissue sections of macaque intestine and lung and tissue from other species. Two types of mast cell were identified in sections of macaque small intestine fixed in 10% formol saline and stained with alcian blue-safranin. The majority of mast cells in the lamina propria contained only blue granules, whereas those in the submucosa below the muscularis mucosae contained both blue and brown granules. Both types of cell were found at the basal level of the lamina propria. This distribution of cells was similar to that found for sections of rat and human bowel fixed with either 10% formol saline or Carnoy's fixative. Although Carnoy's fixative often made the alcian blue staining more prominent, it tended to affect cytology adversely and was not therefore used for routine histology. To investigate the nature of the proteoglycan in the granules of the mast cell, sections were stained with the fluorescent dye berberine sulphate. Many cells which showed fluorescence were regarded as containing heparin (9). Almost all mast cells in the submucosa of the macaque small intestine fluoresced when stained with berberine, but fluorescence was associated with only a few mast cells in the lamina propria.

Bronchoalveolar lavage cells from the macaques infected with A. suum were smeared onto slides and fixed with 10% formol saline. When stained with alcian blue safranin the vast majority of the mast cells contained blue granules in the cytosol and did not fluoresce when stained with berberine sulphate. Only a very few cells had brown staining granules with alcian blue/safranin and a similar proportion fluoresced with berberine sulphate. Similar results were obtained with human bronchoalveolar cells. During this study no difference between the mast cells of the macaque small bowel lamina propria and the bronchoalveolar cell from the same species was seen with any combination of staining techniques.

Histamine release from bronchoalveolar cells. The total histamine content of bronchoalveolar lavage leukocytes correlated closely with the number of mast cells, as defined by absolute cell counts using Kimura's stain. For 18 lavages, the total histamine content was 7.4 $\pm$ 0.6 (mean $\pm$ SEM) pg histamine per mast cell. Release of histamine occurred when cells were incubated with goat antibody to human IgE (Miles) or purified extracts of A. suum. Histamine release took place over 30 min, being largely complete within 10 min. Release of histamine by anti-IgE or antigen varied between animals depending on the sensitivity of the individual monkey. Bronchoalveolar lavage cells from non-infected macaques did not respond to challenge with anti-IgE or antigen. Lavage leukocytes from infected animals do not release histamine when challenged with rabbit antibody to monkey IgG or with compound 48/80 (0.1 - 10 μg/ml).

Effect of sodium cromoglycate, nedocromil sodium and a mono-chromone on histamine release. The pyranoquinoline, nedocromil sodium (disodium 4,6–dioxo–9–ethyl–10–propyl–4H,6H–pyrano–[3,2g]–quinoline–2,8–dicarboxylate) inhibited anti-IgE induced histamine release in a dose–dependent manner, with an IC_{30} of 5.2 x 10^{-6}M (n=5). Sodium cromoglycate, however, gave only 30% inhibition at 10^{-3}M (n=5) 190 times less active than nedocromil sodium. These results are in sharp contrast to those from similar experiments using rat peritoneal mast cells stimulated with anti–rat IgE in which nedocromil sodium (IC_{30} = 1.4 x 10^{-6}M; n=3) and sodium cromoglycate (IC_{30} = 0.9 x 10^{-6}M; n=3) had a similar potency. Further, another chromone, FPL 55618 (5–[3–methylbutyloxy]–4–oxo–8–prop–2–enyl–4H–1–benzopyran–2–carboxy-late) was 1200–fold more potent at blocking histamine release from the rat peritoneal mast cell than nedocromil sodium, having an IC_{30} = 1.1 x 10^{-9}M (n=3). However, this compound showed no activity against the macaque cell when tested at concentrations up to 10^{-6}M.

DISCUSSION

The studies reported in this paper show that a large number of viable mast cells can be retrieved by lavage of the lungs of monkeys infected with the nematode A. suum. It has been reported that mast cells can be obtained from rhesus monkeys (Macaca mulatta) by bronchoalveolar lavage (4,5). However, in these studies only 0.5–1% of all lavaged cells were described as having the characteristics of mast cells and/or basophils. The reason for the large percentage of mast cells in the lung lavage of macaques in the present study (mean 23%) is unknown. There may be a species component, but the effect of repeated infection with Ascaris on the bronchoalveolar lavage cell composition has not yet been thoroughly investigated. However, infection of rats with N. brasiliensis has been reported to induce an intestinal (10) and lung mast cell hyperplasia (11).

In recent years it has been appreciated that heterogeneity of mast cells exists, and in the rat at least two types of cells, connective tissue (peritoneal) and intestinal mucosal mast cells, have been described (see reviews 12-14). Studies using techniques similar to those reported in (12) have shown that the cells found in the lung lavage of macaques have morphological characteristics identical with the mast cells found in the mucosa of the small intestine of the macaque. Bronchoalveolar mast cells from the macaque do not contain heparin, as judged by berberine staining, nor do they respond to challenge with compound 48/80. They therefore appear to be similar to the mucosal mast cell described in the intestine of N. brasiliensis infected rats (12,13).

The histamine content of Macaca arctoides lung lavage mast cells was similar to that reported for rhesus monkeys (4) and for mast cells isolated from human lung tissue (15,16). It has been reported previously that bronchoalveolar cells from macaques infected with A. suum release histamine when challenged in vitro with purified antigen or anti-human IgE (2). This has been confirmed with cells obtained from reinfected animals. It has also been reported that transient airway reactivity to Ascaris antigen can be conferred on non-sensitive monkeys by transferring sensitive bronchoalveolar cells to the bronchial lumen of such an animal (17). Studies in monkeys suggest that mediator release from lung epithelial and lumenal mast cells initiates the bronchoconstrictor episode following inhaled antigen (18). It is therefore likely that immunologically sensitive bronchoalveolar mast cells are involved in the antigen-induced bronchoconstriction of A. suum infected macaques (3).

In developing new topical treatments for asthma, consideration should be given to the types of cell found in the lumen of the lung as these are the first cells to interact with both antigen and drug. The importance of selecting the mast cell most relevant to the disease state is underlined by the vastly different rank order of potency of the drugs on the macaque cell compared to that on the rat peritoneal mast cell.

Bronchoalveolar mast cells from macaques were insensitive to inhibition by sodium cromoglycate. This is in agreement with the results from studies on lung lavage cells from rhesus and cynomologus monkeys (7). However, nedocromil sodium did inhibit mediator release from immunologically activated macaque bronchoalveolar mast cells. These results tend to confirm the importance of the luminal mast cell in the initiation of bronchoconstriction, as nedocromil sodium but not sodium cromoglycate protects against bronchoconstriction when the macaques are challenged 'in vivo' with aerosolized Ascaris antigen (19). Nedocromil sodium is currently being evaluated in patients as a new drug for the treatment of asthma.

REFERENCES

1. Weiszer, I., R. Patterson, and J.J. Pruzansky. 1968. Ascaris

hypersensitivity in the rhesus monkey. I. A model for the study of immediate type hypersensitivity in the primate. J. Allergy 41:14.

2. **Pritchard, D.I., R.P. Eady, S.T. Harper, et al.** 1983. Laboratory infection of primates with Ascaris suum to provide a model of allergic bronchoconstriction. Clin. Exp. Immunol. 54:469.

3. **Richards, I.M., R.P. Eady, D.M. Jackson, et al.** 1983. Ascaris-induced bronchoconstriction in primates experimentally infected with Ascaris suum ova. Clin. Exp. Immunol. 54:461.

4. **Patterson, R., Y. Tomita, S.H. Oh, I.M. Suszko, and J.J. Pruzansky.** 1974. Respiratory mast cells and basophiloid cells. I. Evidence that they are secreted into the bronchial lumen, morphology, degranulation and histamine release. Clin. Exp. Immunol. 16:223.

5. **Patterson, R., J.M. McKenna, I.M. Suszko, N.H. Solliday, J.J. Pruzansky, M. Roberts, and T.J. Kehoe.** 1977. Living histamine-containing cells from the bronchial lumens of humans. Description and comparison of histamine content with cells of rhesus monkeys. J. Clin. Invest. 59:217.

6. **Patterson, R., C. Ts'ao, and I.M. Suszko.** 1980. Heterogeneity of bronchial lumen mast cells which are homogeneous by electron microscopy. J. Allergy Clin. Immunol. 65:278.

7. **Butchers, P.R., C.J. Vardey, I.F. Skidmore, A. Wheeldon, and L.E. Boutal.** 1980. Histamine-containing cells from bronchial lavage of macaque monkeys. Time course and inhibition of anaphylactic histamine release. Int. Arch. Allergy Appl. Immunol. 62:205.

8. **Kimura, I., Y. Moritani, Y. Nishizaki, and Y. Tamzaki.** 1968. Clinical significance of basophil leukocytes in bronchial asthma. Acta Med. Okayama 22:203.

9. **Wingren, V., and L. Enerback.** 1983. Mucosal mast cells of the rat intestine: a re-evaluation of fixation and staining properties, with special reference to protein blocking and solubility of the granular glycosaminoglycan. Histochem. J. 15:571.

10. **Miller, H.R.P., and W.F.H. Jarrett.** 1971. Immune reactions in mucous membranes. I. Intestinal mast cell response during helminth expulsion in the rat. Immunology 20:277.

11. **Mayrhofer, G., H. Bazin, and J.L. Gowans.** 1976. Nature of cells binding anti-IgE in rats immunized with Nippostrongylus brasiliensis: IgE synthesis in regional nodes and concentration in mucosal mast cells. Eur. J. Immunol. 6:537.

12. **Enerback, L.** 1981. The gut mucosal mast cell. Monogr. Allergy 17:222.

13. **Bienenstock, J., A.D. Befus, F.L. Pearce, J. Denburg, and R. Goodacre.** 1982. Mast cell heterogeneity: derivation and function with emphasis on the intestine. J. Allergy Clin. Immunol. 70:407.

14. Jarrett, E.E.E., and D.M. Haig. 1984. Mucosal mast cells in vivo and in vitro. Immunol. Today 5:115.
15. Ennis, M. 1982. Histamine release from human pulmonary mast cells. Agents Actions 12:60.
16. Schulman, E.S., D.W. MacGlashin, S.P. Peters, R.P. Schleimer, H.H. Newball, and L.M. Lichtenstein. 1982. Human lung mast cells: purification and characterization. J. Immunol. 129:2662.
17. Patterson, R., I.M. Suszko, and K.E. Harris. 1978. The in vivo transfer of antigen-induced airway reactions by bronchial lumen cells. J. Clin. Invest. 62:519.
18. Hogg, J.C., P.D. Pare, and R.C. Boucher. 1979. Bronchial mucosal permeability. Fed. Proc. 38:197.
19. Eady, R.P., B. Greenwood, D.M. Jackson, T.S.C. Orr, and E. Wells. 1985. The effect of nedocromil sodium and sodium cromoglycate on antigen-induced bronchoconstriction in the Ascaris-sensitive monkey. Br. J. Pharmac. 85:323.

Mast Cell Differentiation and Heterogeneity,
edited by A. D. Befus et al.
Raven Press, New York © 1986.

Characterization and Mycoplasma Induction of a 71K Receptor for IgE on Rat Basophilic Leukemia Cells

A. Froese, B. M. C. Chan, M. Rao, P. A. Roth, and K. McNeill

MRC Group in Allergy Research, Department of Immunology, University of Manitoba, Winnipeg, Manitoba R3E 0W3, Canada

Studies carried out during the last few years (1–4) have demonstrated that rat basophilic leukemia (RBL) cells carry on their plasma membrane two major receptors for immunoglobulins. In the authors' laboratory these have been designated H and R (5) and were found to have apparent molecular weights (M_r) of 55,000 and 45,000 daltons respectively. Both H and R receptors are also present on rat peritoneal mast cells (6). The latter molecule has also been named α by some authors (7) and it is known to be associated with two other polypeptide chains designated β (7,8) and γ (9). For these chains to remain together during isolation, a certain detergent to lipid ratio has to be maintained (10). So far, H has not been found to be associated with other polypeptide chains.

Of these two receptors, R has the higher affinity for IgE (2, 4) and was shown to be involved in mediator release (11). Receptor H also binds to IgE–Sepharose but with a lower affinity than R (α) (4). Both receptors also interact with rat IgG–Sepharose, and H is most likely the molecule which on intact RBL cells was shown to bind oligomeric rat IgG, but not its monomeric form (2,4).

More recently, a minor surface molecule, having an M_r of 71,000 daltons (71K), was detected in detergent extracts of RBL cells maintained in the authors' laboratory (RBL–Wpg) (12). Similar molecules, but of slightly different M_r could also be detected in detergent extracts of other RBL cell lines (3). A

These studies were supported by a grant from the Medical Research Council of Canada.

molecule with an SDS-PAGE mobility similar to that of 71K had been observed, occasionally in earlier studies (1,13), however, it had not been identified as a receptor for IgE. The following will describe the properties of this molecule and demonstrate that it is induced by mycoplasma.

PROPERTIES OF 71K MOLECULES

As shown previously (12), when receptors are isolated from surface iodinated and Nonidet P-40 (NP-40) solubilized RBL cells, three peaks corresponding to the 71K, H and R receptors are seen upon SDS-PAGE analysis. Using protease inhibitors such as phenyl methyl sulfonyl fluoride (PMSF), iodoacetamide and fetal calf serum during solubilization and isolation steps did not significantly alter the receptor profile. These observations suggested that the H and R molecules are not enzymatic degradation products of the larger 71K molecule. Like R, 71K can also be isolated by means of IgE and anti-IgE, while H cannot (1).

Like H and R, 71K is a glycoprotein since its presence could be demonstrated when RBL cells were labelled with tritiated amino acids or sugars. The binding of 71K to IgE-Sepharose, like that of H or R, could be inhibited by adding an excess of free IgE to receptor-containing RBL cell extracts (14) indicating that the binding is specific. Similarly, inhibition of binding to IgE-Sepharose could also be achieved if intact cells were incubated with an excess of free IgE prior to solubilization (12).

Like R, 71K binds to a variety of lectins including concanavalin A, wheat germ agglutinin, caster bean lectin, winged pea lectin and gorse lectin. By contrast, H bound to some of these lectins and not to others (15).

These observations and those which showed that, during isolation, 71K always segregates with R, suggested that these two molecules may be related. The fact that 71K has a higher molecular weight than R and retains this higher M_r in the presence of SDS may indicate that it consists of an R-like molecule covalently bonded to some other molecule. Early attempts to reduce the M_r of 71K by reduction failed (15). However, upon increasing the concentration of 2-mercaptoethanol (2-ME) to 5% when eluting iodinated receptor from IgE-Sepharose in the presence of SDS, the results shown in Figure 1 were obtained. On SDS-PAGE analysis, the peak representing 71K disappeared and that corresponding to R increased in size. The peak representing H was not affected, nor were any new peaks seen. Similarly, when 71K was isolated, reduced and re-analyzed by SDS-PAGE, a single peak corresponding in mobility to R was observed. Details of these experiments will be presented elsewhere (16). From these observations it could be concluded that 71K consists of a polypeptide chain, identical in M_r to R which is disulfide bonded either to itself or to another molecule which is not surface iodinated. To differentiate between these two possibilities, RBL cells were biosynthetically labelled with

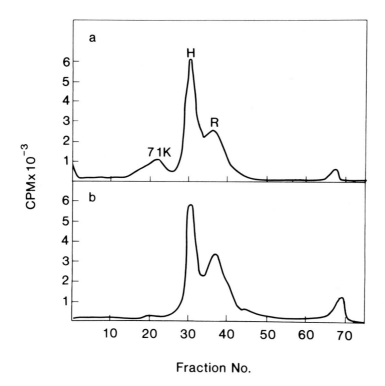

Figure 1. The effect of reduction on the receptors for IgE isolated by means of IgE-Sepharose and eluted by boiling in (a) SDS containing sample buffer in the absence of reducing agent; and (b) SDS sample buffer containing 5% 2-mercaptoethanol. The receptor preparations were subsequently characterized by SDS-PAGE on 10% gels.

^{3}H-leucine and receptors were isolated by repeated IgE-Sepharose chromatography (17) and analyzed by two-dimensional SDS-PAGE, i.e. running the receptor preparation under non-reducing conditions in the first dimension and in the presence of 2-ME in the second (16). This experiment revealed the presence of a ^{3}H-labelled molecule corresponding in mobility to β-chain which was already separated from H and R in the first dimension. Since such a molecule was not detected when surface iodinated RBL cells were used, it most likely represented β non-covalently linked to R (α) (8,12). No ^{3}H-labelled molecule seemed to be released from 71K upon reduction and SDS-PAGE in the second dimension, suggesting that 71K represents an R-like molecule not linked to β by a disulfide bond. This negative result still leaves open the possibility that 71K consists of an R-like molecule linked to, yet another, weakly tritiated molecule, however, it puts more

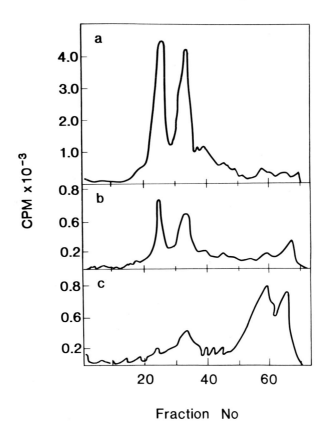

Figure 2.   SDS-PAGE analysis   on 15% gels of tryptic   peptides of the receptors for IgE obtained by digestions of these   receptors in solution, containing some SDS (a) R, (b) 71K and (c) H.

emphasis on the possibility that it represents a disulfide-linked dimer of R-like molecules.
     To obtain more   evidence that   the ^{125}I-labelled component   of 71K consists of peptides   similar or identical   to R, both R   and 71K, and H as a control, were separated by SDS-PAGE in tube gels. Gel slices containing the individual   receptors were minced   and eluted with water.   The concentrated eluates were digested   with trypsin and again analyzed by SDS-PAGE.   The results are shown in Figure 2. As can be seen, while R and 71K yielded very similar patterns, that of H differed significantly. Results using two-dimensional tryptic mapping also showed similarities between R

and 71K but not between these two receptor molecules and H. Differences in tryptic maps between H and R have also been observed by others (18). Details on tryptic digestion and mapping will be published separately (16). Together, these results provide further evidence that R and 71K are closely related.

INDUCTION OF 71K BY MYCOPLASMA

In the course of a routine test of RBL–Wpg cells for mycoplasma using Hoechst dye 33258 (19), it was established that the cell line was contaminated with mycoplasma which upon tests, kindly performed by Dr. G.J. McGarrity of the Institute for Medical Research, Camden, NJ, turned out to be M. hyorhinis. The RBL–Wpg cells were subsequently treated with gentamicin and cloned to eliminate the contaminant. The clone used for subsequent study was designated RBL–CK2. Receptor analysis revealed that while R and H receptors were still present, 71K could no longer be detected (20). Since the absence of 71K could have been the result of cloning, a variant which was free of 71K even in the presence of mycoplasma, another approach was taken. RBL cells were passaged in vivo in Wistar–ICI rats. The tumor cells were harvested, re-adapted to tissue culture and analyzed for surface iodinated receptors. Once again, 71K was absent and the RBL cells were found to be mycoplasma free.

The observations reported so far pointed to mycoplasma as the inducer of 71K, although the evidence was circumstantial. To obtain more direct evidence, the RBL clone was reinfected with mycoplasma. As expected, the receptor profile from ^{125}I–labelled cells once again clearly showed the presence of 71K. Preliminary kinetic experiments revealed that the first signs of 71K can be detected about 10 hr after mycoplasma infection.

The observations on mycoplasma induction of 71K reported so far, and to be published in detail elsewhere (20) have, in fact, provided some explanation why in earlier studies on the FcE receptors of RBL cells 71K was only sporadically observed. At that time, RBL cells were frequently passaged in vivo, a procedure which now has been shown to eliminate mycoplasma and 71K. Re-infection most likely occurred through the fetal calf serum used in the medium. This source of infection must have been absent in more recent years since a cloned line of cells, RBL–3114, has been maintained in culture for a number of years along with RBL–Wpg and was found to be mycoplasma–free and lacking 71K.

The fact that mycoplasma induces an IgE binding surface molecule similar to the FcE receptor (R or α) which is involved in mediator release from some RBL cell lines (21) and rat mast cells (11) suggested that this infectious agent may also affect mediator release properties of RBL cells.

In exploratory experiments, aimed at establishing if mycoplasma can modulate IgE-mediated histamine release from RBL-Wpg cells, it was observed that these cells did not release this mediator, either in the presence or absence of mycoplasma. As a consequence, only the histamine content was examined. Interestingly, the histamine content for infected "cleaned" and reinfected cells was found to be $10.75 + 0.75$ x 10^{-7}, $6.14 + 0.87$ x 10^{-7} and $12.94 + 0.85$ x 10^{-7} g/10^6 cells, respectively (20) showing, once again, a clear influence of mycoplasma. Thus, cells free of mycoplasma have consistently exhibited lower levels of histamine content compared to infected ones.

CONCLUDING REMARKS

The experiments described above have shown that M. hyorhinis is capable of inducing what may turn out to be a very specific modulation of a biologically important cell surface receptor. Other reports on the capability of mycoplasma to modulate biologically important cell functions in vitro exist (22). The biological significance of the phenomena reported here remains to be established. Mycoplasmas are common on mucosal surfaces (23) and thus have access to tissues where they could interact with mast cells. Obviously, no generalizations can be made at this time. It has yet to be explored if other strains of mycoplasma can induce similar changes in RBL cells and if mast cells or basophils of other species can respond to mycoplasma in a fashion comparable to that of RBL cells.

REFERENCES

1. **Conrad, D.H., and A. Froese.** 1978. Characterization of the target cell receptor for IgE. III. Properties of the receptor isolated from rat basophilic leukemia cells by affinity chromatography. J. Immunol. 120:429.
2. **Segal, D.M., S.D. Sharrow, S.F. Jones, and R.P. Siraganian.** 1981. Fc (IgG) receptors on rat basophilic leukemia cells. J. Immunol. 126:138.
3. **Froese, A., R.M. Helm, D.H. Conrad, C. Isersky, T. Ishizaka, and A. Kulczycki Jr.** 1982. Comparison of the receptors for IgE of various rat basophilic leukemia (RBL) cell lines. I. Receptors isolated by IgE-Sepharose and IgE and anti-IgE. Immunology 46:107.
4. **Kepron, M.R., D.H. Conrad, and A. Froese.** 1982. The cross-reactivity of rat IgE and IgG with solubilized receptors of rat basophilic leukemia cells. Mol. Immunol. 19:1631.
5. **Helm, R.M., D.H. Conrad, and A. Froese.** 1979. Lentil-lectin affinity chromatography of surface glycoproteins and the

receptor for IgE from rat basophilic leukemia cells. Int. Arch. Allergy Appl. Immunol. 58:90.

6. **Froese, A.** 1980. The presence of two kinds of receptors for IgE on rat mast cells. J. Immunol. 125:981.

7. **Goetze, A., J. Kanellopoulos, D. Rice, and H. Metzger.** 1981. Enzymatic cleavage of the subunit of the receptor for immunoglobulin E. Biochemistry 20:6341.

8. **Holowka, D., H. Hartmann, J. Kanellopoulos, and H. Metzger.** 1980. Association of the receptor for immunoglobulin E with an endogenous polypeptide on rat basophilic leukemia cells. J. Recept. Res. 1:41.

9. **Perez-Montford, R., J.-P. Kinet, and H. Metzger.** 1983. A previously unrecognized subunit (γ) of the receptor for immunoglobulin E. Biochemistry 22:5722.

10. **Rivnay, B., S.A. Wank, and H. Metzger.** 1982. Phospholipids stabilize the interaction between the alpha and beta subunits of the solubilized receptor for immunoglobulin E. Biochemistry 21:6922.

11. **Ishizaka, T., K. Ishizaka, D.H. Conrad, and A. Froese.** 1978. A new concept of triggering mechanisms of IgE-mediated histamine release. J. Allergy Clin. Immunol. 61:320.

12. **Helm, R.M., and A. Froese.** 1981. The incorporation of tritiated precursors into receptors for IgE of rat basophilic leukemia cells. Immunology 42:629.

13. **Conrad, D.H., A. Froese, T. Ishizaka, and K. Ishizaka.** 1978. Evidence for antibody activity against the receptor for IgE in a rabbit antiserum prepared against IgE-receptor complexes. J. Immunol. 120:507.

14. **Helm, R.M., and A. Froese.** Unpublished observations.

15. **Helm, R.M., and A. Froese.** 1981. Binding of the receptors for IgE by various lectins. Int. Arch. Allergy Appl. Immunol. 65:81.

16. **Roth, P.A., M. Rao, and A. Froese.** To be published.

17. **Kulczycki, A. Jr., and C.W. Parker.** 1979. The cell surface receptor for immunoglobulin E. I. The use of repetitive affinity chromatography for the purification of a mammalian receptor. J. Biol. Chem. 254:3187.

18. **Pecoud, A.R., and D.H. Conrad.** 1981. Characterization of the IgE receptor by tryptic mapping. J. Immunol. 127:2208.

19. **Macy, M.L.** 1979. Tests for mycoplasmal contamination of cultured cells as applied at the ATCC. TCA Manual 5:1151.

20. **Chan, B.M.C., K. McNeill, I. Berczi, and A. Froese.** To be published.

21. **Barsumian, E.L., C. Isersky, M.G. Petrino, and R.P. Siraganian.** 1981. IgE-induced histamine release from rat basophilic leukemia cell lines. Isolation of releasing and non-releasing clones. Eur. J. Immunol. 11:317.

22. **Wayner, E.A., and C.G. Brooks.** 1984. Induction of NKCF-like activity in mixed lymphocyte-tumor cell culture. Direct

involvement of mycoplasma infection of tumor cells. J. Immunol. 132:2135.

23. Bredt, W., J. Feldner, and I. Kahane. 1981. Adherence of mycoplasmas to cells and inert surfaces: Phenomena, experimental models and possible mechanisms. Isr. J. Med. Sci. 17:586.

Mast Cell Differentiation and Heterogeneity,
edited by A. D. Befus et al.
Raven Press, New York © 1986.

Luminal Mast Cells of the Human Respiratory Tract

R. M. Agius, P. H. Howarth, M. K. Church, C. Robinson, and S. T. Holgate

*Medicine I and Clinical Pharmacology, Southampton General Hospital,
Southampton SO9 4XY, England*

The recognition that bronchial asthma is an inflammatory disease with the physiological expression of increased non-specific bronchial hyperreactivity, widespread reversible airways obstruction, and hyperinflation, has led to the intensive investigation of agents which may initiate the response. There is now little doubt that in patients with asthma associated with atopy, the interaction of inhaled allergens with sensitized mediator secreting cells is an important initiating event (1). The fact that asthma is inherited independently of atopy suggests that a predisposition to bronchial hyperreactivity interacts with atopy, giving rise to an increased expression of asthma (2). Since an interactive relationship has been shown between non-specific bronchial hyperreactivity and the degree of mast cell sensitization with specific IgE in predicting the airway response to inhaled allergen (3-5), the most likely mediator secreting cell which interacts with inhaled allergens to initiate the airway response is the bronchial mast cell.

Mast cells are found throughout the human respiratory tract, both in the airways (6) and in the alveolar walls (7). In the bronchi, mast cells are most abundant superficial to the basement membrane where they may be demonstrated by the characteristic metachromatic staining of their proteoglycan-containing secretory granules, immunohistochemically by their high surface content of cytophilic IgE (8), or by electron microscopy (9). Occasionally, mast cells are found between epithelial cells (where they may abut directly onto the airway lumen). Mast cells located beneath the basement membrane are restricted to the connective tissue stroma and in the walls of small blood vessels. In general, the mast cell density in the airways is inversely related to airway

277

size, with the largest number being found in the terminal five generations (7,10). It is those mast cells adjacent to the bronchial lumen that are likely to play the most important role in initiating allergen-induced bronchoconstriction.

Direct evidence for mast cell mediator secretion as being fundamental to allergen-induced bronchoconstriction has been slow to accumulate. It has long been known that in atopic individuals bronchial provocation with specific inhaled antigens causes both early and late bronchoconstriction which reach maximum 15 min and 6-12 hr after challenge respectively (11). Since pretreatment of individuals with sodium cromoglycate attenuates both the early and late asthmatic reaction (12), and in vitro this drug has mast cell stabilizing activity (13), it has been generally assumed that both early and late reactions have a mast cell origin. However, the recognition that sodium cromoglycate may inhibit bronchoconstrictor responses induced by agents such as aspirin (14), propranolol (15), sulfur dioxide (16), and adenosine (17), which are not considered to produce their effects by stimulating mast cell mediator secretion, diminishes the confidence of using this drug as a tool to implicate mast cells in the pathogenesis of asthma.

Direct evidence of a role for mast cells in the pathogenesis of antigen-induced bronchoconstriction would ideally require the demonstration of mast cell degranulation in bronchial biopsies and release of one or more mast cell mediators into the airway lumen, or circulation following antigen provocation. In the nose of allergic subjects, biopsy specimens of the mucosa stained for chloracetate esterase activity to identify mast cells before and after challenge have shown widespread mast cell degranulation from the surface epithelium deep into the lamina propria (18). Similar studies on the airways have not yet been performed, but inhaled antigen provocation is associated with increases in beta-glucuronidase and the lysoacetylglyceryl-ether-phosphorylcholine (a precursor of PAF-acether) in bronchoalveolar lavage following bronchial provocation with antigen (19). This observation implies the activation of mediator secreting cells, but since both of these products are released by a number of inflammatory cells, this cannot be used in support of mast cell activation. Indeed, while purified human lung mast cells are able to generate PAF-acether upon immunological activation, their capacity to release it as an extracellular inflammatory mediator has not been shown (20).

An alternative approach supporting a mast cell role in antigen-induced bronchoconstriction would be the demonstration of increased levels of granule-derived or newly generated inflammatory mediators specific to mast cells in the peripheral circulation, reflecting an overspill from activated cells within the lung. Early attempts to demonstrate increases in plasma histamine following antigen provocation were fraught with difficulties mainly because the assays used to measure this mediator were both insensitive and non-specific, and the factors controlling its release and metabolism poorly understood (21). With the development of highly sensitive and specific isotopic methyltransferase assays for histamine, it has been possible to resolve resting plasma levels of this mediator in the region of 0.05 – 0.2 ng/ml (23–27). Despite being able to resolve baseline levels, conflicting reports have appeared in the literature concerning the detection of increased circulating levels of histamine preceding antigen-induced broncoconstriction (25–27). We have recently demonstrated that the most likely explanation for these discrepancies is that the patients chosen for study differed in their non-specific bronchial reactivity and, as discussed earlier, this is an important component in determining the airway response to inhaled allergens (3–5). When a population of allergic asthmatic patients were selected with almost normal airways reactivity (PC_{20} methacholine, 3.83 mg/ml), antigen provocation, sufficient to produce a mean fall in specific airways conductance (sGaw) of 52% caused two to four-fold increases in plasma histamine that preceded the onset of maximum bronchoconstriction. In contrast, in asthmatic subjects with greatly enhanced airways reactivity (PC_{20} methacholine, 0.07 mg/ml), antigen provocation that caused the same change in sGaw had no detectable effect on plasma levels of histamine (28). One interpretation of this observation is that only a small amount of mediator secretion is necessary to interact with hyperresponsive airways to produce bronchoconstriction.

High molecular weight glycoprotein neutrophil chemotactic factor (NCF) is an inflammatory mediator which has also been used as an in vivo index of mast cell degranulation, since circulating levels are elevated in parallel with the mast cell granule mediator, histamine, in effluent venous blood from patients with cold (29) and cholinergic urticaria (30) following the appropriate physical provocation. Increased circulating levels of NCF have also been reported in association with antigen-induced bronchoconstriction, and used to argue a mast cell hypothesis for this response (31). While NCF has been shown to be released following IgE-dependent challenge of human lung fragments (32), a mast cell origin for this mediator remains to be proven. Indeed, a recent study has demonstrated the release of both high and low molecular weight NCFs following IgE-dependent challenge of peripheral blood monocytes and lymphocytes, at least one major activity being inseparable from that isolated from serum following provocation in vivo (33). In confirmation of our studies with histamine, we have recently shown that, for the same airway response to inhaled allergen, larger amounts of NCF are released into the peripheral circulation in those patients with less reactive airways compared to those with marked airways non-specific reactivity (28).

Sodium cromoglycate and the beta-2-adrenergic agonists, such as albuterol and fenoterol, are inhibitors of IgE-dependent histamine release from human lung fragments (13) and human dispersed lung mast cells (34,35). In isolated mast cells, salbutamol is approximately 30,000 times more potent that sodium cromoglycate as a mast cell stabilizing agent, though for both drugs their efficacy is inversely related to the strength of immunological stimulation (35). In mildly allergic asthmatic subjects who had almost normal airways reactivity to methacholine, pretreatment with inhaled salbutamol (200 ug) or sodium cromoglycate (20 mg) attenuates both the immediate airway, plasma histamine and serum NCF responses following antigen bronchial provocation (36). This confirms that in vivo a component of these drugs' pharmacological activities may result from inhibition of mast cell mediator release. In contrast, the muscarinic cholinergic antagonist, ipratropium bromide (1 mg), while causing bronchodilatation, had no effect in attenuating either the airway or circulating mediator responses following antigen bronchial provocation (36). At least in mild asthma, this finding suggests that mast cell mediator release is not under cholinergic regulation (37).

In the past, the role of histamine as a bronchoconstrictor in asthma has been underestimated on account of the limitations of available H_1-receptor antagonists to act as pharmacological probes (38). Recently, two highly potent and selective histamine H_1 antagonists have been produced, astemizole (39) and terfenadine (40), which attenuate immediate bronchoconstriction following antigen provocation (41). Pretreatment of allergic asthmatic subjects with astemizole (10 mg daily for 14 days) prior to antigen provocation produced maximum attenuation of bronchoconstriction within the first 5 min of challenge, with subsequent loss of effect at later time points. This might suggest that histamine released from bronchial mast cells is more important as a bronchoconstrictor in the earliest phase of the immediate asthmatic reaction, while other pharmacologically active mediators, which include prostaglandins and leukotrienes, are probably more important later in the response. The finding that this dose of astemizole had no effect on antigen-induced increases in plasma histamine showed that at therapeutic concentrations it does not exert a mast cell stabilizing effect (41).

The role of bronchial mast cells in the pathogenesis of the late asthmatic reaction is more contentious. Recent studies by Durham and colleagues have shown that antigen bronchial provocation of patients with asthma who develop late reactions is accompanied by a second increase in circulating levels of plasma histamine and NCF, in association with the delayed airway response (42). However, in patients who only developed an isolated early bronchoconstrictor response to inhaled antigens, only an early increase in circulating mediators was observed. The precise origin of the second wave of mediator release associated with the late reaction is not known. However, upon nasal provocation with antigen that produced both early and late reactions accompanied by release of histamine, only the early response was associated with release of the mast cell derived prostanoid, prostaglandin D$_2$ (43). In the nose, this might suggest that the second wave of histamine release involves basophils which, in contrast to mast cells, are unable to generate PGD$_2$ uon immunologic activation. Whether basophils play an active role in the pathogenesis of the late asthmatic reaction is not known. However, accompanying this reaction there is evidence to suggest that circulating neutrophils and monocytes (and possibly basophils) become activated as shown by an increased expression of their surface C3b and IgG–Fc receptors (44) and enhanced cytotoxicity.

Inhalation of provocative antigen is likely to activate only those mast cells which abut directly onto, or are closely related to, the bronchial lumen. With a knowledge of the half–life of histamine in venous blood when infused into man (21), the content of histamine in human lung (45) and in each lung mast cell (46), the distribution of mast cells between the airway and pulmonary parenchyma (9,10), and finally the amount and time course of histamine release following antigen bronchial provocation (25,28), it is possible to estimate how many mast cells must degranulate within human lung to account for the rises in histamine observed. Thus, immediately following antigen provocation, the immediate two to three–fold rises in plasma histamine observed 5 min post–challenge is equivalent to the total degranulation of 2.5% of all lung mast cells, or 21% of those mast cells superficial to the bronchial basement membrane (9). This relatively low degree of mast cell activation is important, since the efficacy of drugs which inhibit mast cell mediator release is inversely related to the strength of immunological activation. With mediator release in the region of 4–10%, much lower concentrations of anti–allergic drugs, such as sodium cromoglycate and beta–2–adrenergic agonists, will be required to inhibit mediator release than are commonly employed for in vitro experiments (35).

The calculation made above is based on several assumptions, not least of which is that inhaled allergen activates mast cells only superficial to the basement membrane of the airway. As discussed elsewhere in this volume, in rodents there is morphological, histochemical, biochemical and pharmacological evidence of mast cell heterogeneity between the connective tissue and mucosal surfaces. A similar heterogeneity between connective tissue mast cells and mucosal mast cells has been proposed for man, but at present the amount of data to support this is scant and indirect. Strobel and co–workers have demonstrated that mast cells of the human gastrointestinal mucosa exhibit metachromatic staining of their granules, which is particularly susceptible to the choice of tissue fixation (47). Since a similar histochemical property has been demonstrated for mast cells of the rat intestine (48), this observation has been taken as evidence for morphological differences between intestinal mucosal and

connective tissue mast cells in man. In human lung tissue we
have been unable to demonstrate any fundamental differences in
the staining characteristics of mast cells, whether located in
the airways superficial to the basement membrane, deep to the
airway, in the alveoli, or adjacent to the pleural surface (49).
In contrast to rodent mast cells, whereas safranin O is able to
'displace' alcian blue staining of mast cells of connective
tissue but not at mucosal surfaces (49), mast cells throughout
human lung and in bronchoalveolar lavage (BAL) maintain their
alcian blue positivity even in the presence of safranin O.

Mast cells recovered in BAL are presumably derived from the
most superficial areas of the alveoli and bronchi. From normal
lung they constitute in the region of 0.08% of recovered
nucleated cells, whether identified by granule metachromasia or
by their cell surface content of IgE (50,51). The proportion of
mast cells recovered in BAL increases in patients with extrinsic
allergic alveolitis, cryptogenic fibrosing alveolitis and
sarcoidosis (51–53). Since all of these diseases exhibit their
major pathology in relation to the alveoli, it is important to
recognize that alveolar mast cells contribute an important
component to those recovered by BAL, and under normal circumstances may occupy up to 0.4% of the alveolar surface area (7).
In the experimental model of bleomycin-induced lung injury in
rats, up to 10-fold increases in parenchymal mast cells have been
reported, strongly suggesting that mast cells play an important
role in the pathogenesis of interstitial lung disease (54).

Increased numbers of mast cells have also been recovered from
BAL fluid of patients with mild allergic asthma (50,55), but not
in patients who are receiving regular corticosteroids (51). The
recent demonstration that corticosteroids produce a rapid (4–6
hr) decrease in mucosal mast cell populations of the rat gastrointestinal tract (56), might suggest that these drugs not only
suppress the inflammatory response which occurs in the airways of
patients with asthma, but also diminish the number of superficial
mast cells which are activated by inhaled allergens. The presence
of increased numbers of mast cells in BAL implies that the mast
cell population in the lung is not fixed, but may be enhanced in
certain disease states. It is tempting to speculate that this
induction involves the release of macrophage and/or fibroblast
growth factors in the case of parenchymal lung disease (57), and
antigen-induced, T-lymphocyte growth factors (e.g. interleukin-3)
in the case of allergic asthma (58), but at present these hypotheses are purely speculative.

Mast cells recovered by BAL from patients with a variety of
lung diseases show differing contents of the secretory granule
marker, histamine. Using logarithmic regression analysis it can
be calculated that each mast cell contains between 8 and 12 pg of
histamine per cell (51). In patients with bronchial asthma and
cryptogenic fibrosing alveolitis, there is a suggestion that
lower mast cell histamine contents are associated with increased
amounts of histamine recovered in the BAL fluid phase, implying

the presence of active mast cell secretion. Activation of BAL mast cells by cross-linkage of their cytophilic IgE or with calcium ionophore A23187 produces a dose-dependent secretion of histamine with a time course similar to that described for mast cells dispersed from whole lung tissue (59). There is a suggestion that BAL mast cells are more sensitive to immunological stimulation when compared to enzymatically or mechanically dispersed lung mast cells (60). Whether this represents true heterogeneity or differences in the way the mast cells are treated during their recovery remains to be established.

In addition to releasing histamine as an index of degranulation, both immunologic- and calcium-dependent activation of BAL cells generate substantial quantities of PGD_2 (61) with smaller amounts of thromboxane A_2 and PGF_2. Within human lung PGD_2 originates almost solely from mast cells (62,63), though it has been reported that alveolar macrophages are also able to generate small quantities of this eicosanoid (64). In BAL a close positive correlation between PGD_2 generation and histamine release induced by calcium-dependent cell activation with a regression line that passes through the origin implies that the majority of this PGD_2 is derived from mast cells (61). This is particularly important since, on immunologic activation, cultured mouse bone marrow-derived mast cells (which may be precursors of intestinal mucosal mast cells) release only small amounts of PGD_2, but much larger amounts of the 5-lipoxygenase product, LTC_4 (64). In contrast, mast cells of the rat and mouse peritoneal cavity release substantial quantities of PGD_2, but only small amounts of the 5-lipoxygenase products. It has been suggested that differences in the oxidative metabolism of arachidonic acid by the two mast cell subpopulations is an important functional activity of these cells. Our finding of mast cells from BAL generating up to three times as much PGD_2 as mast cells dispersed from human lung parenchyma implies that a simple degree of functional mast cell heterogeneity, as observed in rodents, cannot be applied to human BAL mast cells. However, this does not preclude the possibility of mast cell heterogeneity occurring at a more subtle level in human lung (60).

In conclusion, current knowledge provides strong evidence that antigen bronchial provocation in patients with allergic asthma releases mast cell-dependent inflammatory mediators during the early, and possibly during late, asthmatic reactions. For the immediate airway response, it is likely that those mast cells superficial to the bronchial basement membrane become activated. In asthma there is a suggestion that bronchial mucosal hyperplasia occurs, but whether these mast cells are fundamentally different from mast cells found elsehwere in human lung tissue remains to be established. The hypothesis of mast cell heterogeneity as applicable to the pathogenesis of human disease is an attractive one, although further careful work is necessary to clarify its pathological, physiological and pharmacological significance.

REFERENCES

1. **Holgate, S.T.** 1983. The human lung mast cell: morphology, biochemistry and role in allergic asthma. Adv. Med. 19:287.
2. **McFadden, E.R.** 1984. Pathogenesis of asthma. J. Allergy Clin. Immunol. 73:413.
3. **Bryant, D.H., and M.W. Burns.** 1976. Bronchial histamine reactivity: its relationship to the reactivity of the bronchi to allergens. Clin. Allergy 6:523.
4. **Cockcroft, D.W., R.E. Ruffin, P.A. Frith, A. Cartier, E.F. Juniper, J. Dolovich, and F.E. Hargreave.** 1982. Determinants of allergen-induced asthma: dose of allergen circulating IgE antibody concentrations and bronchial responsiveness to inhaled histamine. Am. Rev. Respir. Dis. 120:1053.
5. **Holgate, S.T., R.C. Benyon, P.H. Howarth, R. Agius, C. Hardy, C. Robinson, S.R. Durham, A.B. Kay, and M.K. Church.** 1985. Relationship between mediator release from human lung mast cells in vitro and in vivo. Int. Arch. Allergy Appl. Immunol. 77:47.
6. **Jeffrey, P., and B. Corrin.** 1984. Structural analysis of the respiratory tract. In: Immunology of the Lung and Upper Respiratory Tract. Edited by Bienenstock, J., McGraw-Hill, New York and Toronto, p.1.
7. **Fox, B., T.B. Bull, and A. Guz.** 1981. Mast cells in the human alveolar wall: an electronmicroscope study. J. Clin. Pathol. 34:1333.
8. **Agius, R.M., D. Jones, and S.T. Holgate.** 1985. Staining characteristics of human pulmonary mast cells. Thorax (In press).
9. **Lamb, D., and A. Lumsden.** 1982. Intra-epithelial mast cells in human airway epithelium: evidence for smoking-induced changes in their frequency. Thorax 37:334.
10. **Guerzon, G.M., P.D. Pare, M.-C. Michoud, and J.C. Hogg.** 1979. The number and distribution of mast cells in monkey lungs. Am. Rev. Respir. Dis. 119:59.
11. **Pepys, J.** 1967. Hypersensitivity to inhaled organic antigen. J. Roy. Coll. Phys. Lond. 2:42.
12. **Booij-Noord, H., N.G. Orie, and K. de Vries.** 1971. Immediate and late bronchial obstructive reactions to inhalation of house dust and protective effects of disodium cromoglycate and prednisolone. J. Allergy Clin. Immunol. 48:344.
13. **Church, M.K., and K.D. Young.** 1983. The characteristics of inhibition of mediator release from human lung fragments by sodium cromoglycate, salbutamol and chlorpromazine. Br. J. Pharmacol. 78:671.
14. **Martelli, N.A., and G. Vsandivaras.** 1977. Inhibition of aspirin-induced bronchoconstriction by sodium cromoglycate inhalation. Thorax 32:684.
15. **Koeter, G.H., H. Meurs, J.G.R. Monchy, and K. de Vries.** 1982. Protective efect of disodium cromoglycate on propranolol challenge. Allergy 37:587.

16. Harries, M.G., P.E.G. Parker, and M.H. Lessoff. 1981. Role of bronchial irritant receptors in asthma. Lancet 1:5.

17. Cushley, M.J., and S.T. Holgate. 1985. Adenosine-induced bronchoconstriction in asthma: role of mast cell mediator release. J. Allergy Clin. Immunol. (In press).

18. Corrado, O.J., E. Gomez, D.L. Baldwin, A.R. Swanston, and R.J. Davies. 1985. Direct evidence for mast cell involvement in Type 1 allergic reactions in man. Thorax. (In press).

19. Tonnel, A., M. Joseph, P.H. Gosset, E. Fournier, and A. Capron. 1983. Stimulation of alveolar macrophages in asthmatic patients after local provocation test. Lancet i:1406.

20. Schleimer, R.P., D.W. MacGlashan, S.P. Peters, R. Naclerio, D. Proud, N.F. Adkinson, and L.M. Lichtenstein. 1984. Inflammatory mediators and mechanisms of release from purified human basophils and mast cells. J. Allergy Clin. Immunol. 74:473.

21. Ind, P.W., M.J. Brown, F.J.M. Lhoste, I. Macquin, and C.T. Dollery. 1982. Determination of histamine and its metabolites. Concentration effect relationships of infused histamine in normal volunteers. Agents Actions 12:12.

22. Brown, M.J., P.W. Ind, P.J. Barnes, D.A. Jenner, and C.T. Dollery. 1980. A sensitive and specific radiometric method for the measurement of plasma histamine in normal individuals. Anal. Biochem. 109:142.

23. Howarth, P.H., G.J.-K. Pao, M.K. Church, and S.T. Holgate. 1984. Exercise and isocapnic hyperventilation-induced bronchoconstriction in asthma: relevance of circulating basophils to measurements of plasma histamine. J. Allergy Clin. Immunol. 73:391.

24. Atkins, P.C., P.-C. Bedard, B. Zweiman, J. Dyer, and M.A. Kaliner. 1984. Increased antigen-induced local and systemic mediator release in rhinitis with pulmonary symptoms in the pollen season. J. Allergy Clin. Immunol. 73:341.

25. Lee, T.H., M.J. Brown, R. Causon, M.J. Walport, and A.B. Kay. 1982. Exercise-induced release of histamine and neutrophil chemotactic factor in atopic asthmatics. J. Allergy Clin. Immunol. 77:73.

26. Moodley, I., D.J.R. Morgan, and R.J. Davies. 1983. The measurement of histamine during allergen-induced asthma. Clin. Sci. 64:13.

27. Hegardt, B., O. Lowhagen, N. Svedmyr, and G. Granerus. 1982. The property of equipotent bronchodilating doses of inhaled KW2131 and terbutaline against allergen-induced bronchospasm. Allergy 37:407.

28. Howarth, P.H., S.R. Durham, A.B. Kay, and S.T. Holgate. 1985. The relationship between mast cell mediator release and bronchial reactivity in allergic asthma. Thorax. (In press).

29. Wasserman, S.I., N.A. Soter, D.M. Center, and K.F. Austen. 1977. Cold urticaria - recognition and characterization of

a neutrophil chemotactic factor which appears in serum during experimental cold challenge.

30. Soter, N.A., S.I. Wasserman, K.F. Ausgent, and E.R. McFadden. 1980. Release of mast cell mediators and alterations in lung function in patients with cholinergic urticaria. N. Eng. J. Med. 302:604.

31. Atkins, P.C., M.E. Norman, and B. Zweiman. 1978. Antigen-induced neutrophil chemotactic factor in man: correlation with bronchospasm and inhibition by disodium cromoglycate. J. Allergy Clin. Immunol. 62:149.

32. O'Driscoll, B.R., T.H. Lee, O. Cromwell, and A.B. Kay. 1983. Immunological release of neutrophil chemotactic activity from human lung tissue. J. Allergy Clin. Immunol. 72:695.

33. Cundell, D.R., J.L. Devalia, and R.J. Davies. 1984. Is NCA-chemotactic activity released solely from mast cells. Proc. Societas European. Physiologiae Respiratoriea, Barcelona. (In press).

34. Peters, S.P., E.S. Schulman, R.P. Schleimer, D.W. MacGlashan, H.H. Newball, and L.M. Lichtenstein. 1982. Dispersed human lung mast cells: pharmacologic aspects and comparison with human lung fragments. Am. Rev. Respir. Dis. 126:1034.

35. Church, M.K., G.J.-K. Pao, and S.T. Holgate. 1985. Inhibition of IgE-dependent histamine release from dispersed human lung mast cells by albuterol and cromolyn sodium. Am. Rev. respir. Dis. (In press).

36. Howarth, P.H., S.R. Durham, A.B. Kay, M.K. Church, and S.T. Holgate. 1985. Influence of albuterol, cromolyn sodium and ipratropium bromide on the airway and circulating mediator responses to antigen bronchial provocation in asthma. Am. Rev. Respir. Dis. (In press).

37. Kaliner, M., R.P. Orange, and K.F. Austen. 1982. Immuno-logical release of histamine and slow reacting substance of anaphylaxis from human lung. IV. Enhancement by cholinergic and alpha adrenergic stimulation. J. Exp. Med. 136:556.

38. White, J., and N.M. Eiser. 1983. The role of histamine and its receptors in the pathogenesis of asthma. Br. J. Dis. Chest. 77:215.

39. Howarth, P.H., and S.T. Holgate. 1984. Comparative trial of two non-sedative H_1-antihistamines, terfenadine and astemizole, for hay fever. Thorax 39:668.

40. Patel, K.R. 1984. Terfenadine in exercise-induced asthma. Br. Med. J. 288:1496.

41. Holgate, S.T., M.B. Emanuel, and P.H. Howarth. 1985. Astemizole and other antihistaminic drug treatment of asthma. J. Allergy Clin. Immunol. (In press).

42. Durham, S.R., T.H. Lee, O. Cromwell, R.J. Shaw, T.G. Merrett, J. Merrett, P. Cooper, and A.B. Kay. 1984. Immunologic studies in allergen-induced late-phase asthmatic reactions. J. Allergy Clin. Immunol. 74:49.

43. Norman, P.S., P.S. Laclerio, A. Creticos, A. Togias, and

L.M. Lichtenstein. 1984. Mediator release after allergic and physical nasal challenges. Proc. CIA Meeting, Mexico. Int. Arch. Allergy Appl. Immunol. 77:57.

44. **Durham, S.R., M. Carroll, G.M. Walsh, and A.B. Kay.** 1984. Leukocyte activation in allergen-induced late phase asthmatic reactions. N. Eng. J. Med. 311:1398.

45. **Kaliner, M.A.** 1980. Mast cell derived mediators and bronchial asthma. In: Airway Reactivity, Mechanisms, and Clinical Relevance. Edited by Hargreave, F.E. Astra, Ontario, p. 175.

46. **Schwartz, L.B., and K.F. Austen.** 1984. Structure and function of the chemical mediators of mast cells. Prog. Allergy 34:271.

47. **Strobel, S., H.R.P. Miller, and A. Ferguson.** 1981. Human intestinal mast cells: evaluation of fixation and staining techniques. J. Clin. Pathol. 34:851.

48. **Wingren, V., and L. Enerback.** 1983. Mucosal mast cells of the rat intestine: a re-evaluation of fixation and staining properties with special reference to protein blocking and solubility of the granular glycosaminoglycan. Histochem. J. 15:571.

49. **Enerback, L.** 1966. Mast cells in rat gastrointestinal mucosa. 2. Dye binding and metachromatic properties. Acta Path. et Microbiol. Scand. 66:303.

50. **Tomioka, M., S. Ida, S. Yuriko, T. Ishihara, and T. Takishima.** 1984. Mast cells in bronchoalveolar lumen of patients with bronchial asthma. Am. Rev. Respir. Dis. 129:1000.

51. **Agius, R.M., R.K. Knight, R.C. Godfrey, P.J. Cole, and S.T. Holgate.** 1984. Significance of mast cells in bronchoalveolar lavage. Thorax 39:708.

52. **Haslam, P.L., A. Dewar, P. Butchars, and M. Turner-Warwick.** 1982. Mast cells in bronchoalveolar lavage fluids from patients with extrinsic allergic alveolitis. Am. Rev. Respir. Dis. 125:51.

53. **Flint, K.C., B. Hudspith, J. Brostoff, K.B.P. Leung, F.L. Pearce, D.G. James, and McI.N. Johnson.** 1985. Hyperresponsiveness of bronchoalveolar mast cells in sarcoidosis. Thorax. (In press).

54. **Goto, T., D. Befus, R. Low, and J. Bienenstock.** 1984. Mast cell heterogeneity and hyperplasia in bleomycin-induced pulmonary fibrosis of rats. Am. Rev. Respir. Dis. 130:797.

55. **Flint, K.C., B.N. Hudspith, K.B.P. Leung, F.L. Pearce, J. Brostoff, and McI.N. Johnson.** 1985. Bronchoalveolar mast cells in extrinsic asthma. Clin. Sci. 68(11):33P.

56. **King, S.J., H.R.P. Miller, G.F.J. Newlands, and R.G. Woodbury.** 1985. Depletion of mucosal mast cell protease by corticosteroids: the effect on intestinal anaphylaxis in the rat. Proc. Natl. Acad. Sci. USA 82:1214.

57. **Metcalfe, D.D.** 1983. Effector cell heterogeneity in immediate hypersensitivity reactions. Clin. Rev. Allergy 1:311.

58. Razin, E., J.N. Ihle, D. Seldin, J.-M. Mencia-Huerta, H.R. Katz, P.A. Leblanc, A. Hein, J.P. Caulfield, K.F. Austen, and R.L. Stevens. 1984. Interleukin 3: a differentiation and growth factor for the mouse mast cell that contains chondroitin sulfate E proteoglycan. J. Immunol. 132:1479.

59. Church, M.K., G.J.-K. Pao, and S.T. Holgate. 1982. Characterization of histamine secretion from mechanically dispersed human lung mast cells: effects of anti-IgE, calcium ionophore A23187, compound 48/80 and basic polypeptides. J. Immunol. 129:2116.

60. Leung, K.B.P., K.C. Flint, F.L. Pearce, B. Hudspith, J. Brostoff, and McI.N. Johnson. 1985. A comparison of histamine secretion from human basophils and from human lung mast cells obtained by bronchoalveolar lavage (BAL) and from dispersion of lung fragments. Clin. Sci. 68:11):7P.

61. Agius, R.M., C. Robinson, and S.T. Holgate. 1985. Release of histamine and newly generated mediators from human broncho-alveolar lavage cells. Thorax. (In press).

62. Lewis, R.A., N.A. Soter, P.T. Diamond, K.F. Austen, J.A. Oates, and J.L. Roberts II. 1982. Prostaglandin D_2 genera-tion after activation of rat and human mast cells with anti-IgE. J. Immunol. 129:1627.

63. Holgate, S.T., G.B. Burns, C. Robinson, and M.K. Church. 1984. Anaphylactic- and calcium-dependent generation of prostaglandin D_2 (PGD_2), thromboxane B_2, and other cyclo-oxygenase products of arachidonic acid by dispersed human lung cells and relationship to histamine release. J. Immunol. 133:2138.

64. MacDermot, J., C.R. Kelsey, K.A. Waddell, R. Richmond, R.K. Knight, P.J. Cole, C.T. Dollery, D.N. London, and I.A. Blair. 1984. Synthesis of leukotriene B_4 and prostanoids by human alveolar macrophages analysed by gas chromatography/ mass spectrometry. Prostaglandins 27:163.

Mast Cell Differentiation and Heterogeneity,
edited by A. D. Befus et al.
Raven Press, New York © 1986.

Flow Cytofluorometric Studies of IgE Binding Sites on Rat Peritoneal Mast Cells

D. Lagunoff, A. Rickard, *D. Peizner, J. Cichon, and M. Zarka

Department of Pathology, St. Louis University Medical Center, St. Louis, Missouri 63104

The capacity to bind IgE at specific cell surface receptors is one of the prominent characteristics of mast cells and basophils. The kinetics of association and dissociation from the cell receptors and the average number of receptors per cell in populations have been thoroughly studied over the past 15 years (1–10). These quantitative studies have utilized human basophils, cultured RBL cells or isolated peritoneal mast cells. Myeloma IgEs labelled with ^{125}I have largely been used in these studies.

Flow cytofluorometry allows the study of IgE binding to mast cells and basophils under a wider range of circumstances than does measurement of radioactivity, particularly the expansion of such studies to measurements on a cell by cell basis. With flow cytofluorometry, it is possible to examine binding to cells in mixed populations without prior isolation and thus to define the heterogeneity of binding independently of any characteristics used to isolate cells (11). Relatively small numbers of cells, of the order of 5×10^4, can be used effectively.

Using flow cytofluorometric technology, we have carried out a series of studies of IgE binding to rat peritoneal mast cells designed to assess the contribution of cell size and secretory activity to the heterogeneity in the number of IgE binding sites per cell.

Supported by grant HL25402 from NIH NHLBI and *Short Term Training Grant HL30572 from NIH.

METHODS

IgE was prepared from rat IgE myeloma IR 162 (12) ascites generously provided by A. Kulczycki. The IgE was precipitated with ammonium sulfate and gel filtered over Sepharose 6B according to Isersky, Kulczycki and Metzger (13). The fractions containing IgE were combined, dialyzed against Tris-phosphate buffer, pH 8, for 48 h and applied to a DEAE column from which the IgE was eluted in the same buffer. The resulting IgE pool was then labelled with fluorescein isothiocyanate for 16 h at 4°C, and the adduct (F-IgE) dialyzed against Tris-phosphate buffer for 24-48 h and passed over Sephadex G-100 to remove residual free fluorescein. F-IgE was eluted from the Sephadex in phosphate buffered saline. The protein content of the F-IgE preparation was estimated with the Lowry reaction, and the fluorescence intensity per mg protein measured at 488 nm excitation and 515 nm emission on a Perkin-Elmer 650-10S spectro-fluorometer. F-IgE was frozen in aliquots and kept at -15°C. Thawed samples were kept up to 1 wk at 4°C.

Rat mast cells were washed from the peritoneal cavity of Sprague-Dawley rats as previously described and separated as indicated from other cell types by centrifugation through Percoll (14). Mast cells comprised 85-95% of the cells after separation on Percoll. In the experiments on young rats and following secretion in vivo, the entire peritoneal fluid cell population was used. Binding of F-IgE to mast cells was standardly carried out at 25°C for 60 min. Following incubation with F-IgE, cells were sedimented at 200 x g for 5 min and the supernatant removed. The cells in a small volume of medium were kept at 0°C until being suspended in balanced salt solution containing 0.1% albumin and analyzed using a Spectrum III flow cytofluorometer (Ortho Diagnostic Systems, Inc., MA). Fluorescence in the instrument is excited by incident light at 488 nm from an argon laser. Right angle scattered light, forward angle (12°) scattered light and right angle fluorescence emission were measured using linear amplification. In general, under the circumstances of measure-ment, the amplified photomultiplier output for fluorescence measurements was near linear over a range of 120 channels as determined by the coincidence method of Watson (15) using 1.8 u microspheres labelled with coumarin (Fluoresbrite carboxylate microspheres, Polyscience, PA). Flow cytometric measurements were obtained as mean channel $\pm$ standard deviation. The number of cells measured per sample varied between 10^3 and 10^4. (The flow cytometer draws-up 0.4 ml of which the cells in a 20 μl sample are measured.)

RESULTS

The high affinity for monomeric IgE of the binding sites on mast cells allowed for good separation of mast cells and other cells by measurement of fluorescence intensity under the

conditions of concentration and time selected for binding (Fig.
1).   That the mast cells and only the mast cells were positive
was   confirmed by measurements on isolated mast cells   and
peritoneal cells depleted of mast cells.   The high intensity of
right angle scatter of the fluorescent cells was consistent with

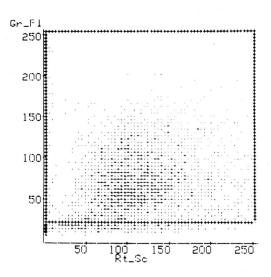

Figure 1. Fluorescence intensity of mast cells and non-mast
cells. Peritoneal cells were isolated from month old rats and
separated on Percoll.   The mast cells were 95% pure.   The
non-mast cells were centrifuged in Percoll twice after which only
0.1% mast cells were present as determined by staining with
acidic toluidine blue. Cells were incubated with 15 μg/ml F-IgE
for 60 min at 25°C and then assayed for fluorescence in the flow
cytometer. The histogram shows cell number on the ordinate and
fluorescence intensity on the abscissa. A. Mast cells.   B. Non-
mast cells.

exclusive heavy labelling of normal mast cells. Studies of IgE
binding site number on mast cells were standardly performed at a
concentration of 15 μg/ml F-IgE calculated to achieve 90% or
greater saturation of the sites after 1 h of binding. Fluorescent
mast cells distributed in a unimodal fashion; the range of fluorescence intensity expressed by isolated mast cells was consider-

able with a coefficient of variation ranging from .3 to .5. One
possible basis for the broad distribution of IgE binding sites
per cell is variation in the surface area of mast cells. If this
were the case, then fluorescence intensity should correlate with
a flow cytometric variable correlated with cell surface   area.

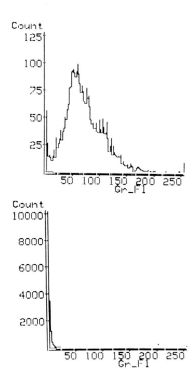

Figure 2. Lack of correlation between fluorescence intensity
and right scatter in a population of mast cells from the same age
rats. Isolated mast cells from the experiment described in Fig.
1 were assessed for right scatter and fluorescence intensity.
The scattergram has fluorescence on the ordinate and right
scatter on the abscissa. The darkness of the printed symbols in
the 64 x 64 matrix is indicative of the number of mast cells
falling in a single cell of the matrix.   The coefficient of
correlation for right scatter and fluorescence for the cells in
the outlined box was .08.

Forward angle light scatter has been shown for other cell types
to correlate with cell volume in a number of instances (16,17)
and, in general, surface area would be expected to increase with
cell volume.   Scattergrams of forward scatter vs fluorescence

exhibited no evidence of correlation. Study of the association of forward and right angle light scatter with direct measurements of mast cell diameter (see below) indicated that in fact the intensity of low angle light scattering did not correlate with mast cell diameters but intensity of right angle scatter did. When right angle scatter was plotted against fluorescence intensity, for collection of mast cells, no correlation was evident (Fig. 2).

TABLE I. Correlation coefficients for mast cell variables from rats of differing age

Variables		Correlation Coefficient
1	2	
Right scatter	(Diameter) 2	.89
Forward Scatter	(Diameter) 2	.08
Fluorescence	Right scatter	.08

Peritoneal cells were collected from rats at varying ages from 3 d postpartum. Diameters of fixed, acid toluidine blue stained mast cells were measured using a calibrated micrometer eyepiece. Twenty-five cells were measured at 400 x for each of 10 different ages. Right angle scatter, forward angle scatter and fluorescence were measured on at least 1000 cells at each of the 10 ages. Correlation coefficients were calculated using the means of the four variables at each age.

Mean volumes of peritoneal mast cells from rats have been demonstrated to exhibit a strong dependence on the age of the animals (18). We therefore examined cell forward angle light scatter, right angle light scatter and IgE binding sites of peritoneal mast cells harvested from rats of increasing age from 3 days post-partum. Mean cell diameter squared and mean right angle light scattering intensity had a high correlation coefficient (Table I). Low angle light scatter intensity showed essentially no correlation with the square of cell diameter. Mean IgE binding sites per cell estimated from fluorescence intensity at near saturating concentrations of F-IgE did not correlate with right angle scattered light in this experiment either.

TABLE II. IgE binding sites on cells from rats 1 month
and 3 months of age

| Age | IgE Binding | |
	Control	Acid Stripped
1 month	90.9 ± 32.1	79.2 ± 32.2
3 month	79.4 ± 32.0	61.5 ± 31.0

Mast cells were obtained from 1 month old and 3 month old rats and isolated on Percoll. Binding sites were assessed in terms of the mean channel of fluorescent intensity for mast cells exposed for 1 h to 15 μg/ml of F-IgE at 25°C. To eliminate any effect of prior bound IgE, mast cells were stripped of IgE in pH 3.2 glycine buffer at 2°C for 10 min, rinsed and then exposed to F-IgE. Values are the mean channel ± SD.

We next examined the possibility that our inability to discern an increase in IgE binding sites corresponding to an increased cell size was attributable to the presence of a greater proportion of already occupied binding sites on the older, larger mast cells. Estimation of spontaneously bound IgE on the mast cell surface using goat-anti-IgE antibody and a second fluorescein-labelled antibody to goat IgG indicated the presence of a small amount of pre-bound IgE, but the method could not be used reliably to measure the number of binding sites because of a relatively high background of goat IgG binding to the rat mast cells and a broad distribution of fluorescence intensity by this indirect method when IgE was added to the cells. We therefore attempted to strip the IgE spontaneously bound to mast cells by treatment with low pH buffer (19) and measure the difference between F-IgE that could be bound to mast cells before and after the bound IgE also reduced the mean number of mast cell binding sites for IgE as measured by re-exposure of the stripped cells to F-IgE, so that this method could not be used to determine reliably the number of occupied binding sites by the difference between binding sites before and after stripping. We did measure binding sites on mast cells from 1 month and 3 month old rats after stripping in one experiment and found no evidence of an increased number of occupied receptors on the cells from the older animals (Table II).

Following stimulation of secretion by peritoneal mast cells a significant decrease in the mean volume of the cells associated with the loss of secretory granules has been described (20). Endocytosis which has been reported to occur in conjunction with secretion in RBL cells (21) might also contribute to a loss of IgE binding sites at the mast cell surface. We therefore

TABLE III. Effect of secretion in vivo on IgE binding sites

| Time after polymyxin B | Percent of Control | | | |
	Histamine	Right Scatter	Fluorescence	Cell No.
2 h	14 ± 3	74 ± 3	87 ± 7	62 ± 14
24 h	12 ± 1	68	86 ± 9	55 ± 12
48 h	24 ± 4	66	102 ± 13	78

Rats were administered 1.5 mg/gm body weight polymyxin B in phosphate buffered saline intraperitoneally and killed at 2, 24 or 48 h after polymyxin B was given. The whole peritoneal cell population was assessed for fluorescence after exposure to 15 µg/ml F-IgE for 1 h at 25°C. Fluorescent cells were counted, and right scatter was determined for the fluorescent cells. Histamine was determined on a sample of cells by the OPT method after extraction in 5% TCA. The values are presented as the mean percent ± SE of the control cells taken from control animals given phosphate buffered saline intraperitoneally but no polymyxin B.

TABLE IV. IgE binding sites on mast cells following secretion induced in vitro

| Time after polymyxin B | Percent of Control | | |
	Histamine	Right scatter	Fluorescence
10 min	46 ± 2	67	81 ± 3
120 min	---	58	76 ± 8

Mast cells were collected, isolated and subjected to secretion induced with 2 µg/ml at 37°C for 10 min polymyxin B. At the indicated time after the addition of polymyxin B, the cells were labelled with 15 µg/ml F-IgE for 60 min at 25°C. Fluorescence and right scatter were assessed 10 and 120 min after adding polymyxin B by flow cytometry. Histamine release was determined by measuring histamine in the pellet and supernatant 10 min after introduction of polymyxin B. All values are given as percent of control cells similarly treated but without polymyxin B. Histamine and fluorescence are means of duplicates in at least four experiments. Right scatter was only determined for two experiments at each time period.

examined the effects of secretion of F-IgE binding to mast cells. Secretion was induced by polymyxin B either in vitro using isolated peritoneal mast cells or in vivo in the peritoneal cavity. Secretion as expected was accompanied by a decrease in mast cell diameter and right angle scatter intensity. A modest decrease in fluorescent intensity of between 15% and 20% was associated in vitro and in vivo (Tables III and IV).

DISCUSSION

In the 15 years since Ishizaka, Tomioka and Ishizaka (22) first unambiguously demonstrated the presence of IgE on the surface of human basophils, an impressive literature has developed characterizing the specific receptors for IgE on mast cells of different species and the kinetics of association of IgE with its receptor on the cells and in isolated form (1,4,5,7-9,23-25). Virtually all of the work on the interaction of IgE and its cell receptors has utilized radioactive iodine as the label. The use of radioactivity provides high sensitivity but is exceedingly cumbersome to apply to quantitative studies of IgE binding to individual cells. Since the number of receptors on mast cells and basophils is high, generally over 10^5 per cell, and the affinity is great, 10^9 M^{-1} to 10^{10} M^{-1}, the use of the IgE labelled with fluorescein in binding studies is relatively straightforward, and flow cytofluorometry permits the quantitative assessment of IgE binding on a cell by cell basis. We have used this technology to examine the distribution of IgE binding sites among rat peritoneal mast cells and to study the effect of cell size, maturational stage and secretory history on mast cell IgE binding sites.

Our flow cytofluorometric studies demonstrate that there is considerable heterogeneity in the number of IgE binding sites per peritoneal fluid mast cell. The variation in number of binding sites per cell is not correlated with the variation in size among peritoneal fluid mast cells. The range of cell size over which the number of IgE binding sites per cell is relatively constant is surprising. Mast cells varying by as much as 3-fold in estimated surface area have essentially the same number of IgE binding sites. Since transmission electron microscopy does not reveal any marked difference in surface anatomy of young and old cells, the estimates of surface area from diameter are probably reasonable. The heterogeneity in the number of IgE binding sites per cell cannot be attributed to the presence of a range of maturational states of mast cells in the peritoneal cavity of adult rats, since previous studies indicate a good correlation between maturational state and size (18,26). We have also shown that cells from very young rats with a high proportion of small immature mast cells (26) have essentially the same mean number of binding sites per cell and the same extent of heterogeneity of number of binding sites as very large mast cells from old rats with almost exclusively fully differentiated mast cells in their peritoneal fluid (27).

Our evidence indicates that the range of the number of IgE binding sites per cell expressed by peritoneal mast cells is not dependent on the secretory history of the cells. About 15% of surface IgE binding sites are lost acutely after secretion, and the original mean number of binding sites per cell is rapidly restored over a period of several days in vivo despite a rather desultory restoration of the fully differentiated mast cell phenotype following extensive secretion of the specific mast cell granules (20). We have not identified the fate of the lost receptors but preliminary evidence indicates that as much as 1/3 of the missing receptors may be endocytosed.

How early in the differentiation program of mast cells IgE binding sites are first expressed is not known. In view of Yong et al.'s (26) observation that peritoneal mast cells from 5 day old rats are virtually exclusively Comb's Type I (2), the earliest form histochemically identifiable as a mast cell, the presence of a full complement of IgE binding sites on cells from 3 day old rats supports the contention that the appearance of IgE receptors is an early event in mast cell differentiation.

The evidence from in vitro studies is very strong that massive secretion is not toxic to mast cells. However, from 12 to 48 h after extensive mast cell secretion is induced in the peritoneal cavity, it is difficult to identify and count mast cells as a consequence of their substantial granule loss. F–IgE binding identifies a substantial, albeit reduced, population of mast cells after in vivo secretion. The presence of cells rich in IgE binding sites but poor in granules adds some additional substance to previous speculation that mast cells extensively degranulated in vivo are capable of surviving and reinitiating their maturational program (20). That it is possible to identify peritoneal cells following extensive mast cell secretion in vivo with only a slightly reduced number of IgE receptors and as high affinity as the mast cell population prior to secretion, indicates that fluorescence activated cell sorting can be used to isolate and characterize the post–secretory mast cell after in vivo induction of secretion. Such studies can be expected to contribute to our understanding of the sequence of mast cell differentiation, its control and its contribution to the heterogeneity of mast cells in vivo.

REFERENCES

1. **Coleman, J.W., and R.C. Godfrey.** 1981. The number and affinity of IgE receptors on dispersed human lung mast cells. Immunology 44:859.
2. **Conrad, D.H., H. Bazin, A.H. Sehon, and A. Froese.** 1975. Binding parameters of the interaction between rat IgE and rat mast cell receptors. J. Immunol. 114:1688.
3. **Conrad, D.H., J.R. Wingard, and T. Ishizaka.** 1983. The interaction of human and rodent IgE with the human basophil

IgE receptor. J. Immunol. 130:327.
4. **Ishizaka, T., C.S. Soto, and K. Ishizaka.** 1973. Mechanisms of passive sensitization. III. Number of IgE molecules and their receptor sites on human basophil granulocytes. J. Immunol. 111:500.
5. **Kulczycki, A. Jr., and H. Metzger.** 1974. The interaction of IgE with rat basophilic leukemia cells. II. Quantitative aspects of the binding reaction. J. Exp. Med. 140:1676.
6. **Lee, T.D.G., A. Sterk, T. Ishizaka, J. Bienenstock, and A.D. Befus.** 1985. Number and affinity of receptors for IgE on enriched populations of isolated rat intestinal mast cells. Immunology 55:363.
7. **Mendoza, G.R., and H. Metzger.** 1976. Disparity of IgE binding between normal and tumor mouse mast cells. J. Immunol. 117:1573.
8. **Metzger, H., B. Rivnay, M. Henkart, B. Kanner, J.-P. Kinet, and R. Perez-Montfort.** 1984. Analysis of the structure and function of the receptor for immunoglobulin E. Molec. Immunol. 21:1167.
9. **Sterk, A.R., and T. Ishizaka.** 1981. Binding properties of IgE receptors on normal mouse mast cells. J. Immunol. 128:838.
10. **Wank, S.A., C. DeLisi, and H. Metzger.** 1983. Analysis of the rate-limiting step in a ligand-cell receptor interaction: the immunoglobulin E system. Biochemistry 22:954.
11. **Titus, J.A., S.O. Sharrow, J.M. Connolly, and D.M. Segal.** 1981. Fc(IgG) receptor distributions in homogenous and heterogenous cell populations by flow microfluorometry. Proc. Natl. Acad. Sci. USA 78:519.
12. **Bazin, H., and R. Pauwels.** 1982. IgE and IgG2a isotypes in the rat. In: Prog. Allergy. Edited by K. Ishizaka. Basel, S. Karger, p.52.
13. **Isersky, C., A. Kulczycki, and H. Metzgter.** 1974. Isolation of IgE from reaginic rat serum. J. Immunol. 112:1909.
14. **Bauza, M.T., and D. Lagunoff.** 1983. Histidine uptake by isolated rat peritoneal mast cells. Effect of inhibition of histamine decarboxylase by α-fluoromethylhistidine. Biochem. Pharmacol. 32:59.
15. **Watson, J.V.** 1977. Fluorescence calibration in flow cyto-fluorometry. Br. J. Cancer 36:396.
16. **Mullaney, P.F., M.A. Van Dilla, J.R. Coulter, et al.** 1969. Cell sizing: a light scattering photometer for rapid volume determination. Rev. Sci. Instrum. 40:1029.
17. **Steinkamp, J.A.** 1984. Flow cytometry. Rev. Sci. Instrum. 55:1375.
18. **Kruger, P.G., and Lagunoff, D.** 1981. Effect of age on mast cell granules. Int. Arch. Allergy Appl. Immunol. 65:291.
19. **Ishizaka, T., and K. Ishizaka.** 1974. Mechanism of passive sensitization. IV. Dissociation of IgE molecules from basophil receptors at acid pH. J. Immunol. 112:1078.
20. **Kruger, P.G., and D. Lagunoff.** 1981. Mast cell restoration.

A study of the rat peritoneal mast cells after depletion with polymyxin B. Int. Arch. Allergy Appl. Immunol. 65:278.

21. **Isersky, C., J. Rivera, S. Mims, and R.J. Triche.** 1979. The fate of IgE bound to rat basophilic leukemia cells. J. Immunol. 133:1926.

22. **Ishizaka, K., H. Tomioka, and T. Ishizaka.** 1970. Mechanisms of passive sensitization. I. Presence of IgE and IgG molecules on human leukocytes. J. Immunol. 105:1459.

23. **Carson, D.A., and H. Metzger.** 1974. Interaction of IgE with rat basophilic leukemia cells. IV. Antibody-induced redistribution of IgE receptors. J. Immunol. 113:1271.

24. **Froese, A.** 1984. Receptors for IgE on mast cells and basophils. In: Prog. Allergy. Edited by K. Ishizaka. Basel, Karger, p.142.

25. **Malveaux, F.J., M.C. Conroy, N.F. Adkinson, and L.M. Lichtenstein.** 1978. IgE receptors on human basophils. Relationship to serum IgE concentrations. J. Clin. Invest. 62:176.

26. **Yong, L.C., S.G. Watkins, and D.L. Wilhelm.** 1977. The mast cell. II. Distribution and maturation in the peritoneal cavity of the young rat. Pathology 9:221.

27. **Yong, L.C., S. Watkins, and D.L. Wilhelm.** 1975. The mast cell: Distribution and maturation in the peritoneal cavity of the adult rat. Pathology 7:307.

28. **Combs, J.W., D. Lagunoff, and E.P. Benditt.** 1965. Differentiation and proliferation of embryonic mast cells of the rat. J. Cell. Biol. 25:577.

Mast Cell Differentiation and Heterogeneity,
edited by A. D. Befus et al.
Raven Press, New York © 1986.

Transmembrane Signaling in Basophils: Ion Conductance Measurements on Planar Bilayers Reconstituted with Purified Fcε Receptor and the Cromolyn Binding Protein

*Israel Pecht, *Vjekoslav Dulic, **Benjamin Rivnay, and **Ayus Corcia

*Departments of *Chemical Immunology and **Membrane Research, The Weizmann Institute of Science, Rehovot 76100, Israel*

Changes in the rather low cytosolic concentration of free Ca^{2+} ions convey a wide range of intracellular messages (1,2). One important way in which this Ca^{2+} concentration can be rapidly increased is by opening Ca^{2+} selective influx pathways in the cells' plasma membrane. Such transient permeability increase of mast cells and basophils to Ca^{2+} ions has been considered to constitute one key element in the process of coupling stimulus to secretion in these cells (3,4). This ion influx has been assigned to the opening of channels which allow the flow of the Ca^{2+} ions down their large concentration gradient (4). Identification and isolation of the membrane components involved in formation and control of such channels along with the resolution of their mode of operation are our main research interest.

In many types of cells, depolarization of the plasma membrane leads to ion influx due to opening of specific channels. Thus, regulation of such ion channels occurs by alteration of the electric potential of the membrane (5). We have therefore examined the relation between a stimulus imposed on the RBL–2H3 cells and a change in their membrane potential (6,7). To that

---

Acknowledgements. Support to I.P. from the Hermann and Lilly Stiftung fuer Medizinische Forschung is acknowledged gratefully. B.R. is an incumbent of the Alan Dixon Career Development Chair in Cancer Research, established by the Chicago Committee for the Weizmann Institute of Science.

end we employed a lipophilic cation (tetraphenyl phosphonium ion) which distributes across cells' membrane in response to their potential (8). Though a marked depolarization was observed in RBL–2H3 cells upon cross–linking their membrane–bound IgE and degranulation (6), the detailed examination of this process led to the conclusion that the induced change in potential is not the cause of ion influx into these cells but rather its result (7). Hence, the channel opening in this system is most probably a ligand activated event.

An important starting point in the search for membrane elements potentially involved in ligand activated channels is naturally the Fc_ε receptor identified and characterized in mast cells and basophils (9). Another direction emerged from a useful tool which we employed in investigation of this system, namely the anti–asthmatic drug cromolyn (10,11). This drug, capable of chelating Ca^{2+} ions in media of low dielectric constant (12), was shown to bind, specifically and in a Ca^{2+}–dependent fashion, to the membranes of mast cells and basophils (13). Furthermore, this binding was shown to block Ca^{2+} influx (14) and degranulation (15). The membrane protein component responsible for binding the drug (cromolyn binding protein = CBP) has been recently isolated (16). Its functional role has been investigated along two parallel lines: on the whole cell level and on the molecular one, in model planar lipid bilayers. Along the first line, variants of the RBL–2H3 cells defective in their capacity of cromolyn binding were isolated and grown (17). These variants were shown to be non-responsive to an immunological stimulus as monitored by Ca^{2+} uptake and mediator release (17). However, upon implanting the cromolyn binding protein (CBP) into the plasma membrane of these variants, their response capacity to the above stimulus had been restored (18). These results provided the first clear evidence for the crucial role played by CBP in coupling the immunological stimulus to secretion by allowing and/or controlling Ca^{2+} ion influx (18).

Direct conductance measurements of planar lipid bilayers containing unfractionated plasma membranes of the RBL–2H3 cells were then initiated. These studies have shown that Ca^{2+} ion conductance could be elicited in the above membranes by cross–linking its Fc_ε receptors with a specific antigen (19). Moreover, conductance was also elicited by directly cross–linking the CBP in these membranes with a monoclonal antibody specific to it (19). Single channel recordings of conductance induced by each of the above agents revealed that marked similarity exists among all of them irrespective of the inducer (19). These results suggested that cross–linking the Fc_ε receptor in the cellular plasma membrane also brings about aggregation of the CBP and thereby leads to channel formation (19). To examine this hypothesis we have produced planar bilayers at micropipette tips, which were reconstituted either with purified Fc_ε receptor or with CBP or with both of these components. Ion conductance of these membranes as affected by different specific reagents was investigated.

MATERIALS AND METHODS

Reagents. Monoclonal anti-CBP antibodies (IgM class) of line 25/94, shown to cause degranulation of RBL-2H3 cells, were generated as described (15). DNP conjugates of bovine serum albumin were prepared. An average of 8 or 16 DNP groups bound per protein molecule were determined from its absorption at 280 and 360 nm, respectively (20). Derivatization of serum albumin with cromolyn was done according to a modification of a published procedure (16). Monoclonal, DNP-specific IgE was kindly supplied by Dr. Z. Eshhar. Cromolyn was kindly supplied by Fisons (Loughborough, England).

Cells. RBL-2H3 cells were grown in stationary flask cultures as described (16).

Quin 2 loading and fluorescence measurements. The RBL-2H3 cells were incubated in batches (1×10^7 cells/ml each) after being saturated with the monoclonal IgE (1 hr, 37°C) and washed three times with Tyrode's buffer. Cells were incubated in the presence of 10 μM quin 2-AM (Calbiochem lot 393105) for 1 hr at 37°C. To the incubation medium an esterase inhibitor was added (BW 284 C51, Wellcome Reagents Ltd.) to minimize extracellular hydrolysis of the Quin 2-AM. After incubation, aliquots of 2×10^6 cell suspensions were centrifuged, supernatants discarded and the pellet resuspended in 2 ml of phosphate-free medium in a 1 cm optical path quartz cuvette. An electrically driven mini-stirrer was inserted into the cuvette ensuring continuous cell suspension without affecting the fluorescence measurement. A Perkin-Elmer MPF-44A spectrofluorometer with a thermostated cell holder (37°C) was employed (excitation at 339 nm, emission at 492 nm). Uptake of Quin 2-AM and hydrolysis of the acetoxymethyl ester were monitored by following the changes in emission spectra of both the cells and supernatants.

Isolation of CBP and Fc$_\varepsilon$ receptor. The CBP was isolated under non-denaturing conditions in nanomolar quantities from RBL-2H3 cells (16). Briefly, ~ 5×10^9 RBL-2H3 cells were incubated with a bovine serum albumin-(DNP)$_8$-(cromolyn)$_9$ conjugate in minimal essential medium supplemented with 10% fetal calf serum, 2 mM CaCl$_2$. The cells were then lysed by n-octylglucoside (40 mM). The lysate was passed through a DNP-specific immunoaffinity column onto which the CBP-Ca^{2+}-cromolyn conjugate had been absorbed. Elution of the CBP from the affinity column was carried out using 5 mM EDTA. The eluate was dialyzed against buffer (0.2M boric acid, 10 mM Hepes, 0.16M NaCl, 2 mM CaCl$_2$, pH 7.4) containing 20 mM n-octylglucoside and concentrated under suction in ultrathimbles (UH 100/25 obtained from Schleicher & Schull). Yields of protein were approximately 100 μg, as measured by the Bradford technique. The protein (at a final concentration of 100 μg/ml in the above buffer and 20 mM n-octylglucoside) was divided into 100 μl aliquots and stored frozen at -70°C until use. The purified CBP appeared as a major band of approximately 70,000 daltons on silver-stained NaDodSO$_4$ polyacrylamide gel. Possible contamination of the CBP, by components of Fc$_\varepsilon$ receptor, was checked both by isoelectric focusing and by cross-reactivity in radioimmunoassay and was found to be <1%. Fc$_\varepsilon$ receptor was isolated and purified according to the method of Rivnay et al. (21,22).

Preparation of lipid-protein vesicles. Two mg of carrier lipids (soybean lipids, phosphatidylcholine and cholesterol in weight/ratio 8:11:1) from a solution of 3.5 mg lipids per ml of chloroform/methanol were dried under nitrogen and then overnight under vacuum. The standard solution contained 145 mM NaCl, 1.8 mM CaCl$_2$ and 10 mM Hepes, pH 7.6. The dry lipid film was resuspended in a volume of 770 µl of this solution that also contained 50 µg of RBL-2H3 phospholipids and when appropriate, either 1.5 µg of CBP and/or 15 nM of Fc$_\varepsilon$-receptor. After sonication, 75 µl of this mixture were added to a 6 mm plastic well (microtiter plates – Nunclone) containing 100 µl of the standard solution.

Bilayer formation. Micropipettes were pulled by the two step method (23) from borosilicate capillary glass tubing (o.d. 1.5 mm; i.d. 0.86 mm; A-M Systems, Everett, WA) in a vertical pipette puller (Model 700C, David Kopf Inst., Tujunga, CA) and filled with the same standard solution used to resuspend the lipid film. Their tip diameters were between 0.8 – 1.5 µm and their tip resistance ranged from 10 to 50 Mohm when the tips were immersed in the standard solution. No polishing or coating of the pipette was necessary.

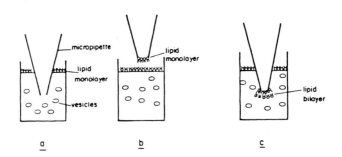

Figure 1. Schematic description of the steps involved in forming a lipid bilayer at the tip of a micropipette. In a) the pipette tip is introduced into the standard solution before the lipid vesicles are added. Pulling it out after a monolayer has formed at the water-air interface produces one monolayer and passing it back through the interface adds the second monolayer.

The method of lipid bilayers formation at the tip of micropipettes is a variation of previously published methods (24, see Fig. 1). Briefly, a micropipette was introduced in a plastic well containing 100 μl of the standard solution. The resistance of the pipette was monitored continuously by applying pulses of 100 μV. 75 μl of the solution containing the lipid vesicles were then added to the well and enough time allowed (up to 20 min) for a lipid monolayer to form at the air–liquid interface. Then the pipette was pulled out of the solution and reintroduced with the help of a three–dimensional hydraulic micromanipulator (Narishiga, Model MO–103). Formation of a bilayer was monitored by measuring the changes in the micropipette resistance with an EPC–5 clamp system (List Electronics). Once a bilayer was formed, it was clamped at different potentials with the patch clamp system. The currents obtained at each potential were low–pass filtered at 1 KHz by the clamp system, monitored with an oscilloscope and recorded on FM tape (HP 3964A).

For measurements of relatively high currents (1 pA or more), the data were replayed from the tape recorder without further filtering, digitized by an Apple–II microcomputer equipped with an 8–bit analog–to–digital converter (Mountain Computing Inc., CA) and plotted on a digital plotter (model DMP–3, Houston Instruments, TX). In these cases, the sampling rate was 3 KHz. For smaller currents, the data were either replayed at lower rate and plotted on an analog recorder with relatively low response time (20 Hz) or filtered through a home–made Bessel–filter (cut–off frequency 10 Hz) and then digitized and plotted as before. In the latter case, the 8–bit analog–to–digital converter is the limiting factor in the resolution of the current values, while both the Bessel–filter and the analog recorder introduce a limitation on the time resolution (see ref. 25 for a comprehensive discussion).

RESULTS AND DISCUSSION

Earlier studies aimed at monitoring calcium influx into triggered mast cells and basophils relied on tracer experiments employing $^{45}Ca^{2+}$. The recent development of the fluorescent calcium indicator Quin–2 provided a new approach (26). This indicator is taken into the cells as the membrane soluble, acetoxy methyl ester and upon hydrolysis by intracellular esterases is converted into the fluorescent chelator–probe which directly monitors Ca^{2+} concentrations. As illustrated in Fig. 2A, stimulation of RBL cells by antigen causes a graded increase in the intracellular free Ca^{2+} ion concentration as reflected by the Quin–2 emission. This process requires the presence of external Ca^{2+}, supporting the notion that Ca^{2+} influx into the cells has indeed occurred. The presence of cromolyn clearly blocks this process at <100 μM concentration (Fig. 2B–D).

As described above, specific cromolyn binding to the surface membranes of RBL–2H3 cells has been shown (13), and this eventually led to the isolation of the CBP (16). Direct conductance measurements of planar lipid bilayers formed at micropipette tips and containing purified components of the RBL-2H3 plasma membranes were carried out as detailed below.

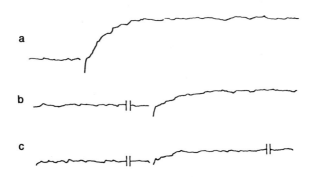

Figure 2. Emission changes due to the Ca^{2+} indicator Quin-2 in the interior of RBL-2H3 cells as effected by antigen stimulation and by the anti-allergic drug cromolyn. Excitation at 339 $\pm$ 1 nm; emission monitored at 492 $\pm$ 3 nm: a) increase in cytosolic-free Ca^{2+} upon adition of 5 ng/ml of antigen (DNP$_{16}$-BSA). b-c) Same as above in the presence of 20, 100 μM of cromolyn.

The procedure employed throughout the following experiments was designed to avoid the possibility of inducing irrelevant conductance changes through the lipid bilayer. As reported in more detail elsewhere (27), we have observed that when the lipid bilayers were formed from vesicles containing pure lipids only, it was possible, by applying voltages above a certain threshold value, to induce conductance changes that appeared very much like conventional channel activity. The voltage threshold for such induction was dependent on the resistance of the lipid bilayer. In order to avoid such spurious conductance changes, only those bilayers which showed at least 10 Gohms resistance and were able to withstand the experimental clamp potential without allowing any conductance change in the absence of applied agonist, were employed.

Conductance induced in membranes containing Fc$_\varepsilon$ and CBP. Vesicles containing both the Fc$_\varepsilon$ receptor and CBP were added to the standard solution and lipid bilayers formed at the tip of the micropipettes. Channel activity can be triggered in such bilayers both by specific antigen and by anti-CBP. Fig. 3 shows examples of the elicited responses. In panel 3A, the clamping potential was 50 mV (interior of the pipette negative). The record starts about 1 min after the addition of antigen (DNP$_{16}$-BSA) to the external solution. At this time, channel activity

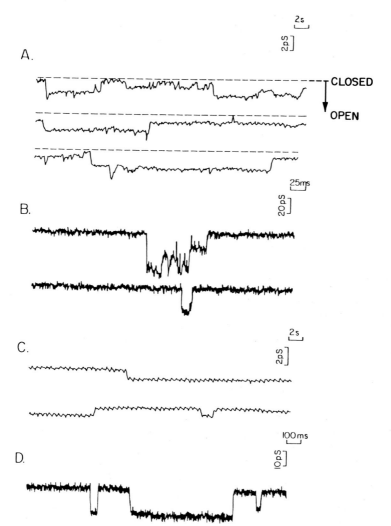

Figure 3. Illustrations of conductance records measured in the micropipette supported bilayer membranes containing both Fcε receptor and CBP. Panels A and B are consecutive records from the same experiment in which antigen (DNP_{16}–BSA) was used as a trigger. Panels C and D are records from different experiments in which the trigger was anti–CBP. All the experiments were initially filtered at 1 KHz cut-off frequency before recording on tape. Panels A and C were then replayed at 1/8th the recording rate and displayed on an analog recorder with a low time response (20 Hz). Panels B and D were digitized without further filtering by an 8-bit analog-to-digital converter. Sampling rate was 10 KHz in B and 5 KHz in D.

can be observed with a basic conductance amplitude of approximately 1.3 pS. Multiplets of this basic amplitude, e.g. doublets or triplets, conceivably representing two or three channels opened simultaneously, are clearly seen in this trace.

Fig. 3B shows the another stretch the record presented in panel A. In addition to the small conductance channel described above, events of relatively large conductance (1050 pS) are clearly seen. Whether this kind of conductance is due to aggregation of the small basic channel units or from a different conductance state of these basic elements is being investigated. Evidence against the latter possibility is that the large conductances were never observed in those cases where only the single basic channel was present yet not multiplets.

The same type of behaviour described above is also observed when anti-CBP antibodies were used to trigger channel activity in a bilayer containing both the Fc$_\varepsilon$ receptor and CBP. Fig. 3C shows a continuous record of an experiment of this kind in which only one type of channel is observed. As in the case of the conductance induced by antigen, the basic channel unit induced by anti-CBP has a very low conductance (approx. 1.3 pS). In the illustrated case, only one channel was observed to open at any given time.

Fig. 3D shows an illustration of the second type of channel activity that has also been observed upon triggering with anti-CBP. In the experiment shown (panel D), anti-CBP induced relatively large conductance (13-17 pS) with short open times (100-700 ms).

The specificity of the observed conductance is further corroborated by their inhibition with cromolyn. The susceptibility to such inhibition depends on the triggering mode. Thus, while conductance induced by antigen is completely blocked by cromolyn (100 μM), that triggered by anti-CBP was not inhibited. Since the monoclonal anti-CBP antibody has been selected on the basis of its competition with cromolyn for binding to CBP (15), the latter result is not surprising. The anti-CBP induces aggregation and channel activity by binding CBP in a way which competes well with cromolyn binding in the given concentration range.

Bilayers containing either Fc$_\varepsilon$ receptors or CBP. In experiments where only the Fc$_\varepsilon$ receptor was introduced into the membrane, neither antigen nor anti-IgE could elicit channel activity. These agonists also failed to induce channel opening in membranes containing CBP only. In contrast, channel activity was readily observed upon adding the monoclonal anti-CBP. Thus, when the only protein incorporated into the membrane was the Fc$_\varepsilon$-receptor, no channel activity could be induced by any of the reagents. However, CBP by itself when aggregated, exhibits channel activity in the absence of Fc$_\varepsilon$ receptor.

In membranes containing CBP alone, the electrical conductances induced by anti-CBP were rather similar to those observed in membranes containing both Fc$_\varepsilon$ receptor and CBP. In most experiments, only channels with small conductances of approximately 1.3 pS were observed. However, in some cases large conductances appeared. Fig. 4 illustrates such an experiment. A bilayer containing only CBP was clamped at 60 mV. Upon addition of anti-CBP, several channels with conductances ranging from 20 to 60 pS appear randomly. Both types of conductances induced in such CBP containing membranes were not inhibited by cromolyn even at 1 mM concentration if added after triggering.

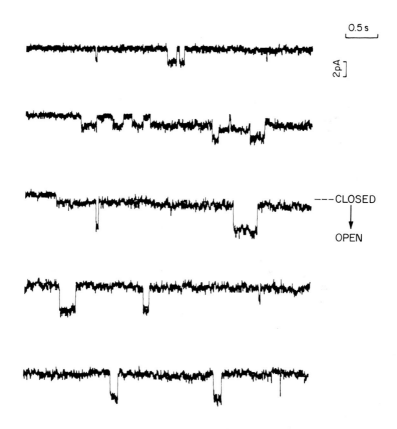

Figure 4. Channel activity induced by anti-CBP in a bilayer containing CBP only. The bilayer was clamped at 60 mV (pipette interior negative), the signal was filtered at 1 KHz before recording it, sampling rate 3 KHz.

CONCLUDING REMARKS

Identification of a membrane protein involved in ion channel formation in RBL–2H3 cells provides a significant advance in defining the early biochemical events which link the immunological stimulus to secretion. We have defined the minimal requirement of two plasma membrane components for eliciting controlled ion conductance through planar lipid bilayers by a specific antibody–antigen reaction: the Fc$_\varepsilon$ receptor and the cromolyn binding protein. Cross–linking the IgE residing in the Fc$_\varepsilon$ receptor most probably leads to aggregation of the CBP which becomes the ion conducting component. CBP aggregation can also be effected directly, by specific monoclonal antibodies. Several major questions regarding the nature of the involved proteins and their mechanism of operation and control are being investigated: (1) their chemical properties, structure and stoichiometry; (2) mutual interactions between the different subunits of Fc$_\varepsilon$ and CBP; (3) the nature of the ion channel formed by CBP, its activation and modulation.

REFERENCES

1. **Campbell, A.B.** 1983. Intracellular calcium, its role as regulator. John Wiley, Chicester.
2. **Douglas, W.W.** 1978. Stimuls–secretion coupling: variations on the theme of Ca^{2+}–activated exocytosis involving cellular and extracellular sources of calcium. Ciba Found. Symp. 54:61.
3. **Foreman, J.C., M.B. Hallet, and J.L. Mongar.** 1977. The relationship between histamine secretion and ^{45}Ca^{2+} uptake by mast cells. J. Physiol. 271:193.
4. **Foreman, J.C., L.G. Garland, and J.L. Mongar.** 1976. In: Calcium in Biological Systems. Edited by J.C. Dwean. Cambridge, Cambridge University Press, p.69.
5. **Tsien, R.W.** 1983. Calcium channels in excitable cell membranes. Ann. Rev. Physiol. 45:341.
6. **Sagi–Eisenberg, R., and I. Pecht.** 1983. Membrane potential changes during IgE mediated histamine release from RBL–2H3 cells. J. Memb. Biol. 75:97.
7. **Sagi–Eisenberg, R., and I. Pecht.** 1984. Resolution of cellular compartments involved in membrane potential changes IgE–mediated degranulation of RBL cells. EMBO J. 3:497.
8. **Lichtshein, D., H.R. Kaback, and A.J. Blume.** 1979. Use of lipophilic cation for determination of membrane potential in neuroblastoma–glioma hybrid cells suspension. Proc. Nat. Acad. Sci. USA 76:650.
9. **Froese, A.** 1984. Receptors for IgE on mast cells and basophils. In: Mast Cell Activation and Mediator Release. Edited by K. Ishizaka. Prog. Allergy 34:188.
10. **Cox, J.S.G., J.E. Beach, A. Blair, A.J. Clarke, J. King, T.B. Lee, D. Loveday, G.F. Moss, T.S.C. Orr, and J.T.A.**

Ritchie. 1970. Disodium cromoglycate. Adv. Drug Res. 5:115.

11. **Altounyan, R.E.C.** 1980. Review of the clinical activity and mode of action of sodium cromoglycate. Clin. Allergy 10:481.

12. **Mazurek, N., C. Geller-Bernstein, and I. Pecht.** 1980. Affinity of Ca^{2+} ions to the antiallergic drug, disodium cromoglycate. FEBS Lett. 111:194.

13. **Mazurek, N., G. Berger, and I. Pecht.** 1980. A binding site on mast cells and basophils for the antiallergic drug-disodium cromoglycate. Nature 286:722.

14. **Pecht, I., and R. Sagi-Eisenberg.** 1985. Calcium channels formation and modulation in secreting basophils and mast cells. In: Calcium, Neural Function and Transmitter Release. Edited by B. Katz and R. Rahaminoff. Boston, Martinus Nijhoff.

15. **Mazurek, N.** 1983. From a Ca^{2+} binding drug to a Ca^{2+} binding protein: Studies on transmembrane signaling in mast cells and basophils. Ph.D. Thesis. The Weizmann Institute of Science, Rehovot, Israel.

16. **Mazurek, N., P. Bashkin, and I. Pecht.** 1982. Isolation of a basophilic membrane protein binding the antiallergic drug-cromolyn. EMBO J. 1:585.

17. **Mazurek, N., P. Bashkin, A. Petrank, and I. Pecht.** 1983. Basophil variants with impaired cromoglycate binding do not respond to an immunological stimulus. Nature 303:528.

18. **Mazurek, N., P. Bashkin, A. Loyter, and I. Pecht.** 1983. Restoration of Ca^{2+} influx and degranulation capacity of variant RBL-2H3 cells upon implantation of isolated cromolyn binding protein. Proc. Nat. Acad. Sci. USA 80:6014.

19. **Mazurek, N., H. Schindler, T. Schurholz, and I. Pecht.** 1984. The cromolyn binding protein constitutes the Ca^{2+} channel of basophils opening upon immunological stimulus. Proc. Natl. Acad. Sci. USA 81:6841.

20. **Dulic, V., N. Mazurek, P. Bashkin, and I. Pecht.** 1985. Preparative isolation of a basophilic Ca^{2+} channel – the cromolyn binding protein. (In preparation).

21. **Rivnay, B., and H. Metzger.** 1982. Reconstitution of the receptor for IgE into liposomes: conditions for incorporation of the receptor into vesicles. J. Biol. Chem. 257:12800.

22. **Rivnay, B., G. Rossi, M. Henkart, and H. Metzger.** 1984. Reconstitution of the receptor for IgE into liposomes, reincorporation of purified receptors. J. Biol. Chem. 259:1212.

23. **Hamill, O.P., A. Marty, E. Neher, B. Sakmann, and F.J. Sigwarth.** 1981. Improved patch-clamp technique for high resolution current recording from cells and cell-free membrane patches. Pflugers Arch. 391:85.

24. **Coronado, R., and R. Latore.** 1983. Phospholipid bilayers made from monolayers on patch-clamp pipettes. Biophys. J. 43:231.

25. **Colquhonn, D., and F.J. Sigwarth.** 1983. Fitting and statistical analysis of single channel records. In: Single Channel Recordings. Edited by B. Sackmann and E. Neher. Plenum Press, p.191.

26. **Tsien, R.Y., T. Pozzan, and T.J. Rink.** 1982. Ca^{2+} homeostasis in intact lymphocytes: cytoplasmic free Ca^{2+} monitored with a new intracellular trapped fluorescent indicator. J. Cell Biol. 94:325.

27. **Corcia, A., I. Pecht, and B. Rivnay.** 1985. (Submitted).

Mast Cell Differentiation and Heterogeneity,
edited by A. D. Befus et al.
Raven Press, New York © 1986.

Histamine Cells in the Canine Fundic Mucosa

*Andrew H. Soll, Mary Toomey, Lovick Thomas,
Fergus Shanahan, and **Michael A. Beaven

*Center for Ulcer Research and Education, Department of Medicine, University of California
at Los Angeles School of Medicine and Medical and Research Services, *Wadsworth
Veterans Administration Hospital Center, Los Angeles, California 90073;
and National Heart, Lung and Blood Institute, **National Institutes of Health,
Bethesda, Maryland 20205*

Histamine and gastric acid secretion are closely linked in the phylogenetic tree; histamine is not found in the gut of species whose stomachs lack the capacity for secreting acid. In the phylogenetic tree, histamine apears in the upper portion of the gut coincident with the development of the capacity for acid secretion (1). Despite the central role for histamine in the regulation of gastric acid secretion, details regarding the cellular localization of histamine and regulation of histamine formation and release within the gastric mucosa remain controversial. The present review focuses upon the identity of the cells that store histamine in the fundic mucosa, and factors that determine histamine delivery to the final target, the parietal cell.

Histamine and the regulation of acid secretion. Considering histamine's role in gastric acid secretion, one faces the complex schematic regulation of acid secretion. Three pathways deliver the chemical messengers that regulate acid secretion: neurocrine (neurotransmitters released from post-ganglionic nerves innervating the fundic mucosa); endocrine (hormones such as gastrin delivered by blood); and paracrine (transmitters that diffuse across the intercellular compartment from local tissue stores).

This study was supported by the National Institute of Arthritis, Diabetes, and Digestive and Kidney Diseases Grants AM19984, AM30444, AM17328, and the Medical Research Service of the Veterans Administration.

In addition to histamine, several other potential chemical transmitters are also stored in cells within the fundic mucosa, including somatostatin, serotonin, and glucagon. Each of these three major pathways is physiologically important (2,3) although, despite the close association of histamine and acid secretion, the importance of histamine was hotly debated (4,5) until Black and co-workers introduced the H_2 receptor antagonists in 1972 (6). H_2 antagonists blocked not only the action of histamine on acid secretion, but also stimulation by food, gastrin and vagal stimuli, thereby guaranteeing a central role for histamine acid secretion (6–8).

This latter point underlines a second complexity of the regulation of acid secretion, an interdependence between these pathways influencing all phases of acid secretion. The mechanism or mechanisms underlying this interplay between the various pathways awaits elucidation. A basic element requiring definition is the regulation of histamine delivery to the parietal cell. Code (4), building upon earlier observations by MacIntosh, hypothesized that histamine was the final chemostimulator at the parietal cell for all pathways stimulating acid secretion; receptors for gastrin and acetylcholine resided on the histamine cell and the parietal cell had receptors only for histamine. However, this view failed to account for findings such as the ability of anti–cholinergic agents to block histamine action and the interdependency between cholinergic pathways and gastrin. The late Morton I. Grossman (7) proposed an alternate theory that the parietal cell had specific receptors for histamine, gastrin and acetylcholine, with the interdependency between the pathways reflecting interaction at the parietal cell itself. The controversy thus centered about whether gastric secretagogues stimulated acid secretion by acting in parallel on the parietal cell, or in series with their final effects being mediated by release of histamine.

Histamine cells of the fundic mucosa. The controversy regarding gastric mucosal histamine stores reflects major differences in histamine cells between species. In the rat, histamine is stored in an enterochromaffin–like (ECL) cell (9–11). These cells lie within the epithelium deep in the gastric glands and not in the lamina propria. The identification of these cells as the site of histamine storage was based upon their parallel distribution with histidine decarboxylase activity and by their fluorescence upon reaction with orthophthalaldehyde (9). These cells also concentrate tritium label after exposure to (5–^{3}H)–hydroxytryptophan (12) and L–[^{3}H]–histidine (13), thus indicating that they possess both APUD (amine precursor uptake and decarboxylation) characteristics and form and store histamine. Histidine decarboxylase activity in the rat fundic mucosa is high, being considerably greater than found in other species. In the rat, gastrin stimulates the formation of histamine by inducing the activity of histidine decarboxylase. Acetylcholine and food, presumably by inducing the release of gastrin, have similar effects (11,14). Histamine release in the rat fundic mucosa may be stimulated by gastrin, but such effects have been demonstrated only by monitoring a decrease in histamine content of the fundic mucosa following gastrin treatment (14). Despite considerable evidence that gastrin interacts with the histamine cell in the rat fundic mucosa, the existence of a causal link between this release of histamine by gastrin, and gastrin stimulation of acid secretion remains controversial (5,15).

In contrast to these findings in the rat fundic mucosa, in dog, man and pig, histamine appears to be stored in a mast-like cell (16–19). The term mast-like is used because these mast cells in the fundic mucosa were reported to have atypical staining properties and to be resistant to histamine release by compound 48/80 (19). In these species, orthophthalaldehyde–induced fluorescence is not found over epithelial cells, but rather only over mast-like cells in the lamina propria, thus providing a sharp contrast to the rat fundic mucosa. Levels of histidine decarboxylase activity are low in these other species (16,17), although some conflicting data exist on this latter point. No firm evidence has been presented to establish that either feeding or gastrin induces histamine formation.

The present work has focused upon the identity of histamine cells in the canine fundic mucosa and elucidation of the factors that regulate histamine release from those cells. Although the theme of this work has not been mast cell heterogeneity (20,21), to gain perspective on the role of histamine cells in the fundic mucosa, it is necessary to contrast fundic cells to non–gastric histamine cells. In the dog, mast cells are present in large numbers in liver; canine hepatic mast cells have been prepared as a control population. Another comparison has been to isolate cells from rat fundic mucosa, a species likely to have fundic mucosal histamine in non–mast cell stores.

Isolation of histamine cells from the canine fundic mucosa. Canine fundic mucosa that was bluntly separated from submucosa was dispersed by sequential exposure to collagenase and EDTA; histamine was present in considerable amounts in these dispersed cells, as were mast cells (22). Other elements from the lamina propria (e.g. endothelial cells) were also present. Separating by elutriation, histamine was found in a cell of small size (22), and mast cells had a similar separation profile by elutriation to histamine. However, this small cell elutriator fraction (SCEF) also contained several other cell types including somatostatin–, glucagon–, and serotonin–containing cells (23,24). Therefore, the SCEF was further separated using a linear density gradient formed from a solution of bovine serum albumin and Ficoll (22). The histamine content of cells was maximal in fractions of high density and correlated with the distribution of morphologically typical mast cells. The mast cells were enriched to about 75% in the fractions of highest density in this gradient (22). In these separations of canine fundic mucosal cells, there was no indication of a second population of histamine-containing cells in the portions of the gradients that contained endocrine markers (glucagon and somatostatin) (24) or endocrine-like cell markers, serotonin and DOPA decarboxylase (23). Despite the degree of enrichment of histamine cells, only trace activities of histidine decarboxylase were found in cell extracts of these fractions. However, using intact cells, histamine formation was detected by both a $^{14}CO_2$ release assay and by isotope dilution (23). Even with intact cells the formation of histamine was low, leaving open the question as to whether histamine was formed primarily in immature mast cells (25), possibly before their migration into the fundic mucosa or whether these in vitro studies under-estimated histamine formation rates by cells in vivo.

Comparison of canine fundic and hepatic mast cells. Identical techniques were used to disperse canine liver, and both histamine and mast cells were present in similar distribution in the smaller cell elutriator fractions, although hepatic mast cells may be slightly larger (23). Fixation of hepatic and fundic mast cells by glutaraldehyde and basic lead acetate, followed by staining with toluidine blue (pH 4.0) yielded comparable numbers of mast cells in fractions from both liver and fundic mucosa. Granular morphology was better preserved with glutaraldehyde fixation. Less than 5% of the mast cells in both populations were apparent after fixation in 10% buffered formalin, and thus formalin sensitivity does not differ between mast cells from these two sites. Both hepatic and fundic mast cells contained about 2.5 pg of histamine per cell. Histidine decarboxylase activity was detected in similar activities in both cell types, but only when studied intact.

Isolation of histamine cells from rat fundic mucosa. When similar techniques to those described above were applied to the rat fundic mucosa, histamine was found in cells of very light density (26). The cellular content of histamine and the activities of histidine decarboxylase and DOPA decarboxylase had similar distributions in the density gradient of rat fundic cells. Cells with granules typical of ECL cells were present in the same region of the density gradient as was histamine. The degree of enrichment of ECL cells was modest, with a maximal content of about 12%. These studies were consistent with the previous findings with intact mucosa noted above, that in the rat, but not canine, fundic mucosa histamine is present in an endocrine-like cell possessing high activities of histidine decaboxylase.

Receptors on parietal and non-parietal cells in the canine fundic mucosa. A brief overview of studies localizing gastrin receptors on parietal and non-parietal cells will provide important background for studies of the regulation of histamine release. Studies with isolated canine parietal cells indicate that histamine, gastrin, and acetylcholine directly activate parietal cell function (27). Potentiating interactions between histamine and both gastrin and cholinergic agents may be an important mechanism underlying the interdependency between secretagogues in vivo. Using the biologically active ligand ^{125}I-[Leu^{15}]-gastrin-17, gastrin receptors have been localized to parietal cells maximally enriched by combined velocity and density separation. However, in the elutriator separation, specific ^{125}I-[Leu^{15}]-gastrin-17 receptors were also found in the SCEF (28). Since the SCEF contains histamine cells plus endocrine and endocrine-like cells, sequential density gradient separation was also performed. A discrete distribution of ^{125}I-[Leu^{15}]-G17 binding was found in fractions of intermediate density; this profile was negatively correlated with the distribution of mast cells, thus indicating that canine fundic mast cells do not have gastrin receptors (29). However, the distribution of ^{125}I-[Leu^{15}]-G17 binding did correlate with that for somatostatin-like immunoreactivity (SLI). Functional evidence that gastrin receptors regulate SLI release was also obtained by studying isolated canine fundic somatostatin cells placed in short term culture (24). SLI release was stimulated by gastrin as well as by epinephrine, the latter acting at a β-receptor.

Muscarinic agents stimulated parietal cell function (27). However, muscarinic agents also influence SLI cell function, serving to inhibit rather than stimulate SLI release (30). This cholinergic effect has a double negative effect; by blocking release of somatostatin, cholinergic agents attenuate the acid inhibitory mechanisms mediated by this peptide, thereby potentially enhancing the secretory response. To localize muscarinic receptors to mast cells, [^{3}H]-QNB binding was performed and QNB receptors in all of the elutriator fractions examined (31). In density gradients performed on the SCEF, [^{3}H]-QNB binding was inversely correlated with histamine content (AH Soll, M Toomey, J Park, D Culp, MA Beaven, manuscript in preparation), suggesting that the histamine cell is not a likely locus of muscarinic regulation.

Regulation of histamine release from fundic mucosal stores. There are some data to indicate that stimulation of acid secretion by pentagastrin causes the release of histamine from dog and human gastric mucosa, although these studies are indirect, contradictory and lack convincing controls (32-36). To study factors directly regulating the release of histamine from canine fundic mast cells, we have placed the SCEF in suspension culture for 24 hr. Mast cells remain viable over this time and accounted for 30% to 60% of the cells surviving at 24 hr. The calcium ionophore A23187 (1 μM), concanavalin A, and antibody to rat IgE stimulate histamine release from these canine mast cells; there was no response to gastrin or cholinomimetics, alone or in combination. These latter negative findings thus agree with the apparent absence of muscarinic and gastrin receptors in the region of the density gradient performed on the small fundic cells that contain the histamine cells. The potential validity of these negative conclusions is reinforced by the findings that gastrin receptors are present on other fundic cells and that these mast cells are capable of releasing histamine in response to other agents. However, negative findings must be interpreted with caution, since it is possible that specific mast cell receptors may be unstable under these conditions. Histamine release in canine fundic mast cells and liver cells was compared. Liver cells appeared to release more histamine in response to the anti-IgE antibody, but this difference was not statistically significant in the experiments performed. Both cells responded to A23187 and concanavalin A and neither responded to compound 48/80 in concentrations of 1 and 10 $\mu g/ml$. Our present data thus do not provide a basis for functionally discriminating hepatic and fundic mast cells as distinct subpopulations. The reason for this apparently negative result may be the possibility that mast cells isolated from the fundic lamina propria contain more than one subtype. We also may not have yet chosen the proper discriminating markers for these potential mast cell subpopulations. Lastly, it is possible that mast cell heterogeneity is more apparent in some species, such as the rat, than in the dog.

Two agents may have important inhibitory effects on mast cell histamine release. Epinephrine inhibited histamine release from fundic and hepatic mast cells over a concentration range from 10 nM to 100 μM, with this effect blocked by propranolol. Furthermore, PGE_2 also inhibited histamine release over a somewhat high concentration range from 0.1 to 100 μM.

Gastrin and acetylcholine activate histamine release from rabbit gastric glands (37) and from amphibian fundic mucosa (38). These findings thus contrast with the studies with canine fundic mast cells discussed above. In light of this apparent difference, it is important to firmly establish the identity of the cell(s) storing histamine in the fundic mucosa of these species. Conclusions regarding the regulation of histamine release based upon studies in rat, and possibly in rabbit and frog, species in which histamine is stored in endocrine-like cells, may not be valid for regulation of histamine formation and release in species in which fundic mucosal histamine is stored only in mast cells.

Our present data obtained from studies with canine fundic cells indicate a complex model wherein both gastrin and acetylcholine have at least dual sets of receptors that are potentially involved in regulating acid-secretory function. Gastrin interacts with receptors on the parietal cell and on the somatostatin cell, and via these receptors has opposing acid-stimulatory and acid-inhibitory actions. This wiring appears to be a short circuit, but several elements, including muscarinic and beta-adrenergic input may modulate these opposing effects of gastrin on different cell types thereby co-ordinating the regulatory process.

There remains a dearth of information regarding the

characterization of histamine cells in the fundic mucosa and the endogenous factos regulating the formation and release of histamine from the fundic mucosa. Our data support the view that canine fundic histamine is stored only in mast cells, and we have not yet been able to discriminate these mast cells from liver or submucosal mast cells. Furthermore, the local paracrine, neuro-crine and hormonal pathways and chemotransmitters stimulating release of mast cell histamine remain to be elucidated. We have found that prostaglandin E_2 in relatively high concentrations and epinephrine acting at a β-receptor inhibit histamine release and may be important factors down-regulating the acid secretory response. There is little doubt that current models and know-ledge will rapidly evolve as systems are developed for studying the cell biology of the fundic mast cell. We anticipate that regulatory pathways of fundic mucosal secretory function will converge at the parietal cell and act in series by modulating the release of paracrine transmitters, such as somatostatin and histamine.

REFERENCES

1. **Reite, O.B.** 1972. Comparative physiology of histamine. Physiol. Rev. 52:778.
2. **Grossman, M.I.** 1981. Regulation of gastric acid secretion. In: Physiology of the Gastrointestinal Tract. Edited by Johnson, L.R. New York, Raven Press, p.659.
3. **Feldman, M.** 1983. Gastric secretion. In; Gastrointestinal Diseases, Third edition. Edited by Fordtran, J.S. Philadelphia, Saunders, p.541.
4. **Code, C.F.** 1965. Histamine and gastric secretion: a later look, 1955–1965. Fed. Proc. 24:1311.
5. **Johnson, L.R.** 1971. Control of gastric secretion: no room for histamine. Gastroenterology 61:106.
6. **Black, J.W., W.A.M. Duncan, C.J. Durant, C.R. Ganellin, and M.E. Parsons.** 1972. Definition and antagonism of histamine H_2-receptors. Nature (Lond.) 236:385.
7. **Grossman, M.I., and S.J. Konturek.** 1974. Inhibition of acid secretion in dog by metiamide, a histamine antagonist acting on H_2 receptors. Gastroenterology 66:517.
8. **Gibson, R., B.I. Hirschowitz, and G. Hutchinson.** 1974. Actions of metiamide, an H_2-histamine receptor antagonist, on gastric H^+ and pepsin secretion in dogs. Gastroenterology 67:93.
9. **Thunberg, R.** 1967. Localization of cells containing and forming histamine in the gastric mucosa of the rat. Exp. Cell. Res. 47:108.
10. **Aures, D., R. Hakanson, and A. Schauer.** 1968. Histidine decarboyxlase and DOPA decarboxylase in the rat stomach. Properties and cellular localization. Eur. J. Pharmacol. 3:217.
11. **Hakanson, R., and G. Liedberg.** 1970. The role of endogenous

gastrin in the activation of gastric histidine decarboxylase in the rat. Effect of antrectomy and vagal denervation. Eur. J. Pharmacol. 12:94.

12. **Rubin, W., and B. Schwartz.** 1979. An electron microscopic radioautographic indentification of the "enterochromaffin-like" APUD cells in murine oxyntic glands. Gastroenterology 76:437.

13. **Rubin, W., and B. Schwartz.** 1979. Electron microscopic radioautographic identification of the ECL cell as the histamine-synthesizing endocrine cell in the rat stomach. Gastroenterology 77:456.

14. **Kahlson, G., and E. Rosengren.** 1968. New approaches to the physiology of histamine. Physiol. Rev. 48:155.

15. **Hakanson, R., G. Liedberg, C.H. Owman, and F. Sundler.** 1973. The cellular localization of gastric histamine and its implications for the concept of histamine as a physiological stimulant of gastric acid secretion. In: Histamine: Mechanisms Regulating the Biogenic Amine Levels in Tissues with Special Regard to Histamine. Edited by Maslinski, C. Stroudsbury, Pa. Dowden, Hutchinson, and Ross, Inc., p.209.

16. **Aures, D., R. Hakanson, C.H. Owman, and B. Sporrong.** 1968. Cellular stores of histamine and monoamines in the dog stomach. Life Sci. 7:1147.

17. **Hakanson, R., B. Lilja, and C.H. Owman.** 1969. Cellular localization of the histamine and monoamines in the gastric mucosa of man. Histochemie 18:74.

18. **Steer, H.W.** 1976. Mast cells of the human stomach. J. Anat. 121:385.

19. **Lorenz, W., A. Schauer, S.T. Heitland, R. Calvoer, and E. Werle.** 1969. Biochemical and histochemical studies on the distribution of histamine in the digestive tract of man, dog and other mammals. Naunyn-Schmiedebergs. Arch. Pharmak. 265:81.

20. **Pearce, F.L.** 1982. Functional heterogeneity of mast cells from different species and tissues. Klin. Wochenschr. 60:954.

21. **Bienenstock, J., A.D. Befus, F. Pearce, J. Denburg, and R. Goodacre.** 1982. Mast cell heterogeneity: Derivation and function, with emphasis on the intestine. J. Allergy Clin. Immunol. 70:407.

22. **Soll, A.H., K. Lewin, and M.A. Beaven.** 1979. Isolation of histamine containing cells from canine fundic mucosa. Gastroenterology 77:1283.

23. **Beaven, M.A., A.H. Soll, and K.J. Lewin.** 1982. Histamine synthesis by intact mast cells from canine fundic mucosa and liver. Gastroenterology 82:254.

24. **Soll, A.H., T. Yamada, J. Park, and L.P. Thomas.** 1984. Release of somatostatin-like immunoreactivity from canine fundic mucosal cells in primary culture. Am. J. Physiol. 247:G558.

25. **Beaven, M.A., D.L. Aiken, E. WoldeMussie, and A.H. Soll.**

1983. Changes in histamine synthetic activity, histamine content, and responsiveness to compound 48/80 during maturation of rat peritoneal mast cells. J. Pharmacol. Exp. Therap. 224:620.

26. Soll, A.H., K.J. Lewin, and M.A. Beaven. 1981. Isolation of histamine-containing cells from rat gastric mucosa: biochemical and morphologic differences from mast cells. Gastroenterology 80:717.

27. Soll, A.H. 1981. Physiology of isolated canine parietal cells: receptors and effectors regulating function. In: Physiology of the Digestive Tract. Edited by Johnson L.R. New York, Raven Press, p.673.

28. Soll, A.H., D.A. Amirian, L.P. Thomas, T.J. Reedy, and J.D. Elashoff. 1984. Gastrin receptors on isolated canine parietal cells. J. Clin. Invest. 73:1434.

29. Soll, A.H., D.A. Amirian, L.P. Thomas, J. Park, M.A. Beaven, T. Yamada. 1984. Gastrin receptors on nonparietal cells isolated from canine fundic mucosa. Am. J. Physiol. 247:G715.

30. Yamada, T., A.H. Soll, J. Park, and J. Elashoff. 1984. Autonomic regulation of somatostatin release: studies with primary cultures of canine fundic mucosal cells. Am. J. Physiol. 247:G567.

31. Culp, D.J., J.M. Wolosin, A.H. Soll, and J.G. Forte. 1983. Muscarinic receptors and guanylate cyclase in mammalian gastric glandular cells. Am. J. Physiol. 245:G641.

32. Mann, W.K., J.H. Saunders, C. Ingoldby, and J. Spencer. 1981. Effect of pentagastrin on histamine output from the stomach in patients with duodenal ulcer. Gut 22:916.

33. Thirlby, R., M. Feldman, M. Tharp, and C. Richardson. 1982. Effect of pentagastrin on gastric mucosal histamine in dogs. Gastroenterology 82:1196.

34. Redfern, S., and Feldman, M. 1985. Am. J. Physiol. (in press).

35. Peden, N.R., E.J.S. Boyd, H. Callachan, D.M. Shepherd, and K.G. Wormsley. 1982. The effects of impromidine and pentagastrin on gastric output of histamine, acid and pepsin in man. Hepato-gastroenterol. 29:30.

36. Lorenz, W., H. Troidl, H. Barth, H. Rohde, S. Schulz, H. Becker, P. Dormann, A. Schmal, J. Dusche, and R. Meyer. 1976. Stimulus-secretion coupling in the human and canine stomach: role of histamine. In: Stimulus-Secretion Coupling in the Gastrointestinal Tract. Edited by Case, R.M., and H. Goebell. MTP Press Limited, p.177.

37. Bergqvist, E., M. Waller, L. Hammar, and K.J. Obrink. 1980. Histamine as the secretory mediator in isolated gastric glands. In: Hydrogen Ion Transport in Epithelia. Edited by Shulz, I., G. Sachs, J.G. Forte, and K.J. Ullrich. Amsterdam, Elsevier/North-Holland Biomedical Press, p.429.

38. Rangachari, P.K. 1975. Histamine release by gastric stimulants. Nature 253:53.

Mast Cell Differentiation and Heterogeneity,
edited by A. D. Befus et al.
Raven Press, New York © 1986.

The Utility of Mouse Cultured Mast Cells in the Analysis of Biochemical Regulation, Mediator Generation, and Drug Effects

Stephen I. Wasserman and Diana L. Marquardt

Division of Allergy, Department of Medicine, University of California at San Diego School of Medicine, University of California at San Diego Medical Center, San Diego, California 92103

The development of non-malignant populations of mast cells which grow and can be maintained in tissue culture has been a major advance in the study of mast cell biology. The mouse bone marrow–derived cultured mast cell (BMMC), grown and differentiated in the presence of the T–cell lymphokine interleukin 3 (1), has proven a particularly useful cell in this regard. It can be grown in large numbers, is fully responsive to IgE–mediated signals, and is stable over long periods of time. Other contributors to this symposium have discussed the identity of this mast cell type, its relationship to the rat intestinal as contrasted with the rat peritoneal mast cell, and have addressed some of the unique functional and structural properties of this cell type. In this latter regard are included its ability to generate platelet activating factor, and to metabolize arachidonic acid by the 5–lipoxygenase pathway. Additionally, the unique sulfated proteoglycans of these mast cells has been utilized to identify them and has been contrasted with heparin found in peritoneal and some other mast cells.

In this discussion our focus will be upon other aspects of this exciting model in cell biology, aspects which include studies of cell surface molecules important in cell activation, the identity of new mediators generated upon cell activation, and finally, the description of models of modulation of mast cell function for which the availability of these cells is crucial.

BIOCHEMICAL REGULATION BY CELL SURFACE ATPase

Several laboratories have previously demonstrated the presence of a Ca/Mg-dependent ATPase on the cell membrane of rat peritoneal mast cells (2,3). Inhibition of this ecto-enzyme by such compounds as ethycrinic acid (2) or the anion channel blocking drug DIDS (4) is associated with parallel degrees of inhibition of IgE-mediated histamine release from these cells. In addition, DIDS also causes parallel, dose-dependent inhibition of IgE-mediated calcium flux, arachidonic acid mobilization and cyclic AMP accumulation, suggesting an early and central role for ecto-ATPase activity in mast cell secretory responses. Studies attempting to further characterize this ecto-ATPase in the rat mast cell are hampered by the high ATPase activity in contaminating peritoneal macrophages, and by the requirement for large numbers of cells for analysis. In this regard, BMMC is particularly useful as large numbers of essentially pure mast cells can be obtained.

Table I. Mouse bone marrow-derived mast cells and rat peritoneal mast cells: ATPase and effect of DIDS

ATPase Type	V_{max} nm/min/10^6 Cells	Kapp μM	I_{50} DIDS M
High Affinity			
Ca^{++} activity, mouse	1.4	0.2	$>10^{-2}$
Ca^{++} activity, rat	0.5	0.17	N.D.
Mg^{++} activity, mouse	0.7	8.9	$>10^{-2}$
Low Affinity			
Ca^{++} activity, mouse	14.0	90	2.5×10^{-3}
Ca^{++} activity, rat	2.2	205	8×10^{-5}
Mg^{++} activity, mouse	14.0	218	$>10^{-2}$

Like the rat serosal mast cell, BMMC possess two ecto-ATPases. One activity displays a high affinity for calcium ion (approximately 0.3 μM), is inhibited by magnesium ions and may be related to the calcium pumping ATPase present in a wide variety of cell types. This enzyme is unaffected by DIDS, but is sensitive to the calmodulin inhibitor trifluoroperazine. As the enzyme is not susceptible to DIDS inhibition, attention was directed instead to a second ecto–ATPase activity. This second enzyme is present on both rat serosal and mouse BMMC (Table I), and it differs from the previously described activity in displaying a much lower affinity for calcium (approximately 0.1 nM) and a nearly ten–fold higher capacity for ATPase hydrolysis. This enzyme is separable from non–specific phosphatase activity as assessed by its inability to hydrolyze p–nitrophenyl phosphate. When extracted in detergent, the enzyme filters on gel filtration columns and migrates in polyacrylamide gels with an apparent molecular weight of 100,000 – 110,000. The enzyme is effectively inhibited by trifluoroperazine and by DIDS at concentrations comparable to those which prevent IgE–mediated release of histamine. In contrast, the enzyme resists inhibition by quercetin, cromolyn, and vanadate ion. The ability to grow the large numbers of BMMC necessary for these biochemical studies has enabled these initial physicochemical analyses, permitted the extraction and separation of the ATPase, and will permit further isolation for physico–chemical and functional analysis.

Mediator Identification

Mouse BMMC were the first mast cell population definitively demonstrated to generate platelet activating factor (5) and to possess chondroitin sulfate E (6), and have been extremely useful in the analysis of the generation of leukotrienes via 5–lipoxygenation reactions.

In concert with attention to ATPase in biochemical regulation of mediator release, our laboratory has addressed the consequences of ATP metabolism in the BMMC. As in rat serosal mast cells, ATP itself is capable of activating the BMMC to release preformed mediators (Fig. 1). More interesting is the finding that IgE–mediated cell activation is accompanied by a rapid and profound fall in ATP concentrations (7). This fall parallels the release of preformed mediators and results in the liberation of large quantities of adenosine. This has been demonstrated utilizing BMMC which are sensitized with IgE anti–DNP and challenged with DNP–BSA. The cells are then extracted with perchloric acid, neutralized with alamine and Freon, and analyzed for ATP utilizing high performance liquid chromatograpy (HPLC) on a Partisil–10 SAX column equilibrated in 10 mM potassium phosphate and eluted with a linear gradient to 25 mM potassium phosphate/0.5M KCl, pH 3.45 (7).

A fall of about 50% in intracellular ATP levels occurred coincident with peak mediator release. At the same time, adenosine, a metabolite of ATP via ATPase and 5′nucleotidase, was released into the supernatant. Both IgE–antigen and calcium ionophore activated systems mediated increases in extracellular adenosine to about twice resting levels. Release of adenosine paralleled that of preformed mediators and, in the presence of an adenosine deaminase inhibitor, remained elevated for at least 5 min. Release was linear with cell number and is maximal in less than 60 sec. Adenosine release was best appreciated when deoxy-coformycin was included in reaction mixtures and was identified, after perchloric acid extraction and neutralization with alamine and Freon, by isocratic elution employing HPLC with a C18 micro Bondapak reverse phase column. Its identity was confirmed by its sensitivity to adenosine deaminase and by its co-elution with authentic adenosine standards. Thus, adenosine joins the mediators histamine, PGD2, leukotrienes C4, D4 and E4, and platelet activating factor as another mast cell spasmogenic-vasoactive mediator.

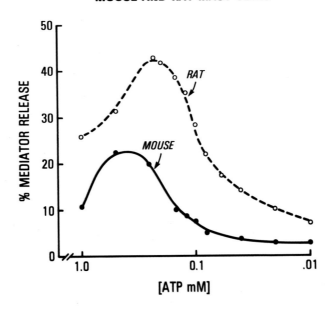

Figure 1. ATP-induced mediator release from cultured mouse, bone marrow-derived, and rat peritoneal mast cells.

MODULATION OF MAST CELL FUNCTION

Having demonstrated ATP hydrolysis and adenosine generation from BMMC, we next assessed the functional consequences of adenosine upon mast cells. Both rat serosal and mouse BMMC demonstrate a marked augmentation of preformed, but not newly generated, mediator release when adenosine is added to various mast cell secretagogues (8,9). These cells possess adenosine receptors upon their surface and the action of adenosine to augment mediator release is inhibited by the adenosine receptor inhibitory drug, theophylline.

Adenosine receptors have been identified utilizing radioligand binding techniques complemented with functional analysis of adenosine modulation of intracellular levels of cAMP. Adenosine and the stable analogues of this compound, L–phenylisopropyl adenosine (L–PIA) and 5'–N–ethycarboxamide adenosine (NECA), have been used to assess binding of ^{3}H–adenosine to mast cell membranes. Membranes were prepared from BMMC by homogenization, pelleted by high speed centrifugation, admixed with deoxycoformycin and ^{3}H–adenosine, and incubated at 4°C for 20 min in the presence or absence of non–radioactive adenosine analog. Reactions were terminated by vacuum filtration on glass fiber filters and bound radioactivity assessed. Specific binding amounted to approximately half of total binding. Binding was linear with membrane protein concentration utilized in the assay, and nearly all eluted ^{3}H–adenosine bound was recovered as authentic adenosine, as assessed by paper chromatography. Binding was rapid, saturable and reversible. By Scatchard analysis, mouse BMMC bound 9.08 $\pm$ 1.23 fmol of adenosine/10^6 cells (about 5000 receptors per cell) with a Kd of 24.4 $\pm$ 3.8 nM. ^{3}H–adenosine binding was not inhibited by the adenosine metabolites, inosine or hypoxanthine. That the adenosine binding was truly with a receptor, was indicated by the ability of adenosine, L–PIA and NECA to enhance intracellular cAMP concentrations and to markedly augment IgE and calcium ionophore stimulated mediator release. These effects occurred at micromolar concentrations of all three agonists with L–PIA being least potent. Increased levels of cAMP could be appreciated within 15 sec of agonist exposure, and persisted for 60 sec when higher concentrations of agonist were utilized.

Theophylline, a phosphodiesterase inhibitor and adenosine receptor blocker, is ineffective in altering IgE–mediated relese of histamine from rat intestinal mucosal mast cells or from mouse BMMC. It is however capable, at high concentrations, of preventing enhanced mediator release induced by adenosine. The ability to maintain BMMC in tissue culture provides an avenue whereby the long–term effects of drugs such as theophylline may be examined. As is true of other antagonist–receptor concentration relationships, BMMC grown in the presence of theophylline display increased numbers of adenosine receptors. This effect of theophylline is dose– and time–dependent and after 6 days in the presence of 100 μM theophylline, BMMC display nearly 8000 (as opposed to 5000) adenosine receptors. The functional consequences of such an increase in adenosine receptors is demonstrated by the fact that such theophylline-grown, adenosine receptor-enriched cells are supra-sensitive to the enhancing effect of adenosine upon IgE-mediated release of preformed, but not newly generated, mediators. An increased enhancement of mediator release of 162% is correlated closely with an increase in receptor number of 156%. The removal of xanthines and continued culture of mast cells in xanthine-free media permits a gradual return to basal levels of cell sensitivity to adenosine. This finding has several important implications: firstly, rapid removal of xanthines might be associated with a clinical hyper-responsiveness to adenosine in such diseases as asthma; and secondly, chronic exposure to a drug may uncover actions unsuspected in acute experiments. Moreover, this model is exceptionally well suited to the analysis of drug effects in which long-term changes are to be expected. Studies of such effects on BMMC of glucocorticoids, cromolyn and other drugs capable of altering adenosine binding are underway.

SUMMARY

The availability of large numbers of stable cultured BMMC has permitted initial steps in isolation and characterization of cell surface molecules present in low concentrations and has enabled identification of new mediators. BMMC provide a uniquely manipulable model for the assessment of long-term effects of therapeutic agents.

REFERENCES

1. **Ihle, J.N., J. Keller, S. Oroszlan, L.E. Henderson, T.D. Copeland, F. Fitch, M.B. Prystowsky, E. Goldwasser, J.W. Schrader, E. Palaszynski, M. Dy, and B. Lebel.** 1983. Biological properties of homogeneous interleukin 3. I. Demonstration of WEHI-3 growth factor activity, mast cell growth factor activity, P-cell stimulating factor activity, colony stimulating factor activity, and histamine producing cell stimulating factor activity. J. Immunol. 131:282.
2. **Magro, A.M.** 1977. Ethacrynic acid inhibitable Ca and Mg activated membrane adenosine triphosphatase in rat mast cells. Clin. Exp. Immunol. 30:160.
3. **Chakravarty, N., and Z. Echetebu.** 1979. Plasma membrane adenosine triphosphatases in rat peritoneal mast cells and macrophages. The relation of the mast cell enzyme to histamine release. Biochem. Pharmacol. 27:1561.
4. **Wasserman, S.I., and J. Kleeman.** 1984. The effect of inhibition of a mast cell ecto-ATPase by DIDS. Fed. Proc. 43:1664.
5. **Mencia-Huerta, J.M., R.A. Lewis, E. Razin, and K.F. Austen.**

1983. Antigen-initiated release of paltelet activating factor from mouse bone marrow-derived mast cells sensitized with monoclonal IgE. J. Immunol. 131:2958.

6. **Razin, E., R.L. Stevens, F. Akiyama, K. Schmid, and K.F. Austen.** 1982. Culture from mouse bone marrow of a subclass of mast cells possessing a distinct chondroitin sulfate proteoglycans with glycosaminoglycan rich in N-acetyl-galactosamine-4.6-disulfate. J. Biol. Chem. 267:7229.

7. **Marquardt, D.L., H.E. Gruber, and S.I. Wasserman.** 1984. Adenosine release from stimulated mast cells. Proc. Natl. Acad. Sci. USA 81:6192.

8. **Marquardt, D.L., D.W. Parker, and T.J. Sullivan.** 1978. Potentiation of mast cell mediator release by adenosine. J. Immunol. 120:871.

9. **Marquardt, D.L., L.L. Walker, and S.I. Wasserman.** Adenosine receptors on mouse bone marrow-derived mast cells: functional significance and regulation by aminophylline. J. Immunol. 133:932.

Mast Cell Differentiation and Heterogeneity,
edited by A. D. Befus et al.
Raven Press, New York © 1986.

Heterogeneity in Human Histamine-Containing Cells

L. M. Lichtenstein, C. C. Fox, R. P. Schleimer, D. Proud, R. M. Naclerio, and A. Kagey-Sobotka

Department of Medicine, Division of Clinical Immunology, The Johns Hopkins University School of Medicine, Good Samaritan Hospital, Baltimore, Maryland 21239

Human mast cells were first described by Ehrlich, who mistakenly suggested that they played a role in nutrition (1). Although there has since been much speculation regarding their function, no known physiological role for these cells has been described; it is indeed possible that they are remnants of man's evolutionary struggle to survive parasitic infestation. On casual evaluation the circulating basophilic leukocyte appears similar to the mast cell; studies of the various mammalian species suggest an inverse quantitative relationship between the two cell types. Rodents, for example, are relatively rich in mast cells but basophils, while they exist, are quite rare (2). In amphibians, on the other hand, up to 65% of the circulating leukocytes are basophils (3). Man, however, has been supplied with a rather generous number of both cell types. Basophils are like mast cells in that they are the only other cells with high affinity IgE receptors, and they are also one of the few cell types to generate and release histamine. It was originally thought that mast cells and basophils would also be functionally similar and a great deal of the early work from this laboratory involved this assumption (4,5). One purpose of this manuscript will be to demonstrate how inaccurate this view was. While there are a number of electron microscopic differences between mast cells and basophils (6), some have felt that both cells arise from the same progenitor and have focused on the morphologic similarities rather than the differences (7).

There are no firm data on the identity of the mast cell precursor in man, although the ability to culture mast cells from the marrow of mastocytosis patients suggests a bone marrow origin (8). Precursors of human basophils have been obtained from umbilical cord blood and characterized as mononuclear, non-adherent cells without surface immunoglobulin or T-cell markers (9,10). The relationship of these cells to mast cell precursors is unclear, and attempts to obtain human mast cells in a similar fashion have apparently been unsuccessful.

In addition to the striking functional differences between basophils and mast cells which will be described, there has recently been reason to question whether there are functional differences between mast cells derived from different tissue sources (11). Elsewhere in this book are descriptions of the morphologic differences between mast cells derived from different loci in the rat. The significance of these morphological differences, however, is put into question by the work of Kitamura, also presented in this book, which demonstrates that one morphologic type may evolve to another simply based on a change in their anatomical location. However useful these morphological criteria may be, there are clearly functional differences between the "mucosal" mast cell and the "connective tissue" mast cell in rodent species (12). Thus far, however, these differences have not been observed in our studies with human mast cells. Similarly, electron microscopic studies by Dvorak on human mast cells from the intestinal mucosa and lung parenchyma demonstrate no differences (13). These studies with human mast cells will constitute the remainder of this manuscript.

Differences and Similarities Between Human Parenchymal Mast Cells, Mucosal Mast Cells and Basophils

As a result of our recent ability to obtain purified human basophils and mast cells and, therefore, to study the biochemistry and function of isolated cells, numerous differences between the two cell types have been appreciated (14). The differences manifest with respect to the mediators which are released, the secretagogues which cause release, the presence of various hormone receptors and the pharmacologic agonists which control the release process. These differences will be itemized and our discussion will close with our initial attempts to investigate these differences in vivo (see Table I).

Mediators. All three cell types, of course, contain histamine. The basophil has on average 1 pg/cell; the range is quite modest from approximately 0.8 to 1.2 pg (15). The lung mast cell contains approximately 3 pg/cell, befitting its increased size and granularity. There is, however, a marked heterogeneity among mast cells (16). A step in the purification of these cells involves countercurrent elutriation and generates fractions of cells ranging from small, which contain approximately 1 pg of histamine cell, to large, with 15 pg/cell. The large cells are also more efficient at releasing this histamine than are the small cells. In the rodent the intestinal mucosal mast cell has distinctly less histamine than does the peritoneal mast cell (11); however, in man the cell obtained from the intestinal mucosa has the same mean and range of histamine content as does the lung cell (13).

	Basophils	Lung Parenchymal Mast Cells	Intestinal Mucosal Mast Cells
Stimuli			
Anti-IgE ($^{Optimal}_{Concentration}$)	0.03 - 0.1 µg/ml	3 - 10 µg/ml	3 - 10 µg/ml
Fmet Peptide	+	−	−
C5a	+	−	?
TPA	+	−	?
AA Products			
LTC_4	90 pm/10^6cells	80 pm/10^6cells	50 pm/10^6cells
PGD_2	−	120 pm/10^6cells	40 pm/10^6cells
AA Metabolism			
Indomethacin	↑	0	0
5HPETE	↑	0	?
Drugs			
Steroids	↓	0	?
Histamine	↓	0	0
Adenosine	↓	↑	?

The picture with respect to the lipid mediators derived from arachidonic acid is not complete. Both basophils and mast cells have an active 5-lipoxygenase pathway and, on average, generate the same quantity of LTC_4, approximately 30 ng/10^6 cells (13,17). However, in both, the range is large: in mast cells from 1 to 300 ng/10^6 cells. LTC_4 production by basophils of different donors is not normally distributed. Most generate <25 ng/10^6 cells while about 10% produce >100 ng. In the cells that have been purified, the lung mast cell and the human basophil, only LTC_4 is produced. In crude preparations of peripheral mono-nuclear cells or in suspensions of lung or gut cells, there is variable metabolism to LTD_4 and LTE_4. This is not complete with respect to the first two cell types, but mucosal preparations rapidly convert all LTC_4 to LTE_4. Mast cells generate 5-HETE and the inactive LTB_4 isomers (18); whether they generate small amounts of chemotactically active LTB_4 remains to be seen. This question has not been addressed in the other cell types. The cyclo-oxygenase pathways of the mast cells produce predominantly prostaglandin D_2 and smaller amounts of thromboxane. While studies are not complete, the intestinal mucosal mast cells appear to generate the same metabolites (13). Thus, the difference found in rodents, with the bone marrow-derived (mucosal) mast cell generating largely leukotrienes and the connective tissue (serosal) cell prostaglandins, is not seen in man (19). Whether basophils have any cyclo–oxygenase product is not yet clear. Attempts by radioimmunoassay have failed to identify PGD_2 or any other prostaglandin. However, in recent studies in which the cells were provided with exogenously labelled arachidonic acid, there appears, on HPLC, to be a product moving in the prostaglandin area, and one would anticipate that such products exist based on the pharmacologic effects of indomethacin, to be described below.

Inasmuch as it has become increasingly evident that platelets are activated during the anaphylactic response, the question of whether mast cells and basophils produce platelet activating factor (AGEPC) is of interest (20). This has been addressed only with lung mast cells and basophils. The former rapidly generate several varieties of AGEPC, the differences being related to the length of the carbon chain in position 1 (21). Interestingly, as in the human polymorphonuclear cell, most or all of the AGEPC generated by this cell is not released (22). The significance of this, and whether there is a pathway in the mast cell whereby a precursor molecule with arachidonate in the 2 position allows the production of both AGEPC and leukotrienes is not yet clear. As earlier reported by Henson, we can find no evidence that the human basophil generate AGEPC (23).

Both the basophil and the lung mast cell contain enzymes with TAME–esterase activity (24,25). A mast cell enzyme, which resembles tryptase, can also generate bradykinin from kininogen (26,27). The relationship of this enzyme to a basophil–derived kininogenase (24) remains to be determined. Although the major protease from rat mast cells is a chymotrypsin–like enzyme (28), the major enzyme from human mast cells is trypsin–like (25). Recently, however, it has been shown that human lung mast cells also contain a chymotryptic protease (perhaps analogous to the rat enzyme) which is capable of converting angiotensin I to angiotensin II) (29). Finally, lung mast cells also contain an enzyme which can cleave Hageman factor (30). These studies are in progress, but it is evident that mast cell proteases can interact with the coagulation pathways.

An interesting question relates to the proteoglycans in these cells. One of the hallmarks which distinguishes the mucosal and connective tissue cell in the rodent is that the latter has heparin while the former has chondroitin sulfate (31). Studies by Galli and co–workers suggest that basophil has chondroitin sulfate (32). We are currently in collaboration with Stevens and Austen seeking to determine the nature of the proteoglycan in purified human lung mast cells. Thus far, in preliminary studies, it would appear that these cells have at least two proteoglycans. This might be taken to indicate that the lung contains a mixture of two types of mast cells, but this speculation is probably premature.

Gleich and his colleagues have demonstrated that the eosinophil major basic protein is found in vivo in man at the site of many different types of allergic inflammation (33,34). They also have evidence that this is an important element in the pathogenesis of chronic asthma and other allergic diseases. In collaboration with his group, we have demonstrated that the basophil also contains major basic protein and, indeed, a variety of other active eosinophil proteins. The lung mast cell, however, does not contain these potent toxic agents.

Thus, with respect to the array of mediators, there are marked differences between basophils and mast cells while the profile in mast cells from gut and lung seems identical. We may expect that IgE-mediated reactions which involve primarily basophils will be quite different from those in which the mast cell plays the major role.

Secretagogues. Again, as with respect to mediators, there are striking differences between basophils and mast cells with regard to the types of secretagogues which are active. Both respond to anti-IgE, but this response is quite different. Basophils manifest optimal release with but 3% of the anti-IgE required to produce this result in the mast cell (16). Both lung mast cells and intestinal mast cells behave identically in this respect (13). It can also be pointed out that anti-IgE induced release in the mast cell occurs at approximately twice the rate as that in basophils. As might be expected, anti-IgE challenge of these cells leads to the whole range of mediators referred to above.

Two peptides of potential biologic importance have been studied, C5a and F-met-leu-phe. The latter is a potent releasing agent from human basophils, but neither the lung nor the intestinal mucosal mast cell is triggered by this peptide. C5a is a potent releaser from basophils, as suggested earlier by Grant (35). Also in accord with his work, we have found that C5a does not generate leukotriene C_4 from human basophils. In recent studies with highly purified C5a obtained from Dr. M. Frank, we found a rather interesting effect caused by D_2O. Heavy water is known to potentiate release caused by many stimuli (36). Its mechanism of action was thought to be through its activity on microtubules, but recent work by Beaven and Metzger indicates that D_2O increases the turnover of phosphatidylinositol and, presumably, the generation of arachidonic acid by this route (37). Addition of D_2O to basophils challenged with C5a, in addition to increasing the level of histamine release, also caused these cells to secrete leukotrienes. There is an interesting parallel to this: chemically linked dimers of IgE cause basophils to release histamine but little or no leukotriene C_4, whereas trimers generate both mediators. If D_2O is added to dimerically challenged cells, however, there is marked leukotriene production.

One of the differences between the two types of rodent mast cells is their response to 48/80 (38). As previously reported by others, 48/80 is generally inactive in both basophils and human mast cells (39). However, another polybasic peptide, poly-arginine, is a potent releaser from all three cell types.

There are three other rather unusual secretagogues which act differently on basophils and mast cells. The first is TPA, which activates protein kinase C, an enzyme felt by most workers in the field to be involved in the release mechanism (40). This secretagogue is a potent releaser from the human basophil and a major component of that process is calcium independent (41).   TPA, however, has little or no effect on the human mast cell. Hyperosmolarity (we used mannitol for this purpose) represents a second interesting stimulus to secretion.   In both the basophil and mast cell the release process is non-cytotoxic, but its mechanism is clearly quite distinct from anti-IgE (42,43).   The importance of this stimulus lies in the fact that it may well be involved in the broncospastic response to the breathing of cold dry air in vivo, which will be referred to below. Though both cells respond to hyperosmolarity, there are differences; the mast cell responds with significant histamine release at much lower increments in osmolarity than does the basophil.   Finally, we have argued that the 5-lipoxygenase pathway is involved in the mechanism of mediator release in the human basophil (44). A part of this argument was that 5-hydroperoxy-eicosatetranoic acid (5-HPETE) augments IgE-mediated release and, in the presence of cytochalasin, can itself cause release (45).   However, in the lung mast cells, 5-HPETE has neither augmenting or releasing properties.

Thus, virtually all secretagogues thus far studied differentiate between basophils and the two kinds of mast cells, while the latter seem to react identically to all of these releasing agents.   The only secretagogue where there are differences between mast cells is morphine, which is well known to degranulate skin mast cells and to cause histamine release in vivo in man.   Morphine in vitro at concentrations up to 0.1 mM causes no release from lung or mucosal mast cells.   Although there is a report that morphine stimulates release from human heart mast cell, our studies in collaboration with Marone failed to confirm this (46).   However, Tharp and Sullivan report that the skin mast cell does respond to this agent (47).   Only this fragmentary evidence suggests that a secretagogue can have a differential effect on mast cells derived from different anatomical locations.

Hormone Receptors

Epinephrine and other β-adrenergic agonists inhibit histamine and leukotriene release from all three cell types (48,49).   There is, however, an interesting difference between the basophil and the mast cell.   For isoproterenol to inhibit basophil histamine release at pharmacologic concentrations, the reaction must be "staged" (50).   That is, there must be interaction between the basophil receptor and the agonist for a brief period in the absence of calcium for inhibition to occur. This does not apply to either the lung or intestinal mucosal mast cell where sub-micromolar concentrations of agonist inhibit in the presence of calcium. Little is known about the effects of α-adrenergic agonists on these cells. There is evidence from experiments with chopped human lung that α-adrenergic stimulation potentiates mediator release (51). This has not been confirmed nor have reports on isolated cells been published, so that one does not know whether this is a tissue effect or is directed to the mast cells. These agonists have no effect on the basophil (52).

The basophil has a histamine type two (H2) receptor which blocks mediator release at low concentrations of histamine or other H2 agonists (53). This receptor is activated at physiologic concentrations of histamine to limit the allergic response (54). However, neither histamine nor the H2 agonist, dimaprit, has any effect on the human lung mast cell, but we have not been able to confirm this finding (55).

The effects of adenosine on these cells is strikingly different. Adenosine, at low concentrations, presumably acting on an A_2 (Ra) receptor, blocks basophil histamine release (56). Work with relatively specific A_1 (Ri) and A_2 agonists confirms that this effect is directed via the A_2 receptor. In the human lung mast cell adenosine has a modest but significant potentiating effect on histamine release. However, adenosine and A_1 and A_2 agonists all have the same effect, so that it does not seem to be mediated via a specific receptor (57). Moreover, theophylline, which is an adenosine antagonist, blocks the inhibition of basophil histamine release by adenosine, but does not have this effect on the mast cell.

There is a report that cholinergic stimulation potentiates histamine release from chopped human lung (51). Methacholine has no effect on basophil histamine release over a large range of concentrations, and cholinergic stimulation of human lung and intestinal mucosal mast cell suspensions had no effect. As with the alpha-adrenergic response, it is difficult to tell whether the chopped lung experiments reflect a tissue effect or an effect directed at the mast cells themselves.

About the only hormone which has a similar effect on basophils and mast cells is the E series of prostaglandins which inhibit at approximately the same concentrations in both cell types (58). It was demonstrated with purified human basophils and mast cells that both prostaglandin and beta-adrenergic agonists cause an increase in intercellular cyclic AMP levels at the concentrations at which inhibition is noted, although the role of cyclic AMP in the release process and its modulation is not yet clear.

Pharmacology

A number of agents used to treat allergic disorders have been studied in rodent and human mast cells and in basophils. The prototype for the development of anti-mediator release drugs, disodium cromoglycate, is active on the peritoneal mast cell of the rat, but does not affect the intestinal mucosal mast cell

(59). While DSCG has a modest effect at very high doses in chopped human lung systems it has little or no effect on the isolated human mast cell (60). A number of investigators have demonstrated that cromolyn has no effect on basophil mediator release (61).

The two drugs which are the mainstays for treatment of asthma are β–adrenergic agonists and theophylline. The therapeutic β–agonists act on human cells in the same way as the endogenous hormone. In the rodent, β–agonists have little effect, whereas theophylline differentially inhibits the peritoneal as opposed to the intestinal mucosal mast cell (62). In humans, theophylline inhibits both basophils and mast cells at similar, relatively high concentrations. The flavonoids, studied extensively by Middleton, seem to be active across the board; they inhibit both types of rodent mast cells, human basophils and mast cells (63–65).

The most striking differences between basophils and mast cells with respect to the pharmacologic agonists concerns those which are thought to be involved in arachidonic acid metabolism. Indomethacin and other NSAID, relatively specific cyclo–oxygenase inhibitors, enhance basophil histamine release (66). They are felt to have this activity by virtue of shunting arachidonic acid metabolism through the lipoxygenase pathway. In accord with this hypothesis, we have found that the release of leukotriene C_4 is greatly enhanced in the presence of pharmacologic concentrations of indomethacin (57). Also, exogenous arachidonic acid increases both histamine release and leukotriene production in the human basophil. In the lung and gut mast cell, however, indomethacin is completely without effect, enhancing neither histamine nor leukotriene production (58). Similarly, exogenous arachidonic acid has no effect with these cell types.

Steroids also have markedly different effects on basophils and mast cells. Pharmacologic concentrations of steroids inhibit basophil histamine and leukotriene production. The time course of this inhibition is very similar to the action of steroids in vivo in man, i.e. a considerable number of hours are required (67). The steroid effect is relatively specific for IgE–mediated release since it does not affect release caused by a number of other secretagogues (68). In the human lung mast cell, however, even high concentrations of steroids have no effect on the production of histamine or on leukotrienes and prostaglandin D_2 production (69).

Thus, the drugs which affect basophils and mast cells are quite different and the question of which cells are involved in different disease processes will have important therapeutic implications. In reviewing all the data described thus far, the possibility that the human basophil plays the same biologic role as the rodent mucosal mast cell seems likely.

IN VIVO CORRELATIONS

Our human in vivo studies involve antigen challenge of the upper airways followed by lavage (70). In the acute response to antigen we can document the release of all mediators referred to in the first section of this paper: histamine, PGD_2, LTC_4 and enzymes with TAME-esterase activity are found, the latter being responsible for the generation of the kinins which are also observed (71). It is highly likely that this response is mast cell related. The late phase reaction of the airways, however, is different. Four to 10 h after the first challenge there is recurrence of symptomatology. Lavage at that time reveals most of the mediators found in the acute response; the exception is prostaglandin D_2 which is never found in the late response (72). Inasmuch as a major difference between basophils and mast cells with respect to mediators is the absence of PGD_2 generation by the basophil, we interpret this result to mean that basophils are involved in the late phase release reaction. The pharmacologic differences between basophils and mast cells also support this conclusion. A study of the effects of oral steroids on this model revealed that the acute phase reaction was not affected by steroids either with respect to symptomatology or to the mediators which could be measured. However, as might have been anticipated, steroids ablated both the cellular and mediator related aspects of the late phase reaction, in accord with the effect of steroids on basophils and not on mast cells. The tricyclic antihistamine, azatadine, inhibits in vitro mediator release from human mast cells if the secretagogue is anti-IgE or antigen. It is, however, inactive against hyperosmolar-induced release (73). In vivo, azatadine was a very effective inhibitor of the antigen-induced release of mediators in the acute phase (74). Hyperosmolar challenge of the upper airways leads to mediator release as does the breathing of cold dry air, the latter accompanied by an increased osmolarity of the nasal secretions (75,76). As predicted from the in vitro studies, azatadine has no effect on the cold dry air response of the upper airways (73). Thus, it seems likely that the differences between mast cells and basophils with regard to meditors released, activity of secretagogues and pharmacologic agonists will have significant implications for the manifestations of allergic disease and for its therapy. The similarities between the in vivo observations and the in vitro experimentation with human basophils and mast cells suggest that the latter will prove to be an important model in developing therapeutic agents for the treatment of human allergic diseases.

REFERENCES

1. **Ehrlich, P.** 1879. Bietrage zur kenntnis der granulierten bindegewebszellen und der eosinophilen leukocyten. Arch. Anat. Physiol. (Liepzig), p.166.

2. **Levy, D.A.** 1974. Histamine and serotonin. In: <u>Mediators of</u> Inflammation. Edited by Weissman, G. New York.

3. **Mead, K.F., M. Borysenko, and S.R. Findlay.** 1983. Naturally abundant basophils in the snapping turtle, <u>Chelydra serpentina</u>, possess cytophylic surface antibody with <u>reaginic function. J. Immunol</u>. 130:334.

4. **Lichtenstein, L.M., and A.G. Osler.** 1964. Studies on the mechanisms of hypersensitivity phenomena. IX. Histamine release from human leukocytes by ragweed antigen. <u>J. Exp.</u> <u>Med</u>. 120:507.

5. **Lichtenstein, L.M., and S. Margolis.** 1968. Histamine release in vivo: Inhibition by catecholamines and methylxanthines. <u>Science</u> 161:902.

6. **Dvorak, A.M., S.J. Galli, E.S. Schulman, L.M. Lichtenstein, and H.F. Dvorak.** 1983. Basophil and mast cell degranulation. Ultrastructural analysis of mechanisms of mediator release. <u>Fed. Proc</u>. 42:2510.

7. **Zucker-Franklin, D.** 1980. Ultrastructural evidence for the common origin of human mast cells and basophils. <u>Blood</u> 56:534.

8. **Horton, M.A., and H.A.W. O'Brien.** 1983. Characterization of human mast cells in long-term culture. <u>Blood</u> 62:1251.

9. **Ogawa, M., T. Nakahata, A.G. Leary, A.R. Sterk, K. Ishizaka, and T. Ishizaka.** 1983. Suspension culture of human mast cells/basophils from umbilical cord blood mononuclear cells. <u>Proc. Natl. Acad. Sci. USA</u> 80:4494.

10. **Ishizaka, T., D.H. Conrad, T.F. Huff, D.D. Metcalfe, R.L. Stevens, and R.A. Lewis.** 1985. Unique features of human basophilic granulocytes developed in <u>in vivo</u> culture. <u>Int. Arch. Allergy Appl. Immunol</u>. 77:137.

11. **Pearce, F.L.** 1982. Functional heterogeneity of mast cells from different species and tissues. <u>Klinishe Wchsr</u>. 60:954.

12. **Befus, A.D., F.L. Pearce, J. Gauldie, P. Horsewood, and J. Bienenstock.** 1982. Mucosal mast cells. I. Isolation and functional characteristics of rat intestinal mast cells. <u>J. Immunol</u>. 128:2475.

13. **Fox, C.C., A.M. Dvorak, S.P. Peters, A. Kagey-Sobotka, and L.M. Lichtenstein.** 1985. Isolation and characterization of human intestinal mucosal mast cell. <u>J. Immunol</u>. 135:483.

14. **Peters, S.P., D.W. MacGlashan Jr., R.P. Schleimer, R.M. Naclerio, D. Proud, W. Kazimierczak, A. Kagey-Sobotka, N.F. Adkinson Jr., and L.M. Lichtenstein.** 1984. IgE-mediated release of inflammatory mediators from human basophils and mast cells <u>in vitro</u> and in <u>in vivo</u>. In: <u>Advances in</u> <u>Inflammation Research</u>. Edited by: Weissman, G. New York, p.227.

15. **MacGlashan, D.W. Jr., and L.M. Lichtenstein.** 1981. The purification of human basophils. <u>J. Immunol</u>. 124:2519.

16. **Schulman, E.S., A. Kagey-Sobotka, D.W. MacGlashan Jr., N. Franklin Adkinson Jr., S.P. Peters, R.P. Schleimer, and L.M. Lichtenstein.** 1983. Heterogeneity of human mast cells. <u>J.</u>

Immunol. 131:1936.
17. MacGlashan, D.W., R.P. Schleimer, S.P. Peters, E.S. Schulman, G.K. Adams III, H.H. Newball, and L.M. Lichtenstein. 1982. Generation of leukotrienes by purified human lung mast cells. J. Clin. Invest. 70:747.
18. Peters, S.P., D.W. MacGlashan Jr., E.S. Schulman, R.P. Schleimer, E.C. Hayes, J. Rokach, N.F. Adkinson Jr., and L.M. Lichtenstein. 1984. Arachidonic acid metabolism in purified human lung mast cells. J. Immunol. 132:1972.
19. Roberts, L.J. II, R.A. Lewis, J.A. Oates, and K.F. Austen. 1979. Prostaglandin, thromboxane and 12-hydroxy-5,8,10,14-eicosatetraenoic acid production by ionophore-stimulated rat serosal mast cells. Biochim. Biophys. Acta 575:185.
20. Hanahan, D.J., C.A. Demopoulos, J. Liehr, and R.N. Pinckard. 1980. Identification of naturally occurring platelet activating factor as acetyl-glyceryl-ether-phosphorycholine (AGEPC). J. Biol. Chem. 255:5514.
21. R.N. Pinckard. Personal communication.
22. Lichtenstein, L.M., R.P. Schleimer, D.W. MacGlashan Jr., S.P. Peters, E.S. Schulman, D. Proud, P.S. Creticos, R.M. Naclerio, and A. Kagey-Sobotka. 1984. In vitro and in vivo studies of mediator release from human mast cells. In: Asthma III: Physiology, Immunopharmacology and Treatment. Edited by: Kay, A.B., K.F. Austen, and L.M. Lichtenstein. London, p.1.
23. Betz, S.J., G.Z. Lotner, and P.M. Henson. 1980. Generation and release of platelet activating factor (PAF) from enriched preparations of rabbit basophils: failure of human basophils to release PAF. J. Immunol. 125:2749.
24. Newball, H.H., R.W. Berninger, R.C. Talamo, and L.M. Lichtenstein. 1979. Anaphylactic release of a basophil kallikrein-like activity. I. Purification and characterization. J. Clin. Invest. 64:457.
25. Schwartz, L.B., R.A. Lewis, D. Seldin, and K.F. Austen. 1981. Acid hydrolases and tryptase from secretory granules of dispersed human lung mast cells. J. Immunol. 126:1290.
26. Proud, D., D.W. MacGlashan Jr., H.H. Newball, E.S. Schulman, and L.M. Lichtenstein. 1985. IgE-mediated release of a kininogenase from purified human lung mast cells. Am. Rev. Respir. Dis. (in press).
27. Proud, D., and L.M. Lichtenstein. 1984. Human lung mast cell kininogenase: apparent identity to tryptase. Fed. Proc. 43:1807.
28. Benditt, E.P., and M. Arase. 1959. An enzyme in mast cells with properties like chymotrypsin. J. Exp. Med. 110:451.
29. Wintroub, B.U., C.E. Kaempfer, N.B. Schechter, and D. Proud. 1984. A human mast cell enzyme converts angiotensin I to angiotensin II by an angiotensin converting enzyme independent pathway. Circulation 70:11.
30. Meier, H.L., L.W. Heck, E.S. Schulman, D.W. MacGlashan Jr., L.M. Lichtenstein, C.G. Cochrane, H.H. Newball, and A.P.

Kaplan. 1983. Lung Hagemen factor cleaving enzyme (LMFA) is a mast cell elastase. Fed. Proc. 42:445.

31. Stevens, R.L., and K.F. Austen. 1981. Proteoglycans of the mast cell. In: Biochemistry of the Acute Allergic Reaction. Edited by Becker, E.L., A.S. Simon, and K.F. Austen. New York, p.69.

32. Galli, S.J., N.S. Orenstein, P.J. Gill, J.E. Silbert, A.M. Dvorak, and H.F. Dvorak. 1979. Sulfated glycosaminoglycans synthesized by basophil-enriched human leukaemic granulocytes. In: The Mast Cell: Its Role in Health and Disease. Edited by Pepys, J., and A.M. Edwards, England, p.842.

33. Frigas, E., D.A. Loegering, G.O. Solley, G.M. Farrow, and G.J. Gleich. 1981. Elevated levels of the eosinophil granule major basic protein in the sputum of patients with bronchial asthma. Mayo Clinic Proc. 56:345.

34. O'Donnell, M.C., S.J. Ackerman, G.J. Gleich, and L.L. Thomas. 1983. Activation of basophil and mast cell histamine release by eosinophil granule major basic protein. J. Exp. Med. 157:1991.

35. Grant, J.A., and L.M. Lichtenstein. 1973. The role of complement in human immediate hypersensitivity: Evidence against involvement of the alternate pathway of complement activation. J. Immunol. 111:733.

36. Gillespie, E., and L.M. Lichtenstein. 1972. Heavy water enhances IgE-mediated histamine release from human leukocytes: evidence for microtubule involvement. Proc. Soc. Exp. Biol. Med. 140:1228.

37. Maeyama, K., R.J. Hohman, H. Metzger, and M.A. Beaven. 1985. IgE-receptor cross-linking, phosphatidyl inositol breakdown and histamine release in rat basophilic leukemia (2H3) cells: enhanced response with D_2O. Fed. Proc. 44:1917 (abstract).

38. Pearce, F.L., A.D. Befus, J. Gauldie, and J. Bienenstock. 1982. Mucosal mast cells. II. Effects of anti-allergic compounds on histamine secretion by isolated intestinal mast cells. J. Immunol. 128:2481.

39. Church, M.K., G.J.-K. Pao, and S.T. Holgate. 1982. Characterization of histamine secretion from mechanically dispersed human lung mast cells: effects of anti-IgE, calcium ionophore, compound 48/80 and basic polypeptides. J. Immunol. 129:2116.

40. Nishizuka, Y. 1984. The role of protein kinase C in cell surface signal transduction and tumor promotion. Nature 308:693.

41. Schleimer, R.P., E. Gillespie, and L.M. Lichtenstein. 1981. Release of histamine from leukocytes stimulated with the tumor-promoting phorbol diesters. I. Characteristics of the response. J. Immunol. 126:570.

42. Findlay, S.R., A.M. Dvorak, A. Kagey-Sobotka, and L.M. Lichtenstein. 1981. Hyperosmolar triggering of histamine

release from human basophils. J. Clin. Invest. 67:1604.

43. **Eggleston, P.A., A. Kagey–Sobotka, R.P. Schleimer, and L.M. Lichtenstein.** 1984. Interaction between hyperosmolar and IgE mediated histamine release from basophils and mast cells. Am. Rev. Respir. Dis. 130:86.

44. **Peters, S.P., M.I. Siegel, A. Kagey–Sobotka, and L.M. Lichtenstein.** 1981. Lipoxygenase products modulate histamine release in human basophils. Nature 292:455.

45. **Schleimer, R.P., S.P. Peters, D.W. MacGlashan Jr., A. Kagey–Sobotka, and L.M. Lichtenstein.** 1985. Arachidonic acid metabolites cause basophil histamine release. In: Dynamic Processes in Membranes and Cells Versus the Drug Receptor Concept. Copenhagen, Munksgaard, (in press).

46. **Marone, G.** Unpublished observations.

47. **Tharp, M.D., M.C. Ray, M.B. Chaker, and T.J. Sullivan.** 1985. Passive sensitization of human cutaneous mast cells with monoclonal murine IgE. Fed. Proc. 44:586 (abstract).

48. **Lichtenstein, L.M., and R. DeBernardo.** 1971. The immediate allergic response: in vitro action of cyclic AMP–active and other drugs on the two stages of histamine release. J. Immunol. 107:1131.

49. **Peters, S.P., E.S. Schulman, R.P. Schleimer, D.W. MacGlashan Jr., H.H. Newball, and L.M. Lichtenstein.** 1982. Dispersed human lung mast cells: pharmacologic aspects and comparison with human lung tissue fragments. Am. Rev. Respir. Dis. 126:1034.

50. **Lichtenstein, L.M., and R. DeBernardo.** 1971. IgE mediated histamine release: in vitro separation into two phases. Int. Arch. Allergy Appl. Immunol. 41:56.

51. **Kaliner, M., R.P. Orange, and K.F. Austen.** 1972. Immunological release of histamine and slow–reacting substance of anaphylaxis from human lung. IV. Enhancement by cholinergic and alpha adrenergic stimulation. J. Exp. Med. 136:556.

52. **Lichtenstein, L.M.** 1973. The control of IgE–mediated histamine release: implications for the study of asthma. In: Asthma: Physiology, Immunopharmacology and Treatment. Edited by: Austen, K.F., and L.M. Lichtenstein. New York, p.91.

53. **Lichtenstein, L.M., and E. Gillespie.** 1975. The effects of the H1 and H2 antihistamines on "allergic" histamine release and its inhibition by histamine. J. Pharm. Exp. Therap. 192:441.

54. **Lichtenstein, L.M., and E. Gillespie.** 1973. Inhibition of histamine release by histamine controlled by H2 receptor. Nature 244:287.

55. **Ting, S., B. Zweiman, R. Larker, and E.H. Dunsky.** 1981. Histamine suppression of in vivo eosinophil accumulation and histamine release in human allergic reaction. J. Allergy Clin. Immunol. 68:71.

56. **Marone, G., S.R. Findlay, and L.M. Lichtenstein.** 1979. Adenosine receptor on human basophils: modulation of

histamine release. J. Immunol. 123:1473.

57. **Kagey-Sobotka, A., G. Marone, S.P. Peters, and L.M. Lichtenstein.** 1985. Differences between human lung mast cells and basophils. Fed. Proc. 44:1917 (abstract).

58. **MacGlashan, D.W. Jr., R.P. Schleimer, S.P. Peters, E.S. Schulman, G.K. Adams, A.K. Sobotka, H.H. Newball, and L.M. Lichtenstein.** 1983. Comparative studies of human basophils and mast cells. Fed. Proc. 42:2504.

59. **Barrett, E., and D. Metcalfe.** 1984. The mucosal mast cell and its role in gastrointestinal disease. Clin. Rev. Allergy 2:39.

60. **Kusner, E.J., B. Dubnick, and D.J. Heriz.** 1973. The inhibition of disodium cromoglycate in vitro of anaphylactically induced histamine release from rat peritoneal mast cells. J. Pharmacol. Exp. Ther. 184:41.

61. **Pearce, F.L., G. Atkinson, M. Ennis, A. Truneh, P.M. Weston, and J.R. White.** 1979. Effect of anti-allergic drugs on histamine release induced by basic agents, antigen and the calcium ionophore A23187. In: The Mast Cell, Its Role in Health and Disease. Edited by: Pepys, J., and A.M. Edwards. England, p.69.

62. **Bienenstock, J., A.D. Befus, F. Pearce, J. Denburg, and R. Goodacre.** 1982. Mast cell heterogeneity: derivation and function, with emphasis on the intestine. J. Allergy Clin. Immunol. 70:407.

63. **Middleton, E. Jr., G. Drzewiecki, and D. Krishnarao.** 1981. Quercetin: An inhibitor of antigen induced human basophil histamine release. J. Immunol. 127:546.

64. **Pearce, F.L., A.D. Befus, and J. Bienenstock.** 1984. Mucosal mast cells. III. Effects of Quercetin and other flavonoids on antigen-induced histamine secretion from rat intestinal mast cells. J. Allergy Clin. Immunol. 73:819.

65. **Wolf, E.J., C.C. Fox, A. Kagey-Sobotka, T.M. Bayless, and L.M. Lichtenstein.** 1985. Inhibition of histamine release from human mucosal mast cells by drugs used in the treatment of inflammatory bowel disease. Gastroenterology 88:1633 (abstract).

66. **Marone, G., A. Kagey-Sobotka, and L.M. Lichtenstein.** 1979. Effects of arachidonic acid and its metabolites on antigen-induced histamine release from human basophils in vitro. J. Immunol. 123:1669.

67. **Schleimer, R.P., L.M. Lichtenstein, and E. Gillespie.** 1981. Inhibition of basophil histamine release by anti-inflammatory steroids. Nature 292:454.

68. **Schleimer, R.P., D.W. MacGlashan Jr., E. Gillespie, and L.M. Lichtenstein.** 1982. Inhibition of basophil histamine release by anti-inflammatory steroids. II. Studies on the mechanism of action. J. Immunol. 129:1632.

69. **Schleimer, R.P., E.S. Schulman, D.W. MacGlashan Jr., S.P. Peters, G. Kenneth Adams III, L.M. Lichtenstein, and N. Franklin Adkinson.** 1983. Effects of dexamethasone on

mediator release from human lung fragments and purified human lung mast cells. J. Clin. Invest. 71:1830.

70. Naclerio, R.M., H.L. Meier, A. Kagey–Sobotka, N.F. Adkinson Jr., D.A. Meyers, P.S. Norman, and L.M. Lichtenstein. 1983. Mediator release after nasal airway challenge with allergen. Am. Rev. Respir. Dis. 128:597

71. Proud, D., A. Togias, R.M. Naclerio, S.A. Crush, P.S. Norman, and L.M. Lichtenstein. 1983. Kinins are generated in vivo following nasal airway challenge of allergic individuals with allergen. J. Clin. Invest. 72:16778.

72. Naclerio, R.M., A.G. Togias, D. Proud, A. Kagey–Sobotka, N.F. Adkinson Jr., P.S. Norman, and L.M. Lichtenstein. 1985. Inflammatory mediators in late antigen–induced rhinitis. N. Engl. J. Med. 313:65.

73. Togias, A.G., R.M. Naclerio, J. Warner, D. Proud, A. Kagey–Sobotka, I. Nimmagadda, P.S. Norman, and L.M. Lichtenstein. 1985. Demonstration of inhibition of mediator release from mast cells by azatadine base: In vivo and in vitro evaluation. JAMA (in press).

74. Naclerio, R.M., H.L. Meier, A. Kagey–Sobotka, P.S. Norman, and L.M. Lichtenstein. 1984. In vivo model for the evaluation of topical anti–allergic medications. Arch. Otolaryngol. 110:25.

75. Silber, G., R. Naclerio, P. Eggleston, D. Proud, A. Togias, L. Lichtenstein, and P.S. Norman. 1985. In vivo release of histamine by hyperosmolar stimuli. J. Allergy Clin. Immunol. 75:176 (abstract).

76. Togias, A.G., R.M. Naclerio, D. Proud, J.E. Fish, N.F. Adkinson Jr., A. Kagey–Sobotka, P.S. Norman, and L.M. Lichtenstein. 1985. Nasal challenge with cold, dry air results in release of inflammatory mediators: possible mast cell involvement. J. Clin. Invest. (in press).

Mast Cell Differentiation and Heterogeneity,
edited by A. D. Befus et al.
Raven Press, New York © 1986.

Mast Cell Heterogeneity: Biological and Clinical Significance

G. J. Gleich, S. J. Ackerman, Ken-ichi Hisamatsu, and K. M. Leiferman

Departments of Immunology, Medicine, and Dermatology, Mayo Medical School and Mayo Clinic and Foundation, Rochester, Minnesota 55905

Recent information indicates that rodents possess clearly distinguishable populations of mast cells in connective tissue and in the mucosa of the gastrointestinal tract. The distinctive properties of intestinal mast cells were initially described by Enerback (1) and confirmed by numerous investigators (2–4). We refer to the mast cells isolated from connective tissue, lung, and serosal cavities as "connective tissue mast cells" and to mast cells derived from the lamina propria of the gastrointestinal tract as "mucosal mast cells". In this overview we will briefly list the characteristics of mast cells and basophils with the aim of illuminating certain of their biological properties and their roles in disease. Table I summarizes the distinctive properties of basophils (5) and two possible types of mast cells.

Origin of mast cells and basophils. First, basophils originate in the bone marrow, circulate in the blood and migrate into tissues to participate in inflammatory and immune reactions

---

Supported by grants from the National Institutes of Health, AI 11483, AI 09728 and AI 15231, and from the Mayo Foundation.

Dr. Ackerman's present address is: Division of Infectious Diseases, Beth Israel Hospital-DA617, 330 Brookline Avenue, Boston, Massachusetts 02215.

Dr. Hisamatsu's present address is: Department of Otorhinolaryngology, Yamanashi College of Medicine, 1110 Shimogato, Tamaho-cho, Nakakoma-gun, Yamanashi-ken 409-39, Japan.

(6). Second, recent studies in mice suggest that precursors of intestinal MC originate in the bone marrow and migrate to the gastrointestinal tract, where they proliferate and differentiate under the influence of local factors and T cells elaborating interleukin 3 (7). Third, evidence derived from cell transfer experiments in mice suggest that bone marrow cells contain MC precursors (8).

TABLE I. Properties of mast cells and basophils

	"Mucosal mast cell"	"Connective tissue mast cell"	Basophil
Location	Intestinal mucosa, intraepithelial within crypts & at base of villi	Respiratory tract, skin, GI tract, bladder, physiologic exudates, fibrous capsules of organs	Differentiate and mature in bone marrow, circulate in blood
Morphology Size	Small, pleomorphic	Large, uniform (9–20 μm)	Large, round (9–12 μm)
Nucleus	Unilobed or bilobed, heavily condensed chromatin	Unilobed	Two or more lobes, heavily condensed chromatin
Granules	Smaller & few	Larger & many	Largest & many, crystalline content
Lifespan	<40 days	>40 days	12 hrs (in circulation)
Proliferation	Thymus-dependent IL-3	Thymus-independent	Possibly T cell directed

Although basophils and mast cells share many similarities, it seems more likely that basophils and eosinophils share a common precursor cell. Eosinophil-associated proteins such as lysophospholipase [Charcot-Leyden crystal (CLC) protein] (9) and eosinophil granule major basic protein (MBP) (10) are present in basophils as well as eosinophils. Although the quantity of MBP present in the basophil is only about 3% of that in the eosinophil, CLC protein is present in approximately the same amount in eosinophils and basophils and both form Charcot-Leyden crystals (9). These findings indicate biochemical similarities between eosinophils and basophils and suggest that they are related. Studies of eosinophil and basophil colony formation have shown the presence of mixed colonies containing both cell types (11-13). Because the mixed colonies appear to be derived from a single cell, these results suggest that eosinophils and basophils share a common precursor. Of interest is the identification of patients who lack both eosinophils and basophils (14,15). Finally, human lung or skin MC, unlike basophils and eosinophils, contain neither MBP nor CLC protein (16). Purified human lung mast cells (17) and cutaneous mast cells present in lesions of patients with urticaria pigmentosa did not show immunofluorescent staining for MBP or CLC; tissue eosinophils in the preparations showed staining for both. The presence of MBP and CLC protein in basophils and not in these MC suggest that MBP and CLC may be used to differentiate these cell types. Whether MBP and CLC protein are present in intestinal mucosal MC remains to be determined.

TABLE II. Cytochemistry of mast cells and basophils

	"Mucosal mast cell"	"Connective tissue mast cell"	Basophils
Formalin sensitivity	Yes	No	Yes
Proteoglycan content	Lower sulfation, chondroitin sulfate E	Higher sulfation, heparin	Heparin, chondroitin sulfates, dermatan sulfate, heparan sulfate
Protease	Mast cell protease Type II	Mast cell protease, Type I (alpha-chymotrypsin-like activity)	Neutral proteases (granule associated trypsin and chymotrypsin-like enzymes, basophil kallikrein of anaphylaxis
Histamine	<1 pg/cell	>15 pg/cell	1 pg/cell
Lipid mediators	? LT[*]	LT-C_4, PGD$_2$	LT-C_4
IgE	Surface & cytoplasmic receptors	Surface receptors	Surface receptors

[*] LT - Leukotriene

Chemical mediators produced by mast cells and basophils.   The
varieties of chemical mediators associated with MC are   reviewed
in detail elsewhere (18).   The chemical mediators associated with
MC from several sites, in several species, and with basophils are
summarized in Table II.

Several differences exist between MC types.   For example,
granule proteoglycans from rat intestinal mucosal MC possess
fewer sulfate groups than those from the peritoneum (19);   the
type of protease in rodent intestinal mucosal MC differs from
that in peritoneal MC (20); and rodent intestinal mucosal MC
contain considerably less histamine than do peritoneal MC (21).

Distribution of mast cells and basophils.   As noted above,
while basophils differentiate and mature in the bone marrow,
circulate in the blood, and then migrate into tissues, MC are
rarely present in the blood and are usually distributed
throughout connective tissues, situated in perivascular locations. Furthermore, though derived from precursors in the bone
marrow and spleen (22), MC can mature in the tissues (8) rather
than in the bone marrow.   Interestingly, the precursors of
peritoneal MC are present in higher concentrations in the
peritoneal cavity than in the bone marrow, and some of these
precursors have characteristics of mast cells themselves (23).

MC have also been identified in the gastrointestinal tract,
especially in the small intestine, although they are also present
in the stomach and the large intestine (24). In the rat, cells
with the histochemical and biochemical properties of intestinal
mucosal MC have not been found outside the gastrointestinal
tract. Evidence for the existence of such MC in humans consists
of identification of a population of mast cells in the gut which
are not revealed by histological procedures used to demonstrate
MC in other sites or in rat peritoneum.   Two recent studies
evaluating conditions for identification of intestinal mucosal MC
in human tissues concluded that fixation in neutral buffered 10%
formalin did not allow identification (25,26); this formalin
sensitivity had previously been found for rat intestinal lamina
propria MC (27).

Figure 1 shows a comparison of MC counts in normal human
jejunum stained with toluidine blue, but treated with different
fixatives (25) and it illustrates the marked differences in the
numbers of MC revealed by the different fixatives. Indeed, even
isolated rat intestinal lamina propria MC require care in their
fixation and staining (21).   A difficulty posed by the use of
Carnoy's fixative is that the tissues are not suitable for
staining of eosinophils (25,28). However, Wingren and Enerback
have reported staining of rat intestinal lamina propria MC in
formalin-fixed tissues by prolonged exposure to 0.5% toluidine
blue for 5-7 days (28). Eosin can then be used as a counterstain;
the resultant tissues are stained for both MC and eosinophils.

Finally, new methods for staining peritoneal or dermal MC in
rodents and man have been introduced based on the marked avidity
of platelet factor 4 (29) and avidin (30) for heparin. We have

confirmed that conjugated avidin stains dermal mast cells from patients with urticaria pigmentosa and even stains mast cell granules after degranulation (16). These fluorescent methods permit rapid enumeration of MC and their granules in formalin-fixed tissues.

The role of mast cells and basophils in homeostasis. In spite of the numerous studies of mast cells and basophils, their function in homeostasis has been elusive. Nonetheless, the ubiquity of mast cells points to their likely role in normal body processes. Also, a role for basophils in host defense has been identified.

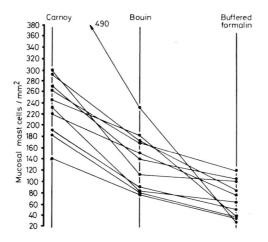

Figure 1. Effects of various fixatives on human intestinal mast cell numbers. MC numbers in tissues fixed in three fixatives including Carnoy's, Bouin's and buffered formalin and stained with toluidine blue. Tissues were operative biopsies of normal human jejunum. The same pattern was seen in all 11 specimens; MC numbers were highest in Carnoy's fixative, were lower in Bouin's fixed preparations, and the same or lower in buffered formalin-fixed preparations. (Used from 25 with permission).

The basophil and resistance to ticks. Evidence suggests that basophils are critical for resistance to ectoparasites. Guinea pigs infected with ticks manifest local inflammatory lesions that are rich in basophils. In these lesions the localization and

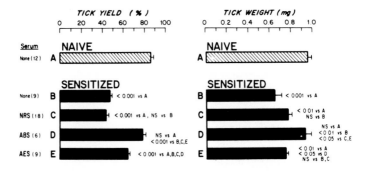

Figure 2. Effects of anti-basophil or anti-eosinophil serum on tick immunity in guinea pigs. Non-sensitized (naive) guinea pigs and guinea pigs sensitized 26 days earlier with a primary infestation of Amblyomma americanum were infested with 100 larval A. americanum ticks placed on the flanks. Sensitized animals were treated with normal rabbit serum, anti-basophil serum, anti-eosinophil serum or no serum. The naive guinea pigs received no serum. Tick yield (number of ticks) and weight of engorged ticks were determined at 90 hours of infestation. Anti-basophil serum (group D) completely ablated resistance; anti-eosinophil serum (group E) partially ablated resistance; normal rabbit serum (group C) had no effect. The total number of animals in each group from 3 separate experiments is shown in parentheses. NS (not significant). (Used from 34 with permission).

degranulation of basophils coincides with tick rejection (31,32). In several species of ticks the degree of resistance correlates with the intensity of the cutaneous basophil response (33). The critical importance of the basophil in tick rejection was demonstrated by administration of specific anti-basophil serum (34). The results of this experiment (Figure 2) indicate that anti-basophil serum totally ablated tick immunity; in the anti-basophil-treated animals both the yield of ticks and their weights were comparable to those in naive animals. Anti-eosinophil serum also reduced tick immunity. In the blood of anti-basophil- and anti-eosinophil-treated guinea pigs the expected effects were observed; the anti-basophil serum caused a disappearance of baso-

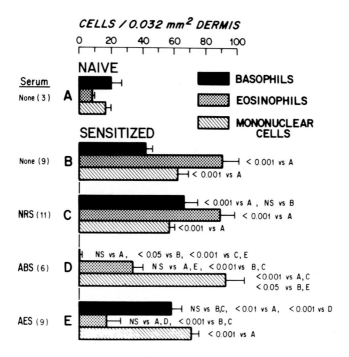

Figure 3. Effects of anti-basophil or anti-eosinophil serum on leukocyte numbers in tick feeding sites. Leukocyte numbers (mean ± SE) were determined in tick feeding sites. Anti-basophil serum guinea pigs (group D) showed no feeding site basophils, reduced eosinophils, and increased mononuclear cell numbers. Anti-eosinophil serum guinea pigs (group E) showed reduced eosinophil numbers but no effect on basophil or mononuclear cell counts. Normal rabbit serum (group C) had no effect. Numbers of neutrophils were not significantly different among the groups. (Used from 34 with permission).

phils (with no effect on other leukocytes, including eosinophils) and the anti-eosinophil serum caused a reduction in eosinophils (with no effect on other cells). However, in the tissues the anti-basophil serum not only abolished the local basophilia, but also caused a marked reduction in the number of eosinophils (compare eosinophils in groups C and D in Figure 3). In contrast, the anti-eosinophil serum while markedly reducing tissue eosinophilia had no effect on the number of basophils. The results of this experiment support a critical role for the basophil in tick immunity and suggest that the basophil expresses this immunity in part through its ability to attract eosinophils to the lesion. In the expression of this reaction, T cells and IgG antibodies are important for the localization of basophils (see 35 and 36 for reviews).

    Mast cells and resistance to helminths. Several lines of evidence point to important roles for mast cells in resistance to helminths. Mast cells adhere to schistosomula of Schistosoma mansoni in the presence of complement or antibody (37,38) and mast cells proliferate in the intestines of rats infected with Nippostrongylus brasiliensis (39). Furthermore, mast cell mediators are able to potentiate eosinophil killing of S. mansoni schistosomula (40, see 41 for review). The results of Dessein et al. show that rats injected with antiserum to IgE do not produce antibodies of the IgE class (42), and when infested with Trichinella spiralis, their mast cells do not release serotonin on exposure to antigens. Furthermore, the anti-IgE-treated rats had fewer tissue eosinophils about the nurse cells (containing T. spiralis larvae), and greater numbers of larvae survived and encysted in muscle. These experiments support a biological role for IgE, mast cells and eosinophils in resistance to helminths.

    The results of IgE suppression studies supporting a role for the MC in resistance to helminths, are opposed by findings in mice deficient in MC and infected with N. brasiliensis; five studies have reported that N. brasiliensis is expelled from such mice in a manner little different from that in normal mice (43-47). Perhaps the best example of these studies is by Crowle who compared the course of N. brasiliensis infection in W/W^v mice, their normal (not MC deficient) littermates and W/W^v mice which were MC-reconstituted by bone marrow or spleen cells (46). Reconstituted animals had numbers of intestinal lamina propria MC comparable to normal littermates, whereas in W/W^v mice intestinal lamina propria MC were totally absent. Following infestation, W/W^v mice had higher peak egg counts than did normal littermates and were slower than littermates to reject the parasite. However, reconstitution with MC did not change the time of parasite rejection or decrease the high peak egg counts even though the numbers of MC in the lamina propria of the small intestine were normal. These results indicate that absence of the MC cannot explain the modest differences between W/W^v mice and normal littermates. Possibly W/W^v mice are deficient in other immunological functions or other cells necessary for normal resistance to N. brasiliensis.

    In another study of the importance of MC in immunity to helminths W/W^v mice were vaccinated with irradiated cercariae of S. mansoni (48). Such vaccinated mice showed resistance to a challenge infection similar to non-deficient littermates; of interest was the finding that SJL mice, which produce IgE antibodies poorly, displayed the same level of resistance to challenge infection as other mice.

One study of the course of T. spiralis infection in W/W^V mice has been reported (49). W/W^V mice, their normal littermates, and MC-reconstituted W/W^V mice were infested with T. spiralis, and the numbers of intestinal lamina propria MC and adult worms were determined. As before (46), W/W^V mice almost totally lacked intestinal lamina propria MC whereas reconstituted animals had normal numbers of these MC. W/W^V mice expelled adult worms more slowly than their normal littermates; MC-reconstituted W/W^V mice were similar to normal littermates. The results suggest that the MC deficiency is the basis for the difference, assuming that bone marrow reconstitution did not repair some as yet unrecognized, defect in W/W^V mice. A possible explanation for the difference between T. spiralis and N. brasiliensis infestation is the observation that N. brasiliensis dwells in the lumen of the gut whereas T. spiralis lies within the cytoplasm of the intestinal mucosal cells (50).

Studies of tissues from humans infected with T. spiralis show a population of intestinal MC not seen in patients infested with Taenia saginata or Giardia lamblia (51) (Table III). These results are illustrative of the studies of intestinal MC in man. In these studies jejunal biopsies were treated with formalin or Carnoy's fixatives and stained with toluidine blue. The numbers of MC in tissues fixed in formalin were increased about two-fold in the biopsies of T. spiralis infected patients; another two-fold increase was seen when the specimens from patients with trichinosis were fixed in Carnoy's. Fixation with Carnoy's permitted identification of small MC not seen after formalin fixation. The finding of a two-fold increase in numbers of MC after fixation in Carnoy's suggests the existence of large MC which are formalin sensitive. They also illustrate certain of the problems in identification of intestinal MC. For example, size in this study is not defined, and the finding of twice as many "large" MC in connective tissue after fixation in Carnoy's is difficult to explain on the basis of current knowledge.

Finally, intraepithelial globule leukocytes (reviewed in 52) may be derived from intestinal MC in sheep, based on cytological and cytochemical similarities (53). On the other hand, intra-epithelial globule leukocytes can develop in nude mice infected with T. spiralis and such nude mice lack intestinal MC (54). The possibility exists that globule leukocytes originate by different mechanisms in different species.

Mast cells and delayed hypersensitivity. Tissue swelling associated with delayed hypersensitivity reactions in mice can be blocked by serotonin antagonists (53). Because the only source of serotonin available for the reaction appeared to be from the MC, it was proposed that MC are needed for the expression of tissue swelling and leukocyte migration in delayed hypersensiti-vity. The later finding that T cells elaborate an antigen-

TABLE III. Effects of fixatives on numbers of mast cells in the mucosa of human jejunal biopsies from patients with trichinellosis, taeniasis and lambliasis[a]

Disease	Mucosal Tissue				Connective Tissue			
	Mast cells[b]		Small mast cells		Mast cells		Small mast cells	
	Formalin	Carnoy's	Formalin	Carnoy's	Formalin	Carnoy's	Formalin	Carnoy's
Trichinellosis n=12[c]	7.9 ± 5.9	14.5 ± 4.9^d	0	2.2 ± 21.3^g	5.8 ± 4.9	12.0 ± 8.7	0	0
Taeniasis n=12[c]	4.6 ± 2.0	4.1 ± 1.4^e	0	0.7 ± 0.9^h	6.1 ± 5.7	6.2 ± 3.6	0	0
Lambliasis n=12[c]	3.0 ± 1.1	3.6 ± 1.5^f	0	0.7 ± 0.3^i	5.3 ± 2.6	7.1 ± 1.4	0	0

a   Modified from (51) with permission

b   Mast cells were enumerated by counting those cells staining metachromatically with toluidine blue (mean $\pm$ SD) in 8 microscopic fields per biopsy using a magnification of 400x. Mast cell counts include both those in connective tissues as well as those at mucosal sites.

c   Of the 12 specimens from 12 patients in each group, 6 were fixed in formalin and 6 were fixed in Carnoy's fixative.

Statistical results: d-e, d-f: p<0.001
                    g-h, g-i: p<0.05

specific factor which sensitizes MC (and is distinct from IgE) provides a mechanism for MC granule secretion in these reactions (56). This concept is reviewed in detail elsewhere (57).

Recently, three groups have tested whether MC deficient W/W^v or Sl/Sl^d mice can express delayed hypersensitivity. Thomas and Schrader found no reduction of contact sensitivity to oxazalone in W/W^v mice (58). Askenase et al. found that W/W^v mice showed a marked reduction in ear swelling 24 hours after challenge with picryl chloride (59). Galli and Hammel found no reduction in contact sensitivity reactions to either picryl chloride or oxazalone in W/W^v or Sl/Sl^d mice (60); in fact, these reactions often exceeded those in littermate controls. Beige mice, whose platelets contain less than 1% of the normal level of serotonin, also showed reactions comparable to littermates. Reserpine blocked the expression of contact hypersensitivity. The potential role of MC in delayed hypersensitivity has recently been reviewed (61). The conclusion is that it is presently not possible to define the role of the MC in delayed hypersensitivity.

Mast cells and fertility. The number of MC in the uterus of hamsters (62) and the ovaries of rats (63) changes during the estrous cycle. In the hamster, the number of mast cells in the myometrium is lowest on the day before ovulation and increased about two-fold on the first and second day after ovulation (62). In pregnant rats, uterine MC, located almost entirely in the myometrium, increase from the time before blastocyst attachment, reach a maximum around the time of blastocyst attachment and then decrease (64). The decrease in MC numbers occurs both at and between implantation sites suggesting that the change is of maternal origin. During early pregnancy in the vole, myometrial MC decrease early and then increase (65). Comparison of the number of MC stained with toluidine blue did not differ when the uterus was fixed with Helly's solution or Carnoy's solution. In the vole, MC are present in equal numbers in the endometrium and myometrium, in contrast to the rat, mouse, guinea pig and hamster, where a large majority of MC are in the myometrium. Injection of histidine decarboxylase into the uterine cavities of the rabbit produces a marked reduction in blastocyst implantation and the viability of the implanted embryos (66). This suggests that rabbit blastocysts have histamine-forming capacity and that blocking this capacity interferes with implantation and embryo development.

Mast cells and tumor immunity. MC and purified MC granules supplemented by H_2O_2 and iodide are cytotoxic to mammalian tumor cells (67). Binding eosinophil peroxidase to mast cell granules markedly increases their ability to damage tumor cells. Another observation supporting a role for MC in tumor immunity comes from methylcholanthrene treated W/W^v mice (68); such mice show an increased tumor incidence. This increase is nullified when they are MC-reconstituted by bone marrow cells. Because W/W^v mice are not deficient in natural killer cell and T cell-mediated cytotoxicity, these results are interpreted as evidence that MC are involved in tumor suppression.

TABLE IV. Evidence suggestive of mast cell involvement in disease

Histologic evidence
    Changes in mast cell numbers in tissues
    Demonstration of mast cell degranulation
    Presence of cells possibly recruited by mast cell mediators
    (e.g. eosinophils, neutrophils)
    Endothelial activation and disconnections

Biochemical evidence
    Elevated levels of mediators in blood or secretions
    Identification of mediators in tissue specimens
    Specific IgE or antigen in blood or tissue

Clinical evidence
    Signs and symptoms reproducible by MC mediators
        Wheal-and-flare skin reaction
        Pruritus
        Bronchospasm
        Rhinorrhea
        Hypotension
        Diarrhea

    Pharmacologic agents that inhibit mast cell mediator release
    or mediator effects reduce symptoms
        Antihistamines
        Theophylline
        Sodium cromolyn
        Beta-adrenergic agonists (?)

*   Modified from (72) with permission

Mast cells and angiogenesis. Studies of angiogenesis in tumors showed that mast cells accumulate at tumor sites before the ingrowth of new capillaries (69). Because of the possibility that MC promote capillary endothelial cell migration, the effect of MC-conditioned media on endothelial cell migration (70) was tested; MC-conditioned medium was prepared by incubation of rat peritoneal MC with medium for 36 hours. The conditioned medium stimulated endothelial cell migration, and subsequent experiments showed that the active factor was heparin; protamine inhibited the effect of heparin on cell migration. In later experiments, protamine inhibited capillary proliferation in several model systems including embryogenesis, tumor angiogenesis, inflammatory angiogenesis, and immune angiogenesis (71).

TABLE V. Pharmacological responsiveness of rat intestinal
lamina propria and peritoneal MC[a]

	Mast Cell	
Agent	Intestinal Lamina Propria	Peritoneal
Secretagogues		
Compound 48/80	0	++
Bee venom 401	0	++
A23187	+	++
Anti-allergic drugs		
Cromoglycate	0	++
Theophylline	0	++
Doxantrazole	++	++
AH 9679	0	++
Secretion enhancing compounds		
Phosphatidyl serine	0	++
Adenosine	++	++
Flavonoids[b]		
Quercetin	++	++
Phloretin	+	++
Neuropeptides[c]		
Somatostatin	0	++
VIP	0	++
Substance P	+	++
Bradykinin	0	++
Neurotensin	0	++

a. Modified from 3 with permission
b. See 74
c. See 75

Implications of Mast Cell Heterogeneity for Disease. Table IV
lists evidence suggestive of MC participation in disease (72).
Certain clinical features such as pruritus, bronchospasm,
rhinorrhea and diarrhea suggest MC involvement. More convincing

is the occurrence of these symptoms along with evidence of specific IgE antibody, either measured by immunoassay or by the occurrence of a positive wheal–and–flare skin reaction. When elevated levels of mediators, such as histamine, are found in blood or secretions, the case for MC or basophil participation becomes even stronger.

Among the pharmacologic criteria listed in Table IV, as suggestive of MC involvement, is a reduction of symptoms after treatment with cromolyn sodium. Because rat intestinal lamina propria MC, unlike peritoneal MC, are not inhibited by cromolyn sodium (73), failure to inhibit mediator secretion by this agent in the presence of other evidence for MC activation could point to the involvement of a specific type of MC. However, it seems premature to make this generalization until studies of isolated cell preparations from particular target organs in humans have been performed. It is of interest that rat intestinal lamina propria MC also differ from peritoneal MC in their responsiveness to several other agents (73); this information is summarized in Table V.

Concerning diseases associated with MC, the literature is replete with lists of these diseases and discussions of their pathophysiology (see 72, 76–80 recent reviews).

Finally, it seems premature to associate a specific type of intestinal or related MC with disease. Indeed, at this point there is a need for studies of the distribution of MC analogous to rat intestinal mucosal mast cells in organs, other than the gut, especially in the lung and the skin during disease to determine whether or not similar MC are present. Unfortunately, at this writing the only criterion that can be readily employed for this identification is the difference in staining of MC after fixation with neutral formalin and other fixatives, such as Carnoy's medium, which permits staining of rat or human intestinal lamina propria MC. Studies of MC staining in the human gastrointestinal tract (Figure 1) show marked differences between the numbers of MC stained with toluidine blue after fixation with neutral formalin and Carnoy's medium (25). Granted the existence of a special type of MC in the gastrointestinal tract, its participation in inflammatory diseases of the bowel seems likely. The recent findings that MC are increased in the jejunum of patients with trichinosis (51) and with untreated coeliac disease (81) strengthen this suggestion.

REFERENCES

1. **Enerback, L.** 1981. The gut mucosal mast cell. Monogr. Allergy 17:222.
2. **Bienenstock, J., A.D. Befus, F. Pearce, J. Denburg, and R. Goodacre.** 1982. Mast cell heterogeneity: derivation and function, with emphasis on the intestine. J. Allergy Clin. Immunol. 70:407.
3. **Shanahan, F., J.A. Denburg, J. Bienenstock, and A.D. Befus.**

1984. Mast cell heterogeneity. Can. J. Physiol. Pharmacol. 62:734.
4. Barrett, K.E., and D.D. Metcalfe. 1984. Mast cell heterogeneity: evidence and implications. J. Clin. Immunol. 4:253.
5. Shulman, E.S., D.W. MacGlashan, R.P. Schleimer, S.P. Peters, A. Kagey-Sobotka, H.H. Newball, and L.M. Lichtenstein. 1983. Purified human basophils and mast cells: current concepts of mediator release. Eur. J. Respir. Dis. 64(128):53.
6. Galli, S.J., A.M. Dvorak, and H.F. Dvorak. 1984. Basophils and mast cells: morphologic insights into their biology, secretory patterns and functions. Prog. Allergy 34:1.
7. Guy-Grand, D., M. Dy, G. Luffau, and P. Vassalli. 1984. Gut mucosal mast cells: origin, traffic and differentiation. J. Exp. Med. 160:12.
8. Kitamura, Y., H. Matsuda, and K. Hatanaka. 1979. Clonal nature of mast cell clusters formed in W/W^v mice after bone marrow transplantation. Nature 281:154.
9. Ackerman, S.J., G.J. Weil, and G.J. Gleich. 1982. Formation of Charcot-Leyden crystals by human basophils. J. Exp. Med. 155:1597.
10. Ackerman, S.J., G.M. Kephart, T.M. Habermann, P.R. Greipp, and G.J. Gleich. 1983. Localization of eosinophil granule major basic protein in human basophils. J. Exp. Med. 158:946.
11. Denburg, J.A., S. Telizyn, M. Richardson, and J. Bienenstock. 1983. Basophil/mast cell precursors in human peripheral blood. Blood 61:775.
12. Leary, A.G., and M. Ogawa. 1984. Identification of pure and mixed basophil colonies in culture of human peripheral blood and marrow cells. Blood 64:78.
13. Denburg, J.A., S. Telizyn, H. Messner, B. Lim, N. Jamal, S.J. Ackerman, G.J. Gleich, and J. Bienenstock. 1985. Heterogeneity of human peripheral blood eosinophil-type colonies: evidence for a common basophil-eosinophil progenitor. Blood 66:312.
14. Juhlin, L., and G. Michaelson. 1977. A new syndrome characterized by absence of eosinophils and basophils. Lancet 1:1233.
15. Mitchell, E.B., T.A.E. Platts-Mills, R.S. Pereira, V. Malkovska, and A.D. Webster. 1983. Acquired basophil and eosinophil deficiency in a patient with hypogammaglobulinaemia associated with thymoma. Clin. Lab. Haemat. 5:253.
16. Leiferman, K.M., G.J. Gleich, H.S. Haugen, G.M. Kephart, D. Proud, L. Lichtenstein, and S.J. Ackerman. 1985. Failure to detect Charcot-Leydon crystal protein and eosinophil granule major basic protein in human mast cells. Fed. Proc. 44:785 (Abstract).
17. Shulman, E.S., A. Kagey-Sobotka, D.W. MacGlashan Jr., N.F. Adkinson Jr., S.P. Peters, R.P. Schleimer, and L.M. Lichtenstein. 1983. Heterogeneity of human mast cells. J.

Immunol. 131:1936.
18. Schwartz, L.B., and K.F. Austen. 1984. Structure and function of the chemical mediators of mast cells. Prog. Allergy 34:271.
19. Tas, J.A., and R.G. Berndsen. 1977. Does heparin occur in mucosal mast cells within the rat small intestine? J. Histochem. Cytochem. 25:1058.
20. Woodbury, R.G., G.M. Gruzenski, and D. Lagunoff. 1978. Immunofluorescent localization of a serine protease in rat small intestine. Proc. Natl. Acad. Sci. USA 75:2785.
21. Befus, A.D., F.L. Pearce, J. Gauldie, P. Horsewood, and J. Bienenstock. 1982. Mucosal mast cells. I. Isolation and functional characteristics of rat intestinal mast cells. J. Immunol. 128:2475.
22. Kitamura, Y., M. Shimada, S. Go, H. Matsuda, K. Hatanaka, and M. Seki. 1979. Distribution of mast-cell precursors in hematopoietic and lymphopoietic tissues of mice. J. Exp. Med. 150:482.
23. Sonoda, T., Y. Kanayama, H. Hara, C. Hayashi, M. Tadokara, T. Yonezawa, and Y. Kitamura. 1984. Proliferation of peritoneal mast cells in the skin of W/W^v mice that genetically lack mast cells. J. Exp. Med. 160:138.
24. Saavedra-Delgado, A.M.P., S. Turpin, and D.D. Metcalfe. 1984. Typical and atypical mast cells of the rat gastrointestinal system: distribution and correlation with tissue histamine. Agents Actions 14:1.
25. Strobel, S., H.R.P. Miller, and A. Ferguson. 1981. Human intestinal mucosal mast cells: evaluation of fixation and staining techniques. J. Clin. Pathol. 34:851.
26. Ruitenberg, E.J., L. Gustowska, A. Elgersma, and H.M. Ruitenberg. 1982. Effect of fixation on the light microscopical visualization of mast cells in the mucosa and connective tissue of the human duodenum. Int. Arch. Allgery Appl. Immunol. 67:233.
27. Enerback, L. 1966. Mast cells in rat gastrointestinal mucosa. I. Effects of fixation. Acta. Pathol. Microbiol. Scand. 66:303.
28. Wingren, U., and L. Enerback. 1983. Mucosal mast cells of the rat intestine: a re-evaluation of fixation and staining properties, with special reference to protein blocking and solubility of the granular glycosaminoglycan. Histochem. J. 15:571.
29. Shaw, S.T. Jr., P.C. Roche, G. Schoner, and L.K. MaCauley. 1982. Immunohistochemical identification of mast cells in paraffin- and epon-embedded tissues using platelet factor 4. J. Histochem. Cytochem. 30;185.
30. Bergstresser, P.R., R.E. Tigelaar, and M.D. Thorp. 1984. Conjugated avidin identified cutaneous rodent and human mast cells. J. Invest. Derm. 83:214.
31. Allen, J.R. 1973. Tick resistance: basophils in skin reactions of resistant guinea pigs. Int. J. Parasitol.

3:195.
32. **Brown, S.J., and F.W. Knapp.** 1981. Response of hypersensitized guinea pigs to the feeding of Amblyomma americanum ticks. Parasitology 83:213.
33. **Brown, S.J., and P.W. Askenase.** 1981. Cutaneous basophil responses and immune resistance of guinea pigs to ticks. Passive transfer with peritoneal exudate cells or serum. J. Immunol. 127:2163.
34. **Brown, S.J., S.J. Galli, G.J. Gleich, and P.W. Askenase.** 1982. Ablation of immunity to Amblyomma americanum by antibasophil serum. Cooperation between basophils and eosinophils in expression of immunity to ectoparasites (ticks) in guinea pigs. J. Immunol. 129:790.
35. **Askenase, P.W.** 1977. The role of basophils, mast cells and vasoamines in hypersensitivity reactions with the delayed time course. Prog. Allergy 23:199.
36. **Askenase, P.W.** 1980. Immunopathology of parasitic disease: involvement of basophils and mast cells. Semin. Immunopathol. 2:417.
37. **Caulfield, J.R., A. Hein, G. Moser, and A. Sher.** 1981. Light and electron microscopic appearance of rat peritoneal mast cells adhering to schistosomula of Schistosoma mansoni by means of complement or antibody. J. Parasitol. 67:776.
38. **MacKenzie, C.D., M. Jungery, P.M. Taylor, and B.M. Ogilvie.** 1981. The in-vitro interaction of eosinophils, neutrophils, macrophages and mast cells with nematode surfaces in the presence of complement or antibodies. J. Pathol. 133:161.
39. **Haig, D.M., E.E.E. Jarrett, and J. Tas.** 1984. In vitro studies on mast cell proliferation in N. brasiliensis infection. Immunology 51:643.
40. **Capron, M., J. Rousseaux, C. Mazingue, H. Bazin, and A. Capron.** 1978. Rat mast cell-eosinophil interaction in antibody-dependent eosinophil cytotoxicity to Schistosoma mansoni schistosomula. J. Immunol. 121:2518.
41. **Gleich, G.J., and D.A. Loegering.** 1984. Immunobiology of eosinophils. Ann. Rev. Immunol. 2:429.
42. **Dessein, A.J., W.L. Parker, S.L. James, and J.R. David.** 1981. IgE antibody and resistance to infection. I. Selective suppression of the IgE antibody response in rats diminishes the resistance and the eosinophil response to Trichinella spiralis infection. J. Exp. Med. 153:423.
43. **Uber, C.L., R.L. Roth, and D.A. Levy.** 1985. Expulsion of Nippostrongylus brasiliensis by mice deficient in mast cells. Nature 287:226.
44. **Kojima, S., Y. Kitamura, and K. Takatsu.** 1980. Prolonged infection of Nippostrongylus brasiliensis in genetically mast-cell depleted W/W^v mice. Immunol. Lett. 2:159.
45. **Crowle, P.K., and N.D. Reed.** 1981. Rejection of the intestinal parasite Nippostrongylus brasiliensis by mast-cell deficient W/W^v anemic mice. Infect. Immun. 33:54.
46. **Crowle, P.K.** 1982. Mucosal mast cell reconstitution and

Nippostrongylus brasiliensis rejection by W/W^v mice. J. Parasitol. 69:66.

47. **Mitchell, L.A., R.B. Wescott, and L.E. Perryman.** 1983. Kinetics of expulsion of the nematode, *Nippostrongylus brasiliensis*, in mast cell deficient W/W^v mice. Parasite Immunol. 5:1.

48. **Sher, A., O.R. Correa, R. Hieny, and R. Hussain.** 1983. Mechanisms of *Schistosoma mansoni* infection in mice vaccinated with irradiated cercariae. IV. Analysis of the role of IgE antibodies and mast cells. J. Immunol. 131:1460.

49. **Ha, T.-Y., N.D. Reed, and P.K. Crowle.** 1983. Delayed expulsion of adult *Trichinella spiralis* by mast cell-deficient W/W^v mice. Infect. Immun. 41:445.

50. **Wright, K.A.** 1979. *Trichinella spiralis*: an intracellular parasite in the intestinal phase. J. Parasitol. 65:441.

51. **Gustowska, L., E.J. Ruitenberg, A. Elgersma, and W. Kociecka.** 1983. Increase of mucosal mast cells in the jejunum of patients infected with *Trichinella spiralis*. Int. Arch. Allergy Appl. Immunol. 71:304.

52. **Gregory, M.W.** 1979. The globule leukocyte and parasitic infection – a brief history. Vet. Bull. 49:821.

53. **Huntley, J.F., G. Newlands, and H.R.P. Miller.** 1984. The isolation and characterization of globule leukocytes: their derivation from mucosal mast cells in parasitized sheep. Parasite Immunol. 6:371.

54. **Ruitenberg, E.J., and A. Elgersma.** 1979. Response of intestinal globule leukocytes in the mouse during a *Trichinella spiralis* infection and its independence of intestinal mast cells. Br. J. Exp. Path. 60:246.

55. **Gershon, R.K., P.W. Askenase, and M.D. Gershon.** 1975. Requirement for vasoactive amines for the production of delayed-type hypersensitivity skin reactions. J. Exp. Med. 142:732.

56. **Askenase, P.W., R.W. Rosenstein, and W. Ptak.** 1983. T-cells produce an antigen binding factor with in vivo activity analogous to IgE antibody. J. Exp. Med. 157:862.

57. **Askenase, P.W., and H. Van Loveren.** 1983. Delayed-type hypersensitivity: activation of mast cells by antigen-specific T-cell fators initiates the cascade of cellular interactions. Immunol. Today 4:259.

58. **Thomas, W.R., and J.W. Schrader.** 1983. Delayed hypersensitivity in mast-cell deficient mice. J. Immunol. 130:2565.

59. **Askenase, P.W., H. Van Loveren, S. Kraeuter-Kops, R. Yacov, R. Meade, T.C. Theoharides, J.J. Nordlund, H. Scovern, M.D. Gershon, and W. Ptak.** 1983. Defective elicitation of delayed-type hypersensitivity in W/W^v and Sl/Sl^d mast cell-deficient mice. J. Immunol. 131:2687.

60. **Galli, S.J., and I. Hammel.** 1984. Unequivocal delayed hypersensitivity in mast cell-deficient (W/W^v and Sl/Sl^d)

and beige mice. Science 226:710.

61. **Galli, S.J., and A.M. Dvorak.** 1984. What do mast cells have to do with delayed hypersensitivity? Lab. Invest. 50:365.

62. **Brandon, J.M., and J.E. Evans.** 1983. Changes in uterine mast cells during the estrous cycle in Syrian hamster. Am. J. Anat. 167:241.

63. **Jones, R.E., D. Duvall, and L.J. Guillette Jr.** 1980. Rat ovarian mast cells: distribution and cyclic changes. Anat. Record 197:489.

64. **Brandon, J.M. and M.C. Bibby.** 1979. A study of changes in uterine mast cells during early pregnancy in the rat. Biol. Reprod. 20:977.

65. **Brandon, J.M., and J.E. Evans.** 1984. Observations on uterine mast cells during early pregnancy in the vole Microtus agrestis. Anat. Record 208:515.

66. **Dey, S.K.** 1981. Role of histamine in implantation: inhibition of histidine decarboxylase induces delayed implantation in the rabbit. Biol. Reprod. 24:867.

67. **Henderson, W.R., E.Y. Chi, E.C. Jong, and S.J. Klebanoff.** 1981. Mast cell-mediated tumor-cell cytotoxicity. J. Exp. Med. 153:520.

68. **Tanoka, H., Y. Kitamura, T. Sato, K. Tanaka, M. Nagase, and S. Kondo.** 1982. Evidence for involvement of mast cells in tumor suppression in mice. J. Natl. Cancer Inst. 69:1305.

69. **Kessler, D.A., R.S. Langer, N.A. Pless, and J. Folkman.** 1976. Mast cells and tumor angiogenesis. Int. J. Cancer 18:703.

70. **Azizkhan, R.K., J.C. Azizkhan, B.R. Zetter, and J. Folkman.** 1980. Mast cell heparin stimulates migration of capillary endothelial cells in vitro. J. Exp. Med. 152:931.

71. **Taylor, S., and J. Folkman.** 1982. Protamine is an inhibitor of angiogenesis. Nature 297:307.

72. **Marquardt, D.L., and S.I. Wasserman.** 1982. Mast cells in allergic diseases and mastocytosis. West. J. Med. 137:195.

73. **Pearce, F.L., A.D. Befus, J. Gauldie, and J. Bienenstock.** 1982. Mucosal mast cells. II. Effects of anti-allergic compounds on histamine secretion by isolated intestinal mast cells. J. Immunol. 128:2481.

74. **Pearce, F.L., A.D. Befus, and J. Bienenstock.** 1984. Mucosal mast cells. II. Effect of quercetin and other flavonoids on antigen-induced histamine secretion from rat intestinal mast cells. J. Allergy Clin. Immunol. 73:819.

75. **Shanahan, F., A.D. Befus, J.A. Denburg, and J. Bienenstock.** 1983. Influence of enteric neuropeptides on intestinal mast cells. Fed. Proc. 42:1343.

76. **Saavedra-Delgado, A.M., and D.D. Metcalfe.** The gastro-intestinal mast cell in food allergy. Ann. Allergy 51:185.

77. **Lemanske, R.F., F.M. Atkins, and D.D. Metcalfe.** 1983. Gastrointestinal mast cells in health and disease. Part I and II. J. Ped. 103:177.

78. **Soter, N.A.** 1983. Mast cells in cutaneous inflammatory

disorders. J. Invest. Derm. 80:22s.

79. **Wasserman, S.I.** 1984. The mast cell and synovial inflammation. Arth. Rheum. 27:841.

80. **Pepys, J., and A.M. Edwards.** 1979. The Mast Cell: Its Role in Health and Disease. Pitman Medical Publishing Co. Ltd., Kent, England.

81. **Strobel, S., A. Busuttil, and A. Ferguson.** 1983. Human intestinal mast cells: expanded population in untreated coeliac disease. Gut 24:222.

Mast Cell Differentiation and Heterogeneity,
edited by A. D. Befus et al.
Raven Press, New York © 1986.

Electron Microscopic Comparison of Human Nasal and Lung Mast Cell Degranulation

*Marc M. Freidman, **Dean D. Metcalfe, and † Michael Kaliner

*Molecular Virology and Immunology, Georgetown University Medical Center, Rockville, Maryland 20852; and **Mast Cell Physiology and †Allergic Diseases Sections, Laboratory of Clinical Investigation, National Institute of Allergy and Infectious Diseases, National Institutes of Health, Bethesda, Maryland 20205*

Mast cells have been shown to play a fundamental role in the generation of immediate hypersensitivity reactions, and in the pathogenesis of allergic diseases. A growing body of evidence indicates that mast cells of different animals, and mast cells residing in different tissues of the same animal, may be distinguished on the basis of morphologic, cytochemical and functional characteristics (1-4). Such heterogeneity may be due to random events such as a stimulus reaching one cell but not another, to differences in the cells' developmental stages, or to normal variability in their response to a common stimulus. Alternatively, morphologic heterogeneity may result from the presence of distinct populations of mast cells within a tissue. To date there are data from only a few studies to suggest that mast cell

---

This research was supported by contract No. 1-AI-22665 from the National Institute of Allergy and Infectious Diseases, NIH.

Acknowledgements: We gratefully acknowledge the assistance of Dr. Gordon D. Raphael for performing the lung incubations; Mr. Thomas L. Brown for preparing specimens for electron microscopy, and for extensive darkroom work; Ms. Cyndi French for secretarial services. Figures 1 and 2a are reprinted by permission of C.V. Mosby Co., from Friedman and Kaliner, J. Allergy Clin. Immunol. 76:70-82, 1985.

heterogeneity, reported in the rat, also exists in man (1,4). In the present study we examined mast cells in the human nasal mucosa and in peripheral lung by light and electron microscopy (LM and EM), for evidence of heterogeneity based on morphologic criteria.

MATERIALS AND METHODS

Human inferior nasal turbinates were obtained from non-atopic patients following turbinectomy for obstructive nasal disease. The mucosae were cut into narrow strips and incubated in rabbit anti-human IgE (1:25), or in ragweed-allergic serum for 2 hr followed by ragweed antigen E (0.5 μg/ml), for 5 or 30 min, or for 1, 2 or 4 hr. Grossly normal human peripheral lung obtained at surgery, generally for cancer resection, was cut into 5 mm cubes. Preincubation in ragweed-allergic serum was followed by incubation either in ragweed antigen E (0.5 μg/ml) or anti-IgE (1:150) for 5 or 30 min. Histamine content of supernatants and tissues was analyzed using a radioenzyme assay (5,6). Specimens were fixed for EM in Karnovsky's fixative (7) diluted with 0.1 M cacodylate buffer. Nasal turbinate mucosae were fixed in a 1:1 fixative:buffer dilution at 4°C for 2 hr, and post-fixed sequentially in osmium tetroxide, thiocarbohydrazide, and again in osmium tetroxide (8). Lung specimens were fixed in a 1:3 fixative:buffer dilution for 30 min at 10°C, and post-fixed in osmium tetroxide alone, or in ferrocyanide-reduced osmium (9). Specimens were dehydrated in acetone and embedded in epoxy resin using routine procedures.

RESULTS

Light Microscopy. Neighboring, unstimulated mast cells in the human nasal mucosa or in lung interalveolar septa may differ from each other not only in the size and shape of their secretory granules, but also in their degree of association with other connective tissue cells. The nasal mast cell at upper-left in Figure 1 is flanked by two mononuclear cells and forms the centre of a cell "islet". The mast cell contains large, polymorphic, densely stained granules similar to those in Figure 2a. The solitary mast cell at lower-right contains smaller granules that appear less prominent. In colour (but not black-and-white) micrographs, lipid droplets are stained a pale green and may therefore be differentiated from the blue or metachromatically-stained secretory granules.

Electron Microscopy: Unstimulated Mast Cells. Unstimulated nasal mast cells typically possessed a round or oval nucleus, lipid droplets, a normal complement of cytoplasmic organelles and filaments, and numerous, dense-staining secretory granules distributed throughout the cytoplasm. In many nasal mast cells

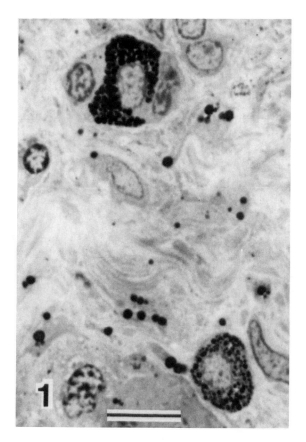

Figure 1. A light micrograph of two nearby mast cells in the propria–submucosa of an inferior nasal turbinate (toluidine blue stain). The cell at upper-left is flanked by two non-granulated mononuclear cells and contains large, polymorphic secretory granules. The cell at lower-right is solitary and contains smaller, less prominent secretory granules. Scale marker = 10 μm.

fixed within 30 min after surgery, the secretory granules contained only a dense-staining, amorphous matrix material that appeared homogeneous even at high EM magnification (Fig. 2a). In other nasal mast cells and in most lung mast cells, the granules contained crystalline "scroll" patterns embedded in an amorphous matrix material (Fig. 2b,c). The matrix material of some granules was comparatively lightly stained, revealing the scroll patterns more clearly and suggesting that the granules might have lost some of their original contents. The scroll patterns,

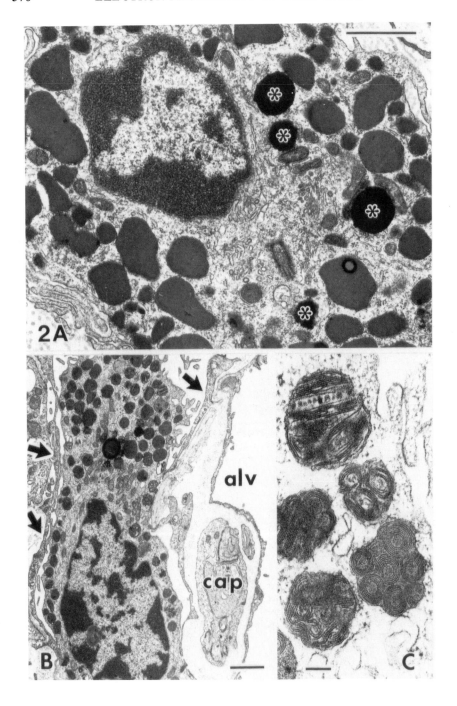

examined in detail by Caulfield et al. (10) appear as whorls when
cut in cross section, and as multi-walled rectangles in longitudinal section.

In one or two regions along the cell surface, the plasma membrane exhibited extensive folding or ruffling and the folds often
interdigitated to form stacks (Figs 2a;3a,b). Within stacks, the
spaces between surface folds often contained collagen fibers and
were therefore continuous with the external, connective tissue
environment. Surface folds on nasal mast cells, measured using
an electronic digitizer, increased the cell surface area, on
average, by almost threefold (8).

Nasal mast cells in the lamina propria-submucosa occurred as
solitary cells, or in multi-cell islets in which they were
closely associated with mononuclear cells. Solitary and islet
mast cells were partly encircled by processes of distinctive,
fibroblast-like stromal cells that appeared to form a webbing
around mast cells and mononuclear cells (Figs 3a,b) (8).

Lung mast cells in the interalveolar septa were solitary, and
rarely formed islets. Attenuated cell processes of fibroblast-like cells, resembling those of stromal cells in the nasal
mucosa, were associated with lung mast cells (Fig. 2b, arrows).

Degranulating Mast Cells. After 5-30 min of stimulation with
ragweed antigen E or with rabbit anti-human IgE, secretory granules in virtually all mast cells exhibited morphologic changes
concomitant with release of up to 29% of total tissue histamine.
The nucleus was often highly indented, or lobulated. Invaginations of the plasma membrane penetrated several micrometers into
the cytoplasm, often to the Golgi zone adjacent to the nucleus
(8). Broader invaginations appeared as large "vesicles" in the
cytoplasm and contained villous membrane projections (Fig. 4a).
Other, narrow invaginations were seen as tortuous channels within
the cytoplasm and were similarly lined with membrane projections
(Fig. 3a). The contents of secretory granules appeared to evolve
in stages from a dense, amorphous material (Fig. 2a), to crystal-

---

Figure 2a. A portion of an unstimulated nasal mast cell,
showing the homogeneous-appearing secretory granules, and
specifically stained lipid droplets (*). Note the interdigitating surface folds at lower-left. Scale marker = 1 μm

Figure 2b. Crystalline scroll patterns are visible in some
secretory granules of an unstimulated lung mast cell in the
interalveolar septum. A capillary (cap) lies to the right of the
mast cell and adjacent to the alveolar air space (alv). Thin
cytoplasmic processes (arrows) of a fibroblast-like cell partly
encircle the mast cell. Scale marker = 1 μm.

Figure 2c. Secretory granules of a lung mast cell, 2 min
after stimulation with anti-IgE. Crystalline, scroll-like
constituents of secretory granules are cut in longitudinal
section (upper left) and in cross-section (lower right). Scale
marker = 0.1 μm.

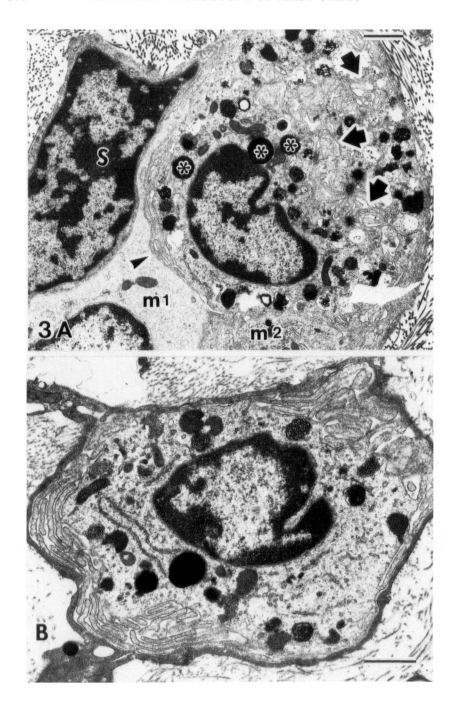

line scroll patterns (Fig. 2c), then to a rope-like material which stained intensely (Figs. 3a,4b) and finally, to a less dense particulate material (Fig. 4a).

After 30-60 min of stimulation, the secretory granules of many nasal and lung mast cells exhibited a much lower staining intensity that was discernible by LM. By EM, some granules contained the intensely stained, rope-like material, while most contained only its apparent remnants – a particulate, almost flocculent material. The amount of stainable material within granules was highly variable.

Under the comparatively gentle conditions of in situ stimulation employed in these studies, only rare nasal mast cells and no lung mast cells exhibited anaphylactic degranulation, as defined by the release of stainable granule contents into the external environment. In such cells, only a portion of the secretory granules were solubilized and undergoing exocytosis (8).

Beginning 30 min after stimulation, mast cells were detected by EM that appeared to be non-granulated, mononuclear cells. The cells possessed a round or oval nucleus, prominent lipid droplets, and often exhibited broad cytoplasmic extensions suggestive of motile cells. Only at high EM magnification were they recognized as degranulated mast cells, due to the presence of small vesicles containing granule remnants (Fig. 4c, arrows). Degranulated mast cells increased in number up to 4 hr after stimulation, the longest incubation employed. Their cytoplasm contained a prominent Golgi apparatus and abundant endoplasmic reticulum, suggesting that these cells may be capable of regranulation.

DISCUSSION

Nasal and lung mast cells gently stimulated in situ exhibited similar patterns of degranulation. The secretory granules of resting mast cells generally comprised a homogeneous, amorphous material often containing embedded, crystalline scroll patterns. Shortly after stimulation, most granules contained prominent

---

Figure 3a. A nasal mast cell forms part of a cell islet in the propria-submucosa. Most secretory granules contain a dense, rope-like material. Adjacent to the nucleus are three lipid droplets (*). A narrow, tortuous system of cytoplasmic channels (arrows) is present to the right of the nucleus and apparently extends to the cell surface. Extensive cell-surface folding is visible (arrowhead). Portions of two mononuclear cells (m1,m2) and a stromal cell (S) are visible. Anti-IgE, 5 min. Scale marker = 1 μm.

Figure 3b. A largely degranulated nasal mast cell is encircled by stromal cell processes. A few secretory granules are present in the cytoplasm, as are lipid droplets. Interdigitating cell-surface folds are prominent. Anti-IgE, 60 min. Scale marker= 1μm.

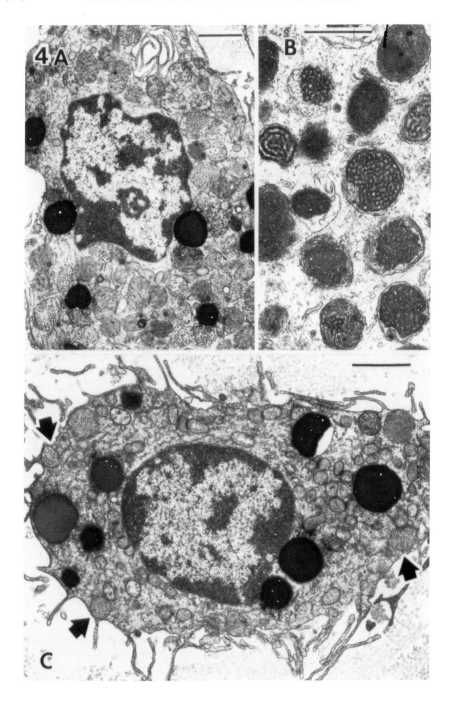

scroll patterns. The amorphous and crystalline granule constituents appeared to evolve to a densely staining, rope-like material and eventually to a less dense, particulate material. The latter material was eventually relegated to small cytoplasmic vacuoles, representing the remnants of secretory granules in degranulated mast cells.

Scroll-like, crystalline material was far more common in the secretory granules of unstimulated lung mast cells than in nasal mast cells. Rather than reflecting a difference between mast cells in these tissues, the increased presence of scroll patterns in "unstimulated" lung mast cell granules may be due to the substantially longer handling time for lung tissue in our (and previous) experiments. We obtained and fixed some of the nasal turbinates within 30 min after their removal, while lung specimens were not received for 2-4 hr. Similarly, the enzymatic digestion and purification procedures employed by previous investigators to isolate lung mast cells (10) may take 1-2 days. We therefore suspect that the greater handling time experienced by lung specimens may have resulted in mast cell stimulation, and the appearance of scroll patterns. Conceivably, variability in the amount of crystalline material present in secretory granules may result from differences between never-before-stimulated mast cells and regranulating mast cells. At present, these mast cells cannot be differentiated.

While the overall process of degranulation appeared to be nearly identical in nasal and lung mast cells, differences in other characteristics were noted. Nasal mast cells commonly formed multi-cell islets with mononuclear cells, were most abundant in the connective tissue below the thick basal lamina, and possessed regions of extensive plasma membrane folding that increased the effective cell surface area by almost three-fold (8). In contrast, lung mast cells seldom formed islets, were

---

Figure 4a. Most secretory granules of this solitary lung mast cell contain either rope-like or particulate material, and some show reduced staining of the background matrix. Remnants of scroll patterns are present in some granules. Lipid droplets are stained black. A broad invagination of the cell surface, with internal, villous projections, is present at upper-right. Anti-IgE, 30 min. Scale marker = 1μm.

Figure 4b. Secretory granules of a degranulating nasal mast cell contain a densely staining, rope-like material and remnants of the crystalline scroll patterns. Anti-IgE, 60 min. Scale marker = 0.5 μm.

Figure 4c. An almost completely degranulated lung mast cell contains the remnants of secretory granules, relegated to small cytoplasmic vacuoles (arrows). Antigen E, 30 min. Scale marker = 1 μm.

typically located in the interalveolar wall within a few μm (or less) of the alveolar air space, and exhibited a more modest increase in cell surface area due to membrane folding.

Anaphylactic degranulation, with exocytosis of stainable, granule-derived material, has been reported in isolated human lung mast cells (1), human skin mast cells in situ (11; present authors, unpublished observations), and rat peritoneal mast cells (12). During this process the granule constituents evolve rapidly to a fibrillar material. Fusion of the perigranular membranes of several granules leads to the formation of broad and smooth-walled labyrinths within the cytoplasm. Subsequent fusion of the labyrinth's lining membrane with the plasma membrane results in the release of stainable, fibrillar granule material into the external environment. Under the gentle stimulation employed in this study, anaphylactic degranulation was observed only rarely (8). Interestingly, only a portion of the secretory granules in such mast cells were involved in anaphylactic degranulation, while the majority remained relatively quiescent and indistinguishable from nearby unstimulated granules.

Both we and previous investigators (13) have failed to observe anaphylactic degranulation of lung mast cells in situ. Anaphylactic degranulation of nasal mast cells in our studies was rare. Nevertheless, significant mediator release (up to 29% of total histamine) occurred. A possible mechanism for effecting mediator release without exocytosis of stainable granule material was proposed by Fernandez, Neher and Gomperts (14). Using a whole-cell patch-pipet technique, they measured changes in cell capacitance that occurred during the degranulation of rat peritoneal mast cells. Large, stepwise increases in cell capacitance were associated with exocytosis and were attributed to the fusion of perigranular membranes (or labyrinths) with the plasma membrane. However, small and rapid capacitance changes, termed "capacitance flicker", were also recorded. The authors postulated that capacitance flicker might be due to the transient, reversible formation of an aqueous pore between individual granule membranes and the plasma membrane. While this hypothesis has yet to be proven, it provides a potential mechanism for mediator release consistent with our observations and those of the previous investigators.

In summary, human nasal and lung mast cells were examined by EM to determine if tissue-specific differences, consistent with the concept of mast cell heterogeneity, could be observed. The observed differences, including the location of mast cells, extent of cell surface folding, and association with other cell types, were insufficient to confirm the presence of distinct mast cell populations. In contrast, human skin mast cells may represent a different population based on the routinely reported pattern of anaphylactic degranulation, characterized by exocytotic release of stainable, granule-derived material into the connective tissue. While morphologic studies in some animals suggest the presence of at least two distinct populations of mast cells, our findings in the human respiratory tract are nevertheless consistent with the concept of tissue-specific variations of a single mast cell population. More extensive comparisons employing respiratory, cutaneous and gastrointestinal mast cells will be required to confirm the existence of mast cell heterogeneity in man.

REFERENCES

1. **Barrett, K.M., and D.D. Metcalfe.** 1984. Mast cell heterogeneity: evidence and implications. J. Clin. Immunol. 4:253.
2. **Kaliner, M.A.** 1980. Is a mast cell a mast cell a mast cell? J. Allergy Clin. Immunol. 66:1.
3. **Bienenstock, J., A.D. Befus, J. Denburg, R. Goodacre, F.L.Pearce, and F. Shanahan.** 1983. Mast cell heterogeneity. Monogr. Allergy 18:124.
4. **Metcalfe, D.D.** 1983. Effector cell heterogeneity in immediate hypersensitivity reactions. Clin. Rev. Allergy 1:311.
5. **Shaff, R.E., and M.A. Beaven.** 1979. Increased sensitivity of the enzymic isotopic assay for tissue histamine: measurement of histamine in plasma and serum. Ann. Biochem. 94:425.
6. **Dyer, J., K. Warren, S. Merlin, D.D. Metcalfe, and M. Kaliner.** 1982. Measurement of plasma histamine: description of an improved method and normal values. J. Allergy Clin. Immunol. 70:82.
7. **Karnovsky, M.J.** 1965. A formaldehyde-glutaraldehyde fixative of high osmolality for use in electron microscopy. J. Cell. Biol. 27:137a (abst.)
8. **Friedman, M.M., and M. Kaliner.** 1985. In situ degranulation of human nasal mucosal mast cells: ultrastructural features and cell-cell associations. J. Allergy Clin. Immunol. 76:70.
9. **Willingham, M.C., and A.V. Rutherford.** 1984. The use of osmium-thiocarbohydrazide-osmium (OTO) and ferrocyanide-reduced osmium methods to enhance membrane contrast in cultured cells. J. Histochem. Cytochem. 32:455.
10. **Caulfield, J.P., R.A. Lewis, A. Hein, and K.F. Austen.** 1980. Secretion in dissociated human pulmonary mast cells. Evidence for solubilization of granule contents before discharge. J. Cell. Biol. 85:299.
11. **Hashimoto, K., B.G. Gross, and W.F. Lever.** 1966. An electron microscopic study of the degranulation of mast cell granules in urticaria pigmentosa. J. Invest. Dermatol. 46:139.
12. **Lagunoff, D.** 1972. Contributions of electron microscopy to the study of mast cells. J. Invest. Dermatol. 58:296.
13. **Kawanami, O., V.J. Ferrans, J.D. Fulmer, and R.G. Crystal.** 1979. Ultrastructure of pulmonary mast cells in patients with fibrotic lung disorders. Lab. Invest. 40:717.
14. **Fernandez, J.M. E. Neher, and B.D. Gomperts.** 1984. Capacitance measurements reveal stepwise fusion events in degranulating mast cells. Nature 312:453.

Mast Cell Differentiation and Heterogeneity,
edited by A. D. Befus et al.
Raven Press, New York © 1986.

Clinical Relevance of Basophil and Mast Cell Growth and Differentiation

*Judah A. Denburg, *Hirokuni Otsuka, **Jerry Dolovich,
† A. Dean Befus, and † John Bienenstock

*Departments of *Medicine, **Pediatrics, and † Pathology, McMaster University,
Hamilton, Ontario L8N 3Z5, Canada*

While it has been recognized that basophils or mast cells participate in a wide range of inflammatory and immune reactions (reviewed in 1), the mechanism whereby these cells accumulate in tissues or gain access to sites of inflammation is poorly understood. Dvorak et al. and Askenase et al. have provided us with an understanding of cutaneous basophil hypersensitivity reactions (CBH) in rodents (1–4) and more recently, Askenase and Galli independently have examined the potential role of mast cells and their principal mediator, serotonin, in delayed hypersensitivity reactions in the mouse with conflicting results (5,6).

In human allergic and atopic disorders, the precise role of basophils or mast cells has not been fully elucidated. Okuda et al have described the functional role of basophils and mast cells in human nasal mucosa in allergic rhinitis (reviewed in 7). Mitchell has demonstrated the accumulation of basophils at skin patch test sites in patients with atopic dermatitis, but the precise mechanism of this cellular reaction was not examined (8). Hirsch has shown that basophil counts in peripheral blood fluctuate before, during and after the pollen season with the IgE levels (9), but the relative contributions of demargination of a

---

Supported by grants from the Medical Research Council and National Cancer Institute of Canada and Fisons Pharmacueticals (UK).
Abbreviations: CFU-c, colony-forming cells in culture; CSA, colony stimulating activity; CBH, cutaneous basophil hypersensitivity; CM, conditioned medium; GM, neutrophil-macrophage; Eo, eosinophil; G6PD, glucose-6-phosphate dehydrogenase; IL, interleukin; NMC, nasal metachromatic cells.

storage pool of mature basophils and bone marrow basophil production have not been clarified.

While it is clear that these cells can and do accumulate both in animal and human models at tissue sites, it is also clear that the number of cells accumulating, for example in the CBH reaction, cannot be explained by migration of the entire circulating pool to a given skin site at any one time (4), suggesting a bone marrow contribution to this process. In the present studies, we have explored whether basophil production may contribute to tissue accumulation of these cells in human allergic states.

MATERIALS AND METHODS

Subjects. Subjects consisted of several groups of comparable normal asymptomatic, skin test-negative individuals and skin test-positive patients with various atopic conditions and for allergic rhinitis, the latter with either seasonal or perennial symptoms as previously described (10). Allergic rhinitis patients' symptoms were assessed by questionnaire in which sneezing, nasal discharge and nasal obstruction were graded on a scale from 0 to 3. The designation 'mild', 'moderate' or 'severe' represents the highest score given to any of these symptoms, as previously described (10).

Cell cultures and assessment of granulocyte colonies. Heparinized venous blood was layered over Ficol-Hypaque (specific gravity 1.077, Pharmacia), and methylcellulose cultures of interface mononuclear cells were performed in supplemented Iscove's modified Dulbecco's medium in the presence of conditioned medium (CM) derived from a human T-cell line (MO, kindly supplied by Dr. D. Golde, UCLA) (11), or from a human keratinocyte cell line (COLO-16) or from antigen- or mitogen-stimulated peripheral blood mononuclear cells obtained from atopic individuals (10,12). Granulocyte colonies of $\geq$ 40 cells were enumerated and classified into neutrophil-macrophage (GM) or eosinophil (Eo) types as previously described (10,13). For each culture of cells from a given subject, 6 Eo-type and 6 GM-type colonies were picked at random from replicate dishes and placed in 400 μl of phosphate-buffered saline for histamine assay; a further 12 Eo-type and 6 GM-type colonies were placed on slides for morphological evaluation and differential counting. All assessments were performed on day 14 in vitro. A colony was considered 'basophil-positive' if it contained $\geq$ 5% metachromatic granule-containing polymorphonuclear or mononuclear cells on differential counting. The progenitor of such a colony was defined as a basophil colony-forming unit in culture (CFU-c). Each colony could be shown to be clonally derived from a single cell using G6PD isoenzyme analysis of single colonies grown from a female G6PD heterozygote (14). The absolute number of basophil CFU-C was calculated by multiplying the total number of Eo- or GM--type colonies (per 10^6 cells plated) by the proportion found to contain $\geq 5\%$ basophils using criteria described below.

Histochemical techniques. Nasal scrapings were taken from anterior surface of the inferior turbinate by a small curette (Nagashima, Japan; 2 x 3 mm) as previously described (26). A total of up to 4 small (approximately 4 mm^2 x 40 μm) scrapings (1-2 from either or both nasal cavities) was smeared thinly on the glass slides. After fixation in Mota's basic lead acetate or 10% buffered formalin (pH 6.9), cell smears were stained with 0.5% toluidine blue (15) in 0.7N HCl at pH 0.5. Total number of metachromatic cells in smears of both sides of nose were counted using a magnification of 500. We have previously demonstrated the reproducibility between nostrils and reliability of replicate NMC counts performed in this way in any given subject (15). To characterize cell morphology and size, 200 metachromatic cells were examined in each specimen under a magnification of 1,250.

For differential counting of cells in individual colonies picked from the methylcellulose cultures, smeared cells were stained with May–Grunwald–Giemsa. To confirm the metachromatic nature of the granules, Hansel's stain (eosin and methylene blue in alcohol) or 0.5% toluidine blue at pH 0.5 was used, after fixation in Mota's basic lead acetate. Cells in which all granules were metachromatic were counted as basophils, thus excluding cells with mixed basophilic-eosinophilic granulation. Direct counting of peripheral blood basophil leukocytes was performed as described by Kimura (16).

Histamine assay. Histamine assay was performed by both isotopic radioenzyme conversion (13,17) and automated spectro–fluorometric methods (18). In our hands, the correlation coefficient between the radioenzyme and automated assays was y = 1.01X + 0.06 (p<0.001), with the reliable limit of sensitivity being 0.12 ng on the radioenzyme assay and 0.20 ng by the automated technique (13,17,19).

Statistical Analysis. For statistical comparisons among groups, F analysis was conducted to determine if variances were equal. If this was true, t-tests were employed; if not, a non–parametric Mann–Whitney U–statistic was used.

RESULTS

Basophil progenitors in atopic patients. Initial observations in a pilot study were that there were increased total CFU–c in methylcellulose and in particular progenitors for basophil or histamine–positive colonies in atopic patients (10). The subjects in these studies included patients with asthma, severe atopic eczema or combinations of allergic rhinitis and the above. Table I shows total CFU–c and frequency of histamine–positive colonies in atopics and other groups; significantly higher values were observed in atopics than in normal controls. Highest values were observed in chronic myeloid leukemia.

TABLE I.   Peripheral blood granulocyte and histamine-positive colonies in atopics and controls

Group (n)	Granulocytes Colonies[a] (per 10^6 cells plated)	Frequency of Histamine-Positive Colonies (%)
Atopics (27)	47 ± 7*	209/528** (40)
Controls (24)	28 ± 6	30/265 (11)

Mean ± SE
a   Total CFU-C in 14 day methylcellulose cultures
* $p < 0.05$, compared to controls
** $p < 0.001$, compared to controls

TABLE II.   Metachromatic properties of cells in colonies grown from circulating progenitors of atopic patients

| Colony No. | Fixative | |
	Mota's (%)	Formalin (%)
1	0/88[a] (0)	0/50 (0)
2	10/145 (6.8)	0/200 (0)
3	3/115 (2.6)	0/200 (0)
4	0/200 (0)	0/156 (0)
5	63/200 (31.5)	0/200 (0)
6	27/200 (13.5)	0/200 (0)
7	0/200 (0)	0/75 (0)
8	0/94 (0)	0/200 (0)
9	0/200 (0)	0/200 (0)
10	1/200 (0.5)	0/200 (0)
11	0/200 (0)	0/200 (0)
12	0/200 (0)	0/200 (0)
13	0/200 (0)	0/200 (0)
14	0/200 (0)	0/200 (0)
15	0/200 (0)	0/150 (0)
16	11/152 (7.2)	0/75 (0)
17	0/171 (0)	0/391 (0)
18	4/184 (2.6)	0/191 (0)
19	9/200 (4.5)	0/136 (0)
20	0/200 (0)	0/136 (0)
21	3/200 (1.5)	0/200 (0)
22	3/153 (1.9)	0/200 (0)
23	0/200 (0)	0/200 (0)
24	0/200 (0)	0/200 (0)
25	24/200 (12)	0/112 (0)
26	29/200 (14.5)	0/200 (0)

a. No. cells staining metachromatically/No. cells counted

Basophil progenitors and nasal allergy. In studies of untreated patients with allergic rhinitis graded according to the numbers of metachromatic cells in the nasal mucosa and/or clinical symptomatology, there was a gradient of the number of basophil CFU-c which was inversely related principally to the number of formalin-sensitive nasal mast cells (19). The increase in granulocyte colonies in allergic patients in both studies was principally due to increase in Eo-type colonies (10), which previously have been shown to include the majority of histamine-positive and metachromatic cell-positive colonies (13).

Staining characteristics of basophils, colony metachromatic cells and nasal metachromatic cells. Peripheral blood basophils and metachromatic cells in colonies both stained after Mota's but not after formalin fixation (Table II). The predominant nasal metachromatic cell in nasal scraping was also a formalin-sensitive cell, the morphology of which resembled mast cells (15).

Basophil progenitors in ragweed nasal allergy. In examining the question of the relation of basophil growth and differentiation to an ongoing IgE response in patients with allergic rhinitis to ragweed only, the same inverse relationship of basophil progenitors to nasal metachromatic cells, symptoms, and to exposure to antigen was observed. The highest number of basophil CFU-C were observed in patients before or after but not during the season, when progenitor levels markedly declined (Figure 1).

Beclomethasone effects on nasal metachromatic cell populations. Effects of beclomethasone in vivo were principally on a nasal metachromatic cell population which was formalin-sensitive. Numbers of these cells, but not of formalin-resistant mast cells, fell significantly on therapy and returned toward normal values following cessation of beclomethasone (20).

Antigen-specific production of basophil and eosinophil colony-stimulating activity. In studies designed to examine the question of the secretion of basophil or eosinophil colony-stimulating factors from lymphomononuclear cells in atopic patients, we found specific antigen-induced elaboration of these factors into supernatants from atopic peripheral blood cells (Table III), or from cells taken from a patient with atopic dermatitis undergoing skin patch testing (8; Figure 2).

Keratinocyte production of eosinophil and basophil colony-stimulating activities. A human keratinocyte cell line (Colo-16) produced a colony-stimulating factor co-eluting with epidermal T-cell activating factor (ETAF) after multi-step purification (21), which was capable of promoting the growth and differentiation of granulocyte colonies, including neutrophil/macrophage (GM-type) and basophil/eosinophil (Eo-type) colonies in methyl-cellulose (22; Table IV). This keratinocyte cell line also produced an IL-3-like activity, as measured by a proliferation assay utilizing a murine IL-3-dependent line, $32Dcl/H_4$ (21,23).

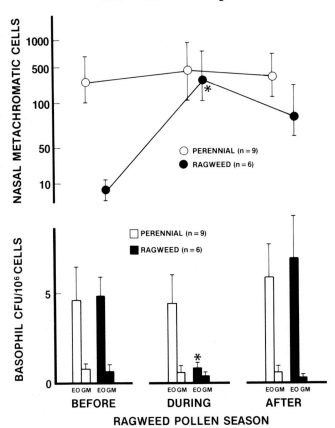

Figure 1. Relationship of basophil colony-forming cells (CFU-c) and nasal metachromatic cells (NMC) to course of seasonal allergic rhinitis. Top graph total NMC counts in nasal scrapings in patients with seasonal allergic rhinitis to ragweed only (closed circles) and in those with perennial allergic rhinitis (open circles); bottom graph, calculated number of circulating basophil CFU-c in methylcellulose cultures from patients with seasonal allergic rhinitis to ragweed only (shaded bars) and in those with perennial allergic rhinitis (open bars); > 1 month before, during or > 1 month after 1984 ragweed pollen season in Ontario, Canada.

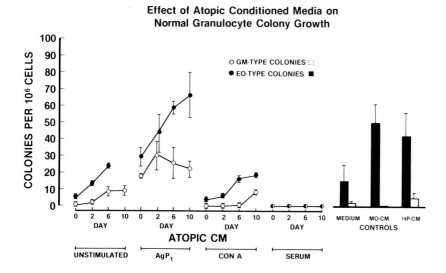

Effect of Atopic Conditioned Media on Normal Granulocyte Colony Growth

Figure 2. Granulocyte colony formation in response to conditioned media (CM) prepared from 3d unstimulated, antigen (house dust, mite, P_1,) - or mitogen (concanavalin A) - stimulated atopic peripheral blood mononuclear cells, or serum from atopic patients on days 0–10 of a chronic antigen skin patch–testing protocol (8). Bar graphs indicate colony formation in presence of medium alone, leukemic T-cell (MO-) CM or human placental (HP-) CM.

TABLE III. Effect of atopic lymphocyte conditioned media on growth of peripheral blood granulocyte colonies

Conditioned Medium	Colonies		
	Total Per 10^6 Cells	Frequency of Histamine–Positive GM	Eo
Unstimulated	31 + 14	–	–
Antigen	68 + 15 *	1/4	3/4
Mitogen	33 + 15	–	–
Positive Control	56 + 5 *	2/23	15/29

Results of 3 expts, mean + SE
* p<0.01, compared to mitogen or unstimulated CM
GM = neutrophil–macrophage–type
Eo = eosinophil–type

TABLE IV.   Human keratinocytes produce granulocyte
colony-stimulating activity

	Colonies $/10^6$ cells plated		
Condition	Colony Type		
	GM	Eo	Total
Negative Control (medium alone)	5.5 ± 1.0	9.2 ± 3.2	14 ± 2.3
Positive controls[a]	22 ± 4.3	24 ± 3.1	46 ± 7.1
COLO-16 20% (v/v) Supernatant	6.3 ± 4.6	7.0 ± 3.8	13 ± 8.4
10%	13 ± 7.0	29 ± 18	42 ± 25
5%	20 ± 11	36 ± 12	56 ± 23
1%	23 ± 7.8	46 ± 10	69 ± 17
0.1%	13 ± 2.9	22 ± 6.0	34 ± 8.6

Mean ± SE, results of 2 experiments
a.   MO-CM or human placental CM, 5% v/v

Production of basophil colony-stimulating activity by and
presence of basophil CFU-c in nasal polyps.   Nasal epithelial
supernatants from nasal polyps supported the growth of normal   or
atopic basophil CFU-c; in addition, nasal polyps contained a   low
frequency of exclusively basophil CFU-c (Table V).

DISCUSSION

Basophil/mast cell progenitors, defined   as those cells   which
give rise   to clonally-derived   metachromatic cell-containing   or
histamine-positive hemopoietic colonies   in vitro   (13,14,24,25),
are present   in   increased   numbers in   circulation   in clinical
situations in which   increases in tissue   numbers of these   cells
are observed   (10,19).   While   it is   not surprising   that   such
progenitors are increased in bone marrow disorders such as CML or
systemic mastocytosis in   which cellular proliferation   involving
these lineages occurs (13,26,27), a similar observation in atopic
states was   less   predictable.   Moreover, that   such   progenitor
levels fluctuate   in   a specific   pattern   in relation   to   local
tissue (nasal)   accumulation of   basophils and mast   cells or   to
symptoms (19;   Figure 1),   including   seasonal allergy,   suggests
strongly that the contribution   of basophil production in   atopic
states is important.   The finding of keratinocyte-derived   hemopoietic factors which promote the growth and differentiation of

granulocyte colonies (21,22; Table V), some of which contain basophils and eosinophils (22), or of such factors produced by peripheral blood mononuclear cells in response to antigen 10,13; Tables III & IV), supports the hypothesis that basophil/mast cell or eosinophil progenitors may be attracted out of circulation by local factors which promote their differentiation in situ in either atopic skin or nose. Finally, our finding of the elaboration of human basophil colony-stimulating activity by and the presence of basophil CFU-c in nasal polyps (Table V) supports this hypothesis. Evidence has been obtained for an analogous role of local environment in murine mast cell proliferation, including studies of mast cell progenitor traffic and frequency in immunodeficient or defective mice and at difference sites in response to immune or inflammatory signals (28-31).

TABLE V. Basophil colony-forming cells and basophil colony-stimulating activity in human nasal polyps

Condition	Basophil CFU-C[a] per 10^6 cells (% Total)[b]
Nasal polyp cells plus MO-CM, 5% v/v (n = 7)	1.0 ± 0.4 (81)
Tonsillar cells plus MO-CM, 5% v/v (n = 5)	0 ± 0 (0)
Normal peripheral blood mononuclear cells plus MO-CM, 5% v/v (n = 4)	5 ± 5 (63)
Normal peripheral blood mononuclear cells plus NE-CM, 5% v/v (n = 4)	20 ± 7 (100)
Normal peripheral blood mononuclear cells plus NE-CM, 2.5% v/v (n = 4)	8 ± 5 (57)

a. Colonies containing $\geq$ 5% basophils
b. Percent of total granulocyte colonies at 14d in methylcellulose cultures.

MO-CM: Leukemic T cell conditioned medium
NE-CM: Nasal polyp epithelial cell conditioned medium

In human allergy, antigen may act not only to promote inflammatory reactions which result in chemotactic signals for mature basophils/mast cells or eosinophils to accumulate locally (1,2,32), but could also stimulate the production of factors from peripheral blood or locally situated progenitor cells to promote their growth and differentiation into effector cells. Traffic studies of circulating progenitors, as well as assays for the presence of hemopoietic factors or progenitors at local sites may help clarify these proposed mechanisms. Finally, the cloning of subpopulations of human basophils and mast cells may be possible by utilizing purified growth factors and/or enriched progenitor populations from appropriate individuals. This clearly could have impact on the management of allergic disorders in the future.

REFERENCES

1. **Askenase, P.W.** 1977. Role of basophils, mast cells, and vasoamines in hypersensitivity reactions with a delayed time course. Prog. Allergy 23:199.
2. **Dvorak, H.F., B.A. Simpson, R.C. Bast, and S. Leskowitz.** 1971. Cutaneous basophil hypersensitivity. III. Participation of the basophil in hypersensitivity to antigen-antibody complexes, delayed hypersensitivity and contact allergy. J. Immunol. 107:138.
3. **Mitchell, E.B., S.J. Brown, and P.W. Askenase.** 1982. IgG1 antibody-dependent mediator release after passive systemic sensitization of basophils arriving at cutaneous basophil hypersensitivity reactions. J. Immunol. 128:1663.
4. **Leonard, E.J., M.A. Lett-Brown, and P.W. Askenase.** 1979. Simultaneous generation of tuberculin-type and cutaneous basophilic hypersensitivity at separate sites in the guinea pig. Int. Arch. Allergy Appl. Immunol. 58:460.
5. **Askenase, P.W., S. Bursztajn, M.D. Gershon, and R.K. Gershon.** 1980. T-cell dependent mast cell degranulation and release of serotonin in murine delayed-type hypersensitivity. J. Exp. Med. 152:1358.
6. **Galli, S.J., and I. Hammel.** 1984. Unequivocal delayed hypersensitivity in mast cell deficient (W/W^v and Sl/Sl^d) and beige mice. Science 226:710.
7. **Okuda, M.** 1977. Mechanisms in nasal allergy. O.R.L. Digest 39:26.
8. **Mitchell, E.B., J. Crow, M.D. Chapman, S.S. Jouhal, F.M. Pope, and T.A.E. Platts-Mills.** 1982. Basophils in allergen-induced patch test sites in atopic dermatitis. Lancet i:127.
9. **Hirsch, S.R., and J.H. Kalbfleisch.** 1976. Circulating basophils in normal subjects and in subjects with hay fever. J. Allergy Clin. Immunol. 58:676.
10. **Denburg, J.A., S. Telizyn, A. Belda, J. Dolovich, and J. Bienenstock.** 1985. Increased numbers of circulating basophil/mast cell progenitors in atopic patients. J.

Allergy Clin. Immunol. 76:466.

11. Golde, D.W., S.G. Quan, and M.J. Cline. 1978. Human T-lymphocyte cell line producing colony-stimulating activity. Blood 52:1068.

12. Ahlstedt, S., I. Hammarstrom, M.-B. Into-Malberg, B. Bjorsten, and J.A. Denburg. 1985. Appearance in vitro of large histamine-containing basophilic cells. Int. Arch. Allergy Appl. Immunol., in press.

13. Denburg, J.A., S. Telizyn, M. Richardson, and J. Bienenstock. 1983. Basophil/mast cell precursors in human peripheral blood. Blood 61:775.

14. Denburg, J.A., H. Messner, B. Lim, N. Jamal, S. Telizyn, and J. Bienenstock. 1985. Clonal origin of human basophil/mast cells from circulating multipotent hemopoietic progenitors. Exp. Hematol. 13:185.

15. Otsuka, H., J.A. Denburg, J. Dolovich, D. Hitch, P. Lapp, R.S. Rajan, J. Bienenstock, and A.D. Befus. 1985. Heterogeneity of metachromatic cells in human nose: Significance of mucosal mast cells. J. Allergy Clin. Immunol., in press.

16. Kimura, I., M. Yoshiakia, and Y. Tanizaki. 1973. Basophils in bronchial asthma with reference to reagin-type allergy. Clin. Allergy 3:195.

17. Denburg, J.A., A.D. Befus, R. Goodacre, and J. Bienenstock. 1981. Basophil production. III. Relation of histamine to guinea pig basophil growth in vitro. Exp. Hematol. 9:214.

18. Siraganian, R.P., and M.J. Brodsky. 1976. Automated histamine analysis for in vitro allergy testing. J. Allergy Clin. Immunol. 57:525.

19. Denburg, J.A., H. Otsuka, S. Telizyn, R. Rajan, P. Lapp, D. Hitch, A.D. Befus, J. Bienenstock, and J. Dolovich. 1985. Basophil precursors, peripheral blood basophils and nasal metachromatic cells in patients with allergic rhinitis. J. Allergy Clin. Immunol. 75:153A.

20. Otsuka, H., J.A. Denburg, A.D. Befus, D. Hitch, P. Lapp, R. Rajan, J. Bienenstock, and J. Dolovich. 1985. Effect of beclomethasone diproprionate on nasal metachromatic cell populations. Clin. Allergy, in press.

21. Denburg, J.A., and D.N. Sauder. 1985. Hemopoietic colony stimulating activity derived from human keratinocytes. Clin. Res. 33:633A.

22. Denburg, J.A., E.B. Mitchell, and D.N. Sauder. 1985. Basophil colony growth stimulated by human keratinocytes: possible relevance in atopic dermatitis. Clin. Res. 33:160A.

23. Denburg, J.A., Y. Tanno, and J. Bienenstock. 1985. Growth and differentiation of human basophils, eosinophils and mast cells. In: Mast Cell Heterogeneity and Differentiation. Edited by A.D. Befus, J.A. Denburg, and J. Bienentsock. New York, Raven Press. This volume.

24. Denburg, J.A., S. Telizyn, H. Messner, B. Lim, N. Jamal, S.J. Ackerman, G.J. Gleich, and J. Bienenstock. 1985.

Heterogeneity of human peripheral blood eosinophil–type colonies: evidence for a common basophil–eosinophil progenitor. Blood 66, in press.

25. **Leary, A.G., and M. Ogawa.** 1984. Identification of pure and mixed basophil colonies in culture of human peripheral blood and marrow cells. Blood 64:78.

26. **Denburg, J.A., W.E.C. Wilson, and J. Bienenstock.** 1982. Basophil production in myeloproliferative disorders: increases during acute blastic transformation of chronic myeloid leukemia. Blood 60:113.

27. **Denburg, J.A., S. Telizyn, S. Ahlstedt, J.H. Olafsson, G. Roupe, and J. Bienenstock.** 1985. Basophil/mast cell precursors in mast cell proliferative disorders. Clin. Invest. Med. 9, in press.

28. **Crapper, R.M., and J.W. Schrader.** 1983. Frequency of mast cell precursors in normal tissues determined by an in vitro assay: antigen induces parallel increases in the frequency of P–cell precursors and mast cells. J. Immunol. 131:923.

29. **Guy–Grand, D., M. Dy, L. Luffau, and P. Vassalli.** 1984. Gut mucosal mast cells: origin, traffic and differentiation. J. Exp. Med. 160:12.

30. **Mayrhofer, G., and H. Bazin.** 1981. Nature of the thymus dependency of mucosal mast cells. III. Mucosal mast cells in nude mice and nude rats, in B rats and in a child with the Di George Syndrome. Int. Arch. Allergy Appl. Immunol. 64:320.

31. **Kitamura, Y., K. Hatanaka, M. Murakami, and H. Shibata.** 1979. Presence of mast cell precursors in peripheral blood of mice demonstrated by parabiosis. Blood 53:1085.

32. **Ward, P.A., H.F. Dvorak, S. Cohen, R. Yoshida, R. Data, and S. Selvaggio.** 1975. Chemotaxis of basophils by lymphocyte-dependent and lymphocyte–independent mechanisms. J. Immunol. 114:1523.

Mast Cell Differentiation and Heterogeneity,
edited by A. D. Befus et al.
Raven Press, New York © 1986.

Epilogue

Mast Cell Heterogeneity: Basic Questions and Clinical Implications

*John Bienenstock, **A. Dean Befus, and † Judah A. Denburg

*Host Resistance Programme, Departments of *Pathology and † Medicine, McMaster University, Hamilton, Ontario L8N 3Z5, Canada; and **Gastrointestinal Research Group, Department of Microbiology and Infectious Diseases, University of Calgary, Calgary, Alberta T2N 4N1, Canada*

Mast cells and basophils are involved through their high affinity IgE receptors in allergic or immediate hypersensitivity type reactions. However, there have been a number of observations suggesting that they are also involved in the maintenance of physiological homeostasis and/or in response to various types of injury. In this paper we will review some of these situations in which there is good experimental evidence for mast cell involvement, offer some mechanisms whereby they may be involved, and through this expose the variety of conditions which may be affected by application of the knowledge of work being done in this field, as outlined in this book.

Mast cell increases in different conditions. It has been known for a number of years that mast cell numbers increase in draining lymph nodes following immunization. Table I lists the many conditions known to be associated with mast cell increases. The massive increase which was shown to occur in thymuses of NZB mice and which led Burnet (1) to postulate that mast cells might be derived from lymphocytes is still completely unexplained. It has been shown, however, that mast cells may be derived from lymphocyte-like cells (2), but it is clear that at least in the intestine they do not arise from cells with a lymphocytic phenotype (3).

Increased numbers of mast cells are also found in most types of fibrous tissue and in association with scars, especially keloids (4). We will review some of the possible relationships of mast cells to fibrosis later. Suffice it to say at this stage that mast cell numbers increase in tissues including the skin in murine chronic graft versus host disease, a model of scleroderma, and also in human scleroderma (5). In this situation they appear often to be degranulated, although the pathogenetic mechanisms

underlying this enigmatic condition are still obscure. In acute
graft versus host disease the intestinal mast cell count is also
elevated (6). In many diseases which affect the gastrointestinal
tract, mast cells are increased drastically in number, the most
classic example being in response to nematode infections (7,8).
In both Crohn's disease and ulcerative colitis (9), significant
increases are found both in number and in histamine content, and
similar increases occur in celiac disease especially in patients
who are not in remission (10).

TABLE I.   Examples of non-"allergic" conditions in which
            increased numbers of mast cells occur

Lymphoid tissues

    Thymus (NZB mice)
    Immunization
    Mastocytosis

Fibrous Tissue

    Scars, especially Keloids

Bone/Joint

    Osteoporosis
    Callus formation
    Rheumatoid synovitis
    Mastocytosis

Nervous System

    Peripheral neuropathy
    Distal portion sectioned nerves
    Neuromata

Lung

    Hypoxia/hypertension
    Immunization
    Interstitial lung disease

    Fibrosis
    Asbestosis, silicosis

Skin

    Urticaria pigmentosa

Intestine

    Celiac disease
    Crohn's disease
    Ulcerative colitis
    Graft vs host disease
    Nematode infection
    Mastocytosis

Tumors

    Many solid tumors

    In the lung, mast cell numbers are considerably elevated as a
response to hypoxia and in association with pulmonary hyperten-
sion (11).   Aerosolized antigen also leads to increases in
parenchymal mast cell numbers (12). Interstitial lung diseases,
most of which lead to pulmonary fibrosis, are associated with
mast cell number increases in the lung tissue as well as in the
bronchoalveolar lavage fluid obtained from such patients (13).
In pulmonary fibrosis itself, mast cell numbers are increased and
they are also elevated in conditions associated with fibrosis
such as asbestosis, silicosis and sarcoidosis (14). Alveolar
macrophages produce a factor which degranulates mast cells (15)
and it is interesting that histamine, through an H2 receptor on

the fibroblast, can cause proliferation or inhibition of growth of fibroblasts depending on the stage of cell cycle reached (16). Furthermore, mast cell granules may be phagocytosed by fibroblasts and affect some of their functions (17).

In various pathological conditions which affect bones and joints mast cell numbers have been shown to be elevated. For example, this occurs in osteoporosis (18), in the lytic bone lesions associated with systemic mastocytosis and even in callus formation as part of the normal repair process (19). It is interesting that long-term heparin administration as part of anticoagulant therapy is known to be associated with osteoporosis (20); the mechanisms involved are obscure. In rheumatoid arthritis, the synovium has recently been shown to be heavily infiltrated with mast cells although it is not clear how this promotes or regulates the inflammatory process in this type of synovitis (21).

Many solid tumors are associated with mast cell hyperplasia, particularly neurofibromata, in which significant increases are seen. It is of interest that amputation neuromata, and nerves acutely injured or involved in diseases causing peripheral neuropathy, have been shown to have a significant increase in mast cell numbers and histamine content (22). For example, acute sectioning of a peripheral nerve results in a few days in significantly increased numbers of mast cells within the nerve. Many of these cells have been shown to be in mitosis, and their number increases to a maximum 12 wk after sectioning, only in the portion of the nerve distal to the injury. Some mast cell-nervous system connections including the formation of specific synapses or boutons, have been described (23). Since nerve growth factor causes mast cell degranulation in vitro (24) as well as mast cell proliferation in vivo (25), and since neuropeptides have a differential effect on intestinal mucosal mast cells and peritoneal mast cells (26), it is not difficult to construct hypotheses involving mast cells in the regulation of inflammation. Interactions between nerves and mast cells are made more plausible by observations suggesting that neuropeptides such as vasointestinal polypeptide (VIP) (27) or substance P are found in many cells which include mast cells, basophils and neutrophils.

Factors which inhibit or promote growth. In an experimental model of pulmonary fibrosis induced by a single administration of bleomycin intra-tracheally to rats, we have previously shown that the fibrosis is accompanied by significant (> 50 fold) increases in mast cells and histamine content in pulmonary parenchyma (14). This mast cell hyperplasia is not controlled by T cells since experiments with two different types of (thymus-less) nude rats obtained from Holland or the National Institutes of Health showed similar findings (28). It is interesting that the nude animals showed, if anything, an exaggerated mast cell response as seen in Table II. These results are in accord with our observations suggesting that from the histochemical and functional point of view, mast cells isolated from the lung of such animals with pulmonary fibrosis were more like cells from the peritoneal cavity than those from the intestinal mucosa. This is the only situation of which we are aware where the peritoneal type of mast cell is undergoing significant proliferation and hyperplasia; this hyperplasia has been shown to be clearly thymus independent. It is well known that peritoneal mast cells are found in nude mice and rats. Whether IL-3 or IL-3-like factors are necessary for the proliferation of all mast cells is not known. Indeed the experiments of Kitamura (29) suggest that peritoneal mast cells may be acted on by IL-3-like factors in his "transdifferentiation" model to promote the differentiation of these cells into cells characteristic of intestinal mucosal mast cells. There are distinct possibilities that IL-3 or IL-3-like products may be derived from many cells not of thymic origin. In this respect it is worth noting that the cell line from which IL-3 was originally isolated, the WEHI-3B is in fact a myelomonocytic cell line with no lymphocytic lineage. Recent observations (30) have suggested that human and murine keratinocytes as well as human epithelial cell-derived tumours may secrete IL-3 or IL-3-like molecules as well as hemopoietic growth factors.

TABLE II. Effect of bleomycin on histamine content of nude rat lung (day 30)

Littermates	Histamine (μg/g wet weight)
Sham	0.91 ± 0.13
Bleomycin	5.24 ± 1.67
Nu/nu	
Sham	2.67 ± 0.56
Bleomycin	13.64 ± 2.47

The lineage of the cells identified as intestinal mucosal mast cell in type is likely to be from the bone marrow (31). In the rat, cultured bone marrow derived mast cells and intestinal mast cells both contain rat mast cell protease II and chondroitin sulfate di-B (32). It is noteworthy that the rat basophil leukemic cell also contains these two markers (32). While the extrapolation may be too far, Otsuka et al. (33) have shown that the frequency of basophil precursors in the blood of patients with allergic rhinitis is inversely proportional to the number of nasal metachromatic, and presumed mucosal type of, cell. It is tempting to speculate that as with Kitamura's transdifferentiation experiments basophil precursors in the presence of epithelial or T cell derived IL-3-like or basophil growth factors in the nose become "mucosal mast cells". However, it is true to say that mouse bone marrow, spleen, mesenteric lymph node, Peyer's patch and intestinal mast cell growth is IL–3–dependent. The same is implied from the experiments with rat bone marrow, mesenteric lymph node and intestine since T cells produce the growth promotion factors (34). What is not known is whether such factors are obligatorily produced by T cells and whether IL–3 promotes proliferation, differentiation or both. In another well studied system, the B cell growth promoting factor appears to be separable from that which promotes differentiation, so that it is distinctly possible that separate factors will be found to be responsible for mast cell proliferation and differentiation. While most of the work on mast cell growth factors is designed to study positive events, there must also be negative or inhibitory signals which regulate growth of these cells. In fact we observed such in mesenteric lymph node derived factors at particular times after infection of rats with Nippostrongylus (35).

Human and guinea pig basophil growth is T–dependent in vitro. Factors which down regulate growth of human basophils have been separated from factors responsible for basophil growth stimulation (36). The presence or absence of basophils, or for that matter mast cells are not known in the human counterpart of the nude animal, the diGeorge syndrome. Human IL–3 like factors from mitogen–stimulated lymphocytes and solid tumour cell lines such as the bladder appear to cause human bone marrow as well as murine IL–3 dependent mast cell growth (37). Such factors probably promote the growth of only one type of human metachromatic cell, possibly mast cells and not basophils. The gene for murine IL–3 does not appear to bear any significant homology to genes for human erythroid potentiating or multipotent hemopoietic factors. Human mast cell colony growth from blood does not absolutely require T–derived factors. Thus the role of mast cell growth factors and their thymic derivation or dependence in the human is an area of significant uncertainty at the present.

Mast cells, cytotoxicity and cell–mediated immunity. Since mast cells are involved in so many non–allergic types of reactions, it may be instructive to look at some of these to derive some clues as to their function. It is well–established by Askenase (38) that a T–cell derived non–IgE antigen specific factor may bind to mast cells and cause them to release and secrete material by a different mechanism of degranulation than that found with the IgE system. It has been suggested that MC's are essential in delayed hypersensitivity through serotonin-dependent mechanisms which are blocked by reserpine and methysergide. However, recent evidence shows that mast cell deficient mice strains W/W^v and Sl/Sl^d produce competent delayed hypersensitivity responses. Thus, there is considerable question whether mast cells are involved in an obligatory fashion in the expression of delayed hypersensitivity (39).

Another type of delayed hypersensitivity reaction is that seen with granuloma formation in the experimental schistosomiasis model in mice in which it has been shown that granulomata may be inhibited by T-suppressor cells which can adoptively transfer this function to a host animal. Mast cells are invariably present in granulomata and intestinal granulomata are smaller than those found in the liver. The involvement of histamine has been shown by the ability of cimetidine to antagonize the suppressor cell effect through inhibition of the H2 receptor. Thus the size of the granulomata appears to be controlled through some involvement of mast cells (40).

Another way in which mast cells can be involved in tissue injury and regulation of parasite infestation is through the observation that mast cells in conjunction with eosinophils can kill schistosomula (41). This has been refined to show that peritoneal mast cell granules together with eosinophil peroxidase can also effect cell death (42). Furthermore, mouse peritoneal cells will kill fibrosarcoma cells induced by methylcholanthrene (43). We have recently shown that such cytotoxicity is extended to the target of the classical natural cytotoxic (NC) cell, the fibrosarcoma WEHI 164 (44). Rat PMC form conjugates with this target, do not kill another natural killer target, the YAC line, and all rat, including intestinal mucosal, and murine, including mouse bone marrow and intestinal derived mast cells in culture, have this cytotoxic activity. It is interesting that the degranulation by 48/80 of rat PMC had no effect on the subsequent cytotoxicity for the WEHI 164 tumour cell. Thus mast cells may be involved directly in host defense through such mechanisms.

Mast cells as regulators of physiological events. Although the heterogeneous effects of histamine are well-known on large numbers of immune effector systems (45), those of proteoglycans such as heparin, for example, are less well-known and are shown in Table III. The purpose of this table is to remind the reader of the variety of effects of these proteoglycans (8) and the fact that the biological effects of over-sulfated chondroitin sulfates such as chondroitin sulfate di-B have yet to be investigated. Proteoglycan functions may serve as important indicators of the role of the rat intestinal mucosal mast cell.

TABLE III.   Effects of heparin

Blocks activated properdin (46)
Blocks mitogen induced blastogenesis (47)
Inhibits phagocytosis by neutrophils (48)
Enhances phagocytosis by Kupffer cels (49)
Enhances pinocytosis by fibroblasts (50)
Blocks some neutrophil enzymatic activity (51)
Blocks tissue damage by eosinophil MBP (52)
Inhibits smooth muscle growth (53)
Causes osteoporosis (20)
Stimulates angiogenesis (54)

The role of mast cell derived substances in the regulation of many physiological circumstances is very broad when we think of the approximately two dozen mediators so far identified in this cell type (55). Castro and colleagues (56) were one of the first to observe that immune factors may have a regulatory effect on epithelial cell function in some classic experiments involving rats originally infected with Trichinella and subsequently re-exposed to the worm. Other experiments recently have extended these observations. For example, Perdue et al. have shown that intestinal anaphylaxis in the rat induced by specific antigen is accompanied by major changes in the intestinal handling of sodium, chloride and water (57). These major changes were inhibited by doxantrazole in the rat, a drug we have previously shown to be an effective anti-allergic for rat intestinal mucosal mast cells (58). Disodium cromoglycate which we had shown to be ineffective on these cells was without effect in this system. A similar type of experiment in guinea pigs sensitized to cow's milk showed that short circuit current epithelial changes could be elicited by β-lactoglobulin and that this was inhibitable by indomethacin at physiological doses (59). The fascinating observation that a psychologically conditioned stimulus could modify histamine release in guinea pigs (60), coupled with those observations quoted above as well as suggesting a nervous system–mast cell relationship, open up the concept of mast cells being involved in reactions to injury and controlled by nervous impulses. Thus mast cells would be part of a homeostatic mechanism which in the event of dysregulation might lead to them being pathogenetically involved in many clinical disease states.

Clinical implications. The clinical implications of research on factors which modulate growth of mast cells or basophils must relate both to the diagnosis of allergic and other conditions described above, and to therapy of allergic, inflammatory or fibrotic disorders. The identification of growth promoting or inhibitory factors should lead to their cloning, or synthesis of active peptides which would be candidates for new drug development. As a consequence of much of the research on mast cell heterogeneity outlined in this book, the screening of anti-allergic drugs by pharmaceutical companies needs to be performed on mast cells from different target tissues. Rats may not be good models for the human since drugs which are potent on rat intestinal mucosal mast cells may not necessarily have any effect in the human. There is an obvious need for development of homogeneous human mast cell lines or clones which could be used in such drug screening. Lastly, it must be asked whether monkey broncho–alveolar lavage derived mast cells are satisfactory test target cells and really equate to the mucosal mast cell found in the human lung mucosa or gastrointestinal tract (61). We can, however, safely assume that the application of this type of research will lead to important new discoveries which will bring totally new approaches to bear on many existing diseases including those in which allergic mechanisms appear to play a role.

REFERENCES

1. **Burnet, F.M.** 1965. Mast cells in the thymus of NZB mice. J. Pathol. Bacteriol. 89:271.
2. **Combs, J.W., D. Lagunoff, and E.P. Benditt.** 1965. Differentiation and proliferation of embryonic mast cells of the rat. J. Cell. Biol. 25:577.
3. **Schrader, J.W., R. Scollay, and F. Battye.** 1983. Intramucosal lymphocytes of the gut: Lyt 2 and Thy 1 phenotype of the granulated cells and evidence for the presence of both T cells and mast cell precursors. J. Immunol. 130:558.
4. **Kischer, C.W., H. Bunce, and M.R. Shetlar.** 1978. Mast cell analyses in hypertrophic scars, hypertrophic scars treated with pressure and mature scars. J. Invest. Dermatol. 70:355.
5. **Claman, H.N.** 1985. Mast cells, T cells and abnormal fibrosis. Immunol. Today 6:192.
6. **Levy, D.A., A.F. Wefald, and W.E. Beschorner.** 1985. Intestinal mast cell proliferation in graft-versus-host reaction in rats. Int. Arch. Allergy Appl. Immunol. 77:186.
7. **Befus, A.D., F.L. Pearce, J. Gauldie, P. Horsewood, and J. Bienenstock.** 1982. Mucosal mast cells. 1. Isolation and functional characteristics of rat intestinal mast cells. J. Immunol. 128:2475.
8. **Befus, D., F. Pearce, and J. Bienenstock.** 1985. Intestinal mast cells in pathology and host resistance. In: Food Allergy and Intolerance. Edited by J. Brostoff and S.J. Challacombe. London, Saunders. (In press).
9. **Lloyd, G., F.H.Y. Green, H. Fox, et al.** 1975. Mast cells and immunoglobulin E in inflammatory bowel disease. Gut 16:861.
10. **Strobel, S., A. Busuttil, and A. Ferguson.** 1983. Human intestinal mucosal mast cells: expanded population in untreated coeliac disease. Gut 24:222.
11. **Williams, A., D. Heath, J.M. Kay, and P. Smith.** 1977. Lung mast cells in rats exposed to acute hypoxia, and chronic hypoxia with recovery. Thorax 32:287.
12. **Ahlstedt, S., B. Bjorksten, H. Nygren, and G. Smedegard.** 1983. Induction of mucosal immunity and pulmonary mast cells in mice and rats after immunization with aerosolized antigen. Immunology 48:247.
13. **Tomioka, M., S. Ida, Y. Shindoh, T. Ishihara, and T. Takishima.** 1984. Mast cells in bronchoalveolar lumen of patients with bronchial asthma. Am. Rev. Resp. Dis. 129:1000.
14. **Goto, T., D. Befus, R. Low, and J. Bienenstock.** 1984. Mast cell heterogeneity and hyperplasia in bleomycin-induced pulmonary fibrosis of rats. Am. Rev. Resp. Dis. 130:797, 1984.
15. **Schulman, E.S., M.C. Liu., D. Proud, D.W. MacGlashan Jr., L.M. Lichtenstein, and M. Plaut.** 1985. Human lung macrophages induce histamine release from basophils and mast

cells. Am. Rev. Respir. Dis. 131:230.
16. Jordana, M., J. Gauldie, A.D. Befus, M. Newhouse, and J. Bienenstock. 1985. Effect of histamine on proliferation of human adult lung fibroblasts. Submitted.
17. Atkins, F.M., M.M. Friedman, P.V.S. Rao, and D.D. Metcalfe. 1985. Interactions between mast cells, fibroblasts and connective tissue components. Int. Arch. Allergy Appl. Immunol. 77:96.
18. Frame, B., and R.K. Nixon. 1968. Bone-marrow mast cells in osteoporosis of aging. N. Eng. J. Med. 279:626.
19. Lindholm. R.V., and T.S. Lindholm. 1970. Mast cells in endosteal and periosteal bone repair. A quantitative study on callus tissue of healing fractures in rabbits. Acta. Orthop. Scandinav. 41:129.
20. Griffith, G.S., G. Nichols Jr., J.G. Asher, and B. Flanagan. 1965. Heparin osteoporosis. J.A.M.A. 193:91.
21. Crisp, A.J., C.M. Chapman, S.E. Kirkham, A.L. Schiller, and S.M. Krane. 1984. Articular mastocytosis in rheumatoid arthritis. Arthr. Rheum. 27:845.
22. Olsson. Y. 1968. Mast cells in the nervous system. Int. Rev. Cytology 24:27.
23. Newson, B., A. Dahlstrom, L. Enerback, and H. Ahlman. 1983. Suggestive evidence for a direct innervation of mucosal mast cells. Neuroscience 10:565.
24. Bruni, A., E. Bigon, E. Boarato, L. Mietto, A. Leon, and G. Toffano. 1982. Interaction between nerve growth factor and lysophosphatidylserine on rat peritoneal mast cells. FEBS Letters 138:190.
25. Aloe, L., and R. Levi-Montalcini. 1977. Mast cells increase in tissues of neonatal rats injected with the nerve growth factor. Brain Res. 133:358.
26. Shanahan, F., J.A. Denburg, J. Fox, J. Bienenstock, and D. Befus. 1985. Mast cell heterogeneity: effects of neuroenteric peptides on histamine release. J. Immunol. 135:1331.
27. Cutz, E., W. Chan., N.S. Track, A. Goth, and S.I. Said. 1978. Release of vasoactive intestinal polypeptide in mast cells by histamine liberators. Nature 275:661.
28. Befus, D., T. Lee, T. Goto, R. Goodacre, F. Shanahan, and J. Bienenstock. Histologic and functional properties of mast cells in rats and humans. In: Mast Cell Differentiation and Heterogeneity. This volume.
29. Kitamura, Y., T. Nakano, T. Sonoda, Y. Kanayama, T. Yamamura, and H. Asai. Probable transdifferentiation between connective-tissue and mucosal mast cells. In: Mast Cell Differentiation and Heterogeneity. This volume.
30. Denburg, J., Y. Tanno, and J. Bienenstock. Growth and differentiation of human basophils, eosinophils and mast cells. In: Mast Cell Differentiation and Heterogeneity. This volume.
31. Haig, D.M., C. McMenamin, C. Gunneberg, R. Woodbury, and

E.E.E. Jarrett. 1983. Stimulation of mucosal mast cell growth in normal and nude rat bone marrow cultures. Proc. Natl. Acad. Sci. USA 80:4499.

32. Stevens, R.L., H.R. Katz, D.C. Seldin, and K.F. Austen. Biochemical characteristics distinguish subclasses of mammalian mast cells. In: Mast Cell Differentiation and Heterogeneity. This volume.

33. Otsuka, H., J. Dolovich, A.D. Befus, S. Telizyn, R. Rajan, J. Bienenstock, and J. Denburg. 1985. Peripheral blood basophils, basophil progenitors and nasal metachromatic cells in allergic rhinitis. Submitted.

34. Jarrett, E.E.E., and D.M. Haig. 1984. Mucosal mast cells in vivo and in vitro. Immunol. Today 5:115, 1984.

35. Denburg, J.A., A.D. Befus, and J. Bienenstock. 1980. Growth and differentiation in vitro of mast cells from mesenteric lymph nodes of Nippostrongylus brasiliensis infected rats. Immunology 41:195.

36. Tanno, Y., J. Bienenstock, M. Richardson, T.D.G. Lee, A.D. Befus, and J. Denburg. 1985. Reciprocal regulation of human basophil and eosinophil differentiation by separate T cell derived factors. Submitted.

37. Stadler, B.M., K. Hirai, T. Schaffner, and A.L. de Weck. A growth factor for human mast cells. In: Mast Cell Differentiation and Heterogeneity. This volume.

38. Askenase, P.W., and H.V. Loveren. 1983. Delayed-type hypersensitivity: activation of mast cells by antigen-specific T-cell factors initiates the cascade of cellular interactions. Immunol. Today 4:259.

39. Galli, S.J., and A.M. Dvorak. 1984. What do mast cells have to do with delayed hypersensitivity? Lab. Invest. 50:365.

40. Weinstock, J.V., S.W. Chensue, and D.L. Boros. 1983. Modulation of granulomatous hypersensitivity: V. Participation of histamine receptor positive and negative lymphocytes in the granulomatous response of Schistosomiasis mansoni-infected mice. J. Immunol. 130:423.

41. Capron, M., J. Rousseaux, C. Mazingue et. al. 1978. Rat mast cell-eosinophil interaction in antibody dependent eosinophil cytotoxicity to Schistosoma mansoni schistosomula. J. Immunol. 121:2518.

42. Henderson, W.R., E.Y. Chi, E.C. Jong, and S.J. Klebanoff. 1981. Mast cell-mediated tumor-cell cytotoxicity. Role of the peroxidase system. J. Exp. Med. 153:520.

43. Farram, E., and D.S. Nelson. 1980. Mouse mast cells as anti-tumor effector cells. Cell. Immunol. 55:294.

44. Ernst, P., T. Lee, A.D. Befus, and J. Bienenstock. 1985. Unpublished observations.

45. Beer, D.J., and Rocklin, R.E. 1984. Histamine-induced suppressor-cell activity. J. Allergy Clin. Immunol. 73:439.

46. Wilson, J.G., D.T. Fearon, R.L. Stevens et al. 1984. Inhibition of the function of activated properdin by squid chondroitin sulfate E glycosaminoglycan and murine bone

marrow–derived mast cell chondroitin sulfate E proteoglycan. J. Immunol. 132:3058.

47. **Frieri, M., and D.D. Metcalfe.** 1983. Analysis of the effect of mast cell granules on lymphocyte blastogenesis in the absence and presence of mitogens: identification of heparin as a granule–associated suppressor factor. J. Immunol. 131:1942.

48. **Victor, M., J. Weiss, and E. Elsbach.** 1981. Heparin inhibits phagocytosis by polymorphonuclear leukocytes. Infect. Immun. 32:295.

49. **Kitchen, A.G., and R. Megirian.** 1971. Heparin enhancement of Kupffer cell phagocytosis in vitro. J. Reticuloendoth. Soc. 9:13.

50. **Dougherty, T.F., G. Scheebeli, and N. Panagiotis.** 1984. Stimulation of motility and pinocytosis by heparin in amoeba proteus and murine fibroblasts. J. Reticuloendothel. Soc. 1:363.

51. **West, B.C., C.H. Dunphy, and C.A. Moore.** 1983. Human neutrophil N–acetyl–β–glucosaminidase: heparin inhibition. Biochem. Med. 29:1.

52. **Kierszenbaum, F., S.J. Ackerman, and G.J. Gleich.** 1982. Inhibition of antibody–dependent eosinophil–mediated cytotoxicity by heparin. J. Immunol. 128:515.

53. **Hoover, R.L., R. Rosenberg, W. Haering, and M.J. Karnovsky.** 1980. Inhibition of rat arterial smooth muscle cell proliferation by heparin. II. In vitro studies. Circ. Res. 47:578.

54. **Azizkhan, R.G., J.C. Azizkhan, B.R. Zetter, and J. Folkman.** 1980. Mast cell heparin stimulates migration of capillary endothelial cells in vitro. J. Exp. Med. 152:931.

55. **Schwartz, L.B., and K.F. Austen.** 1984. Structure and function of the chemical mediators of mast cells. Prog. Allergy 34:271.

56. **Castro, G.A.** 1982. Immunological regulation of epithelial function. Am. J. Physiol. 243:G321.

57. **Perdue, M.H., M. Chung, and D.G. Gall.** 1984. Effect of intestinal anaphylaxis on gut function in the rat. Gastroenterology 86:391.

58. **Pearce, F.L., A.D. Befus, J. Gauldie, and J. Bienenstock.** 1982. Mucosal mast cells. II. Effects of anti–allergic compounds on histamine secretion by isolated intestinal mast cells. J. Immunol. 128:2481.

59. **Cuthbert, A.W., P. McLaughlan, and R.R.A. Coombs.** 1983. Immediate hypersensitivity reaction to β–lactoglobulin in the epithelium lining the colon of guinea pigs fed cow's milk. Int. Arch. Allergy Appl. Immunol. 72:34.

60. **Russell, M., K.A. Dark, R.W. Cummins, E. Ellman, E. Callaway, and H.V.S. Peeke.** 1984. Learned histamine release. Science 225:733.

61. **Eady, R.P., B. Greenwood, S.T. Harper, J. Mann, T.S.C. Orr,**

and E. Wells. Bronchoalveolar mast cells from macaques infected with Ascaris suum. In: <u>Mast Cell Differentiation and Heterogeneity</u>. This volume.

Mast Cell Differentiation and Heterogeneity,
edited by A. D. Befus et al.
Raven Press, New York © 1986.

Nomenclature[1]

*A. Dean Befus, **John Bienenstock, and † Judah A. Denburg

*Host Resistance Programme, Departments of **Pathology and † Medicine, McMaster University, Hamilton, Ontario L8N 3Z5, Canada; and *Gastrointestinal Research Group, Department of Microbiology and Infectious Diseases, University of Calgary, Calgary, Alberta T2N 4N1, Canada*

It was generally agreed that a recommendation for an appropriate nomenclature was neither feasible nor reasonable given the state of development of knowledge at the present time. Nevertheless, there was consensus that workers in this field should abandon the use of such general and potentially confusing descriptive terms as "mucosal", "connective tissue", "typical" or "atypical" mast cells or "P cells" in favour of an indication of derivation, and as full a description as possible in terms of organ source, proteoglycan content or other qualifiers important in the definition of such cells (see below). A suggestion to employ the terms "mast cell-like" or "basophil-like", using the suffix to denote likely relationship, was debated, but not uniformly adopted.

It was agreed that the use of cultured cells to define mast cell type must be done with caution since there were several caveats associated with the effects of culture, which included acquisition or loss of factor dependency, and the effect of culture conditions on selection for certain cell types, not necessarily representing the predominant type present at the outset.

While the issue of "thymus dependency" was important, it was agreed to be a difficult area to currently resolve since some of the factors important for promotion of mast cell growth such as IL-3 were not necessarily only produced by T cells. Furthermore, it was not certain whether different factors were responsible for proliferation as opposed to maturation, and it was still not certain whether peritoneal mast cells may in fact exhibit thymus-dependency.

Notwithstanding the above, there was consensus that in the rat

---

[1]Excerpted from the report of the Millcroft Meeting entitled "Mast Cell Differentiation and Heterogeneity": Immunol. Today, October 1985, with permission from the publishers.

there may be two types of mast cell which are distinguishable on the basis of staining, fixation, functional attributes and factor dependency. The bone rat marrow–derived mast cell is likely phenotypically similar to the intestinal mucosal mast cell, but not to that found in the peritoneal cavity. There was considerable doubt expressed as to evidence for heterogeneity in other species and the evidence, is at best, inconclusive. In the human there is no good evidence yet for two or more functionally distinct types of mast cells; any evidence supporting mast cell heterogeneity in man is circumstantial.

The meeting suggested that all investigators working in this area take the following factors into account when writing about mast cells or basophils and delineate a) species of origin; b) whether cultured and, if so, culture conditions; c) organ source; d) proteoglycan content, where it is chemically defined (in this regard "heparin-containing cells" as a descriptor was considered acceptable); e) where possible, response to secretagogues or, more specifically, secretion of leukotrienes and prostaglandins in response to anti–IgE or antigen; f) histochemistry, especially the conditions of both fixation and staining.

With adoption of these suggestions it might be possible for workers in this field to gather further information and allow this to form the basis for a comparison of results and concepts with a view to meeting at the 6th International Congress in Toronto in 1986 to decide if the accumulated information would then allow for recommendations to be made as to a nomenclature.

It was recommended by Dr. Austen on behalf of the I.U.I.S. Standardization Committee that Dr. Bienenstock chair such a working group to make recommendations as to nomenclature.

Mast Cell Differentiation and Heterogeneity,
edited by A. D. Befus et al.
Raven Press, New York © 1986.

Methods for the Identification and Characterization of Mast Cells by Light Microscopy

*Lennart Enerback, **Hugh R. P. Miller, and † Graham Mayrhofer

*Department of Pathology, Sahlgren Hospital, University of Goteborg, S413 45 Goteborg,
Sweden; **Department of Pathology and Immunology, Moredun Institute, Edinburgh,
Scotland; and † Department of Microbiology and Immunology, The University of Adelaide,
Adelaide 5000, South Australia*

The general discussion at the mast cell symposium revealed some degree of uncertainty, or perhaps even disagreement, about suitable methods for the identification of mast cells in tissues or cell mixtures. It was also questioned to what extent presently available histotechnical or histochemical methods can be used to define mast cell subsets. Several of the participants of the meeting expressed the need from some simple guidelines or advice from investigators in this field. This brief review was written in response to a request by the editors of the symposium, as an attempt to meet this demand.

1. FIXATION AND STAINING OF MAST CELL PROTEOGLYCANS

Methods which are used for the identification of mast cells in situ are directed towards their main components such as proteoglycan, proteinase, or IgE. Proteoglycan is the single most distinctive constituent of the mast cell granules, where its glycosaminoglycan (GAG) component is responsible for the specific metachromatic dye-binding. This discussion will concern dye-binding methods for the visualization of the GAG of mast cell granules. It will deal with methods for fixation and staining of mast cells, as well as ways of defining mast cell subsets with existing methodology, and possible or desirable future developments. A detailed discussion about histochemistry of mast cell GAG will be found in (1) which also contains protocols for recommended methods.

Fixation

The purpose of fixation is to preserve the structure of the granular proteoglycan, leaving the anionic groups of the GAG (sulphates and carboxyls) available for cationic dye-binding. Failures to preserve the dye-binding may be the result of extraction of soluble GAG by the fixative or of blocking of dye-binding groups of the GAG by aldehydes, which are very common components of fixatives. A suitable fixative should therefore have a very low aldehyde concentration (IFAA, see below) or the fixation time should be kept short (paraformaldehyde). A low pH (IFAA) prevents solubilization of the granular proteoglycan by promoting ionic linkages between GAG and protein. Aldehyde blocking of dye-binding as a result of routine fixation in formaldehyde or glutaraldehyde may be overcome by very long staining times (see below). Addition of salts of heavy metals such as lead to fixatives was often recommended in the past as a means of preventing solubilization of GAG. It should be avoided since heavy metals may block the binding of cationic dyes.

Recommended Procedures

IFAA. This is an iso-osmotic solution containing 0.6% formaldehyde and 0.5% acetic acid. It was originally designed for the fixation of gut mucosal mast cells (2), but has since then been successfully used for the fixation of various tissues in a great number of species, including man. It is recommended as a general fixative for tissue mast cells. Small tissue pieces are kept in the fixative at room temperature or at +4°C for 4 to 12 h and then in 70% ethanol for 12 h, followed by routine paraffin embedding. Fixed tissue pieces are soft and should be handled very carefully. Dye-binding of GAG is excellently preserved. Histochemical esterase activity is also well preserved by this fixative. However, the GAG may be partially extracted from sections of fixed tissue by salt solutions at neutral or alkaline pH (3).
Carnoy. This fixative contains chloroform, methanol and acetic acid. It was found to be one of very few of the common fixatives which preserved the dye-binding of gut mucosal mast cells (2) and has been frequently used for the study of these cells. It may be used as a general mast cell fixative. It does not block the dye-binding and probably does not extract GAG to any significant extent. However, structural details are not as well preserved as after fixation in IFAA and the staining of mast cells is often somewhat indistinct.
Ethanol-acetic acid. Ethanolic or methanolic fixatives are recommended for the fixation of cell smears, sediments or imprints. This solution, made up of 3 parts of ethanol and 1 part acetic acid, is recommended for cell smears or sediments,

and is used routinely with a fixation time of 30 min to 2 h by one of the authors for isolated mast cells or imprints of human nasal mucosa.

MFAA. This solution containing 4.0% formaldehyde and 5.0% acetic acid in methanol was selected with the object of preserving epithelial tissue optimally, and for the study of intraepithelial mucosal mast cells and granulated lymphocytes in the rat (4).

This fixative is also suitable for preservation of mast cells in other tissues. The activity of chloroacetate esterase is preserved, and tissue basophils can be distinguished from mast cells in the gut mucosa because they have sparse granules that are chloroacetate esterase-negative under these conditions (Mayrhofer and Pitts, unpublished). Unfortunately, as with Carnoy's, the binding of eosin by eosinophils is abolished by MFAA. Mast cells in mouse (Carroll et al., 1984) and human (Mayrhofer, 1980) small intestinal mucosa are preserved well by MFAA, but comparison with other fixatives has not been made on human tissue.

Paraformaldehyde. The dye-binding of tissue eosinophil granules is not well preserved by special mast cell fixatives such as IFAA and MFAA or by Carnoy fixation. Short time (6 h) fixation in 4% neutral buffered paraformaldehyde was recently recommended for the simultaneous preservation of the dye-binding of mucosal mast cell glycosaminoglycan, esterase activity and tissue eosinophils in rat intestinal mucosa (5). It is well suited for this purpose and preserves excellently tissue structure as well as dye-binding of mucosal mast cells, especially when stained with copper phthalocyanin dyes (see below). Preliminary studies indicate that 4% paraformaldehyde fixation of human intestine for 24-48 h provide excellent preservation for both GAG and esterase staining. However, it cannot be excluded that this fixative, like 4% formaldehyde (3), will extract GAG of mast cell granules in species or sites where the mast cell proteoglycan is highly soluble.

4% formaldehyde. This solution, buffered to neutral pH, is widely used in routine histology and histopathology. It cannot be recommended as a general mast cell fixative due to the fact that it may block the dye-binding of the GAG by its reaction with structurally associated protein, or extract the GAG from mast cell granules (6). However, such an extraction does not appear to occur to a significant extent in mast cells of the mouse, rat or man. In these species, cationic dye-binding of mucosal mast cells of several sites is blocked by the fixative (see below). This blocking of dye-binding may be reduced by long staining times, of the order of 5 to 7 days (6). An important practical consequence of this is that the paraffin blocks of tissue available in the files of laboratories of clinical pathology can be used for the study of human mast cells.

Staining

The principle of the various staining methods is to visualize the GAG by the binding of cationic dyes to its anionic sites (sulphate and carboxyl groups). The specificity of the staining is enhanced by the use of strongly metachromatic dyes such as toluidine blue or azure A, and/or low pH of the dye solutions by which staining of other tissue anions is suppressed.

Mast cells can also be stained with Romanowsky-type dye mixtures such as the Giemsa stain routinely used for the identification of basophils in blood smears. Such dye mixtures contain anionic and cationic dyes. The staining depends partly on a balance between these two components obtained during a "differentiation" step of the staining procedure which is not easily standardized. Mast cell granules contain cationic protein and anionic GAG whose charged groups are available for dye-binding to a variable degree, depending on proteoglycan structure and/or effects of fixation. The results of Giemsa staining may therefore be difficult to interpret, especially when it comes to identification of mast cell subsets differing in proteoglycan structure from the classical connective tissue mast cells. It is instructive in this context to recall that the granules of mucosal mast cells and globule leukocytes were classified in the past as "amphiphil" or "acidophil", based on their staining properties in such dye mixtures (7). We therefore strongly advise against the use of the Giemsa stain as a sole means of identifying mast cells in tissue sections.

Recommended Methods

Toluidine blue. This metachromatic dye stains mast cell granules red or violet and other tissue components blue. Staining of two sections or cell imprints, one at pH 0.5 and the other at pH 4, is often useful (8). At low pH mast cell granules are stained against a virtually unstained background, facilitating their identification and counting, while staining at higher pH helps to identify related cells or tissue components.

Alcian blue. This non-metachromatic copper phthalocyanin dye may be used at low pH (dye dissolved in 0.7N HCl) as a general mast cell stain, alone or in a sequential staining with the red dye safranin to differentiate intestinal mucosal mast cells (staining blue) from connective tissue mast cell of other sites (staining red) in rats and mice (see below).

Histochemical Definition of Mast Cell Subsets

As discussed in the introductory chapter to this volume, intestinal mucosal mast cells can be differentiated from the classical connective tissue mast cells in the rat by their fixation and staining properties. These properties are very well defined in the rat, but much less so in the various tissues of other species, including man. The histochemical differences reflect the fact that the two cell types contain GAG of different structure: macromolecular heparin in the case of connective tissue mast cells (9) and oversulphated chondroitin sulphate in the case of mucosal mast cells (10). So far, detailed information on GAG structure is only available for a few cells or tissues, and it is not known to what extent mast cells in the various species differ with respect to proteoglycan content. Therefore, none of the histochemical criteria used to identify mucosal mast cells in the rat can be applied to other species without further study. On the other hand, it is important that mast cells subjected to different kinds of study are defined as well as possible with the object of disclosing evidence of heterogeneity. However, knowledge in this field is still very limited and only allows a few general recommendations about suitable methods for this purpose.

Blocking of dye-binding by aldehyde. As discussed in the introductory paper of this volume, many mast cells of the human gastrointestinal mucosa are susceptible to fixation by aldehyde to a higher degree than mast cells of other tissue sites, as in rats and mice (3,11-13). The degree of aldehyde blocking of dye-binding may thus be used as a means to define mast cells of a particular species or site. There are at least two techniques which may be used for this purpose:

1) Fixation of adjacent tissue pieces or duplicate slides in 4% neutral buffered formaldehyde and in a mast cell fixative such as IFAA, followed by cell counting in sections of the two specimens stained by toluidine blue or alcian blue.

2) Cell counting in adjacent sections from tissue fixed in 4% neutral buffered formaldehyde, where one of the sections has been stained by the conventioned method (0.5% toluidine blue at pH 0.5 for 30 min) and the other by long toluidine blue staining (5 to 7 days).

Experience so far indicates that these two methods give comparable results. For instance, in normal human skin and mastocytosis lesions, cell counts in sections of tissue fixed in 4% formaldehyde and stained by the conventioned method are 20% to 30% lower than in IFAA fixed tissue, or in tissue fixed in 4% formaldehyde and stained for 5 days (14). Under similar conditions mast cells in the gastrointestinal and some other mucous membranes are 70% to 80% lower (3,11,12,13). Human intestinal mucosal mast cells thus seem to differ from dermal mast cells with respect to aldehyde sensitivity like those of the rat, but to a somewhat lesser degree. Obviously, we cannot decide if the partial blocking of dye-binding is an expression of a distinctive property of one single cell population at each tissue site, or of local heterogeneity, in the sense that mucosal and dermal mast cells are composed of an aldehyde sensitive and a non-aldehyde sensitive population in varying proportions.

Staining methods. As indicated in the foregoing, cationic

(metachromatic) dye–binding at low pH is a property of the family of sulphated GAG (e.g. heparin, chondroitin sulphates, dermatan sulphate and keratan sulphate). It should be made very clear that no staining method has yet proved to specifically stain an individual GAG.

Critical electrolyte concentration (CEC) staining with alcian blue. This staining principle was introduced by Scott and Dorling (15), who demonstrated that alcian blue stains different GAG with increasing selectivity as increasing concentrations of salt (MgCl$_2$) is included in the dye solution. CEC were defined as electrolyte levels above which staining was extinguished and were determined for various tissue sites. The gut mucosal mast cell of the rat has a substantially lower CEC than the dermal mast cell (16), reflecting the different GAG content of these two cell types. The CEC staining principle has the advantage of being well founded in physicochemical dye–binding theory and is equivalent to a similar and well established cetylpyridinium system for polyanion fractionation. It can be recommended as a general method to define the dye–binding of mast cells. However, it should be remembered that the CEC, like other similar staining patterns, may not necessarily depend on GAG stucture alone, but also on other properties of the proteoglycan, such as spatial relations between GAG and protein, resulting in variable degrees of protein blocking of anionic sites.

Berberine. As discussed in the introductory chapter to this volume, fluorescent berberine binding appears to have a certain degree of specificity for heparin. Its absence is a distinctive property of rat intestinal mucosal mast cells (16). However, the exact requirement for the fluorescent binding is not yet known, and experience of staining of mast cells of other species than the rat is so far limited. This dye is of potential interest as a means of distinguishing mast cell subsets but must be further evaluated before any general recommendations can be made.

Alcian blue–safranin. Differences in affinity for alcian blue of mast cells or individual mast cell granules can be visualized in a sequential staining with alcian blue followed by safranin, as first demonstrated by Spicer (17). Mucosal mast cells of rats and mice are distinguished from mast cells of other sites by their stronger affinity for alcian blue in this staining sequence (8). As discussed in detail elsewhere (1) the preference for alcian blue in this staining sequence may be an indiction of a low degree of sulphation, possibly due to the absence of N–sulphate (18), but the staining patterns may also depend on other physicochemical properties of the proteoglycan, unrelated to the structure of the GAG per se (19). Moreover, the staining patterns with the alcian blue–safranin sequence appear to be much less distinctive in other species than in the rat and mouse. Therefore, alcian blue–safranin staining cannot be recommended as a general method for the definition of mast cell subsets.

2. IDENTIFICATION OF IGE AND MAST CELL ANTIGENS

Immunohistochemical localization of mast cell-associated IgE or mast cell antigens can be difficult, especially in tissue sections or on mixed cell suspensions. This is because in order to identify labelled cells, the characteristic mast cell granules must also be preserved and stained appropriately. That is, the fixative used in cell preparation must preserve not only the antigenicity of the relevant moiety but also the staining proper- ties of the mast cell granules. Unfortunately, the fixatives that will preserve particular antigens cannot usually be predict- ed and the range of fixatives that preserves the granules of rat intestinal mucosal mast cells is limited (see above) and on the whole not conducive to retention of antigenicity. Because the requirements for fixation of the connective tissue type of mast cells are less exacting, a correspondingly greater range of less denaturing fixatives should be available for immunohistochemistry on these cells.

Pure Cell Populations

In principle, the simplest approach to this problem is to start with a pure or greatly enriched population of mast cells – identification of stained cells in then no longer necessary and provided the antigen in question is present on the cell surface, fixation is not required. Cultured mast cells and purifed mast cells from mesothelial spaces (20) can be studied in this way by the conventional methods of binding radiolabelled, fluorochrome- labelled or enzyme-labelled ligands to the surfaces of viable cells. Isolation procedures yielding semi-purified mast cell populations from certain tissues have also been described (21- 24). For detection of internal antigens in pure preparations, cell smears can be fixed with mild agents such as ethanol or acetone because preservation of granules is not obligatory. Huntley, Newlands and Miller (23) used these methods to detect surface and internal immunoglobulin in isolated ovine intestinal mucosal mast cells and globule leukocytes.

However, if the mast cell preparation is considered to be heterogeneous and a correlation is sought between morphological and/or histochemical criteria and the expression of an antigen, it will be necessary to fix the cells as described in the next section.

Mixed Cell Populations

In mixed cell suspension, it may be possible to identify mast cells and fluorochrome label simultaneously by flow cytometry because the granular cytoplasm produces 90° high light scatter. Alternatively, cells bearing labelled surface antigen can be sorted and methanol-fixed smears can be stained with Giemsa to determine whether mast cells are included in the sorted popula-

tion (25).   If cell—sorting or other methods of positive
selection (e.g. panning) are not available, it will be necessary
to produce a preparation in which staining for antigen and for
granules can be examined simultaneously.   In this case, the
simplest solution for visualization of surface antigens is to
label the cells prior to preparation of smears. This obviates
the need for special fixatives to preserve granule integrity in
mast cells analogous to rat intestinal mucosal mast cells,
because Giemsa acts not only as a stain on alcohol—fixed cells
but also as an additional "fixative" for mast cell granules (see
below).   With immunoperoxidase methods, formation of reaction
product should be undertaken before preparation of the cell smear
to ensure that proteoglycan is not leached during incubation with
substrate. Giemsa cannot be applied before incubation with substrate because it is bleached by hydrogen peroxide. Furthermore,
the dye cannot be used in combination with immunofluorescence and
it causes positive chemography in autoradiography. If used after
development of silver grains in autoradiographs, the proteoglycan
will already have been leached from mast cells analogous to the
rat intestinal mast cells.
     In the case of internal antigens (e.g. internal IgE; 23,26),
fixation is required both to allow antibody penetration of the
cell membrane and to stabilize the granule structure.   Granules
of mucosal mast cells cannot be identified by alcian blue staining in air—dried cryostat sections of rat gut (unfixed or
ethanol—fixed), presumably because the proteoglycan is soluble
(Mayrhofer, unpublished results).   One would expect therefore
that the granules of rat intestinal mucosal mast cells in
ethanol—fixed smears would also be water-soluble, and this is
certainly the case in smears of granular intraepithelial lymphocytes which also contain non-heparin—containing proteoglycan
(4; Mayrhofer, unpublished results).   This may be surprising,
because the granules of air—dried, alcohol-fixed intestinal
intraepithelial lymphocytes from rabbits and rats can be stained
with Giemsa (27,28).   However, brief previous exposure of the
fixed cells to aqueous buffers completely abolishes the staining.
The difference in staining between the two dyes appears to be
because Giemsa binds rapidly to the granule matrix and
insolubilizes it, whereas the less rapidly binding alcian blue is
unable to prevent loss of the proteoglycan. Once air—dried, it
is no longer possible to prevent dissolution in water of the
granules of either intraepithelial lymphocytes or mucosal mast
cells by fixation with Carnoy of MFAA.   However, there may be
differences in this respect between intestinal mucosal mast cells
of rats and sheep (23).
     There is little information available on which to base advice
about how to fix antigens and mast cell granules for their
simultaneous identification. However, from the above discussion
it seems important that fixation should precede air—drying if the
granules in mast cells of all types are to be identified by a
reasonably specific stain such as alcian blue. Some experience

with simultaneous staining for glycosaminoglycan and granule enzyme activities (Mayrhofer, unpublished results) may illustrate the general strategies that could be employed. Two methods have proven useful in studies on granular intraepithelial lymphocytes and large granular lymphocytes, which also contain glycosamino- glycan. Firstly, washed cell pellets can be fixed (in this case in 90% ethanol, 10% formalin), rehydrated, infiltrated with OCT compound (Tissue-Tek, Miles Scientific, Naperville, Illinois, USA) overnight, frozen in liquid nitrogen-cooled isopentane and cryostat sections prepared for staining with alcian blue. This was followed by demonstration of appropriate enzyme activity. Secondly, the granule staining by alcian blue can be preserved in cell smears, provided that the smear is fixed just as visible fluid is drying but before the cells themselves lose cytoplasmic water. In this study, Carnoy, MFAA, ethanol-formalin and methanol-formalin all preserved granules for staining with alcian blue, but only ethanol-formalin preserved the activities of all three enzymes. It seems likely that experimentation with a range of similar rapid fixaties could yield combinations that would allow localization of various antigens while preventing loss of proteoglycan. For instance, Carnoy appears to be a satisfactory fixative for immunohistochemical demonstration of IgE (26) and mast cell protease (29), while paraformaldehyde (and probably the fixatives mentioned above) can be used to demonstrate immuno- globulin light chain (23). It is worth noting that alcian blue firstly tends to prevent loss of proteoglycan during subsequent incubations and secondly, neither fluoresces itself nor quenches the fluorescence of Fluorescein.

Mast Cells in Tissue Sections

For many purposes it is not possible to dissociate solid tissues and it is desirable to have techniques that can detect in tissue sections cell-associated antigens on morphologically recognizable mast cells. Particularly for intestinal mucosal mast cells, cryostat sections of unfixed tissue are quite unsuitable in the rat, although at least some mast cells in human mucosae can be identified in frozen sections that have been lightly fixed (30,31). Frozen sections of unfixed tissue may also have application for immunohistochemical examination of connective tissue mast cells (30,32). However, one must always be concerned that only a subpopulation of the available mast cells will be preserved by such methods.

There is unlikely to be a universal fixative that will pre- serve mast cell granules reliably and also be suitable for the detection of all antigens. However, Carnoy has been of considera- ble use and has allowed simultaneous demonstrations of mast cells and sheep immunoglobulin (33), rat IgE (26,34), rat mast cell protease (29) and a mast cell-specific surface antigen (34). Recent studies with mouse anti-rat monoclonal antibodies on rat intestine (Mayrhofer and Pitts, this volume; Mayrhofer, unpublished results) have had some success on Carnoy-fixed, paraffin-embedded tissue (26). This fixative produced excellent staining with MRC OX8 (most intraepithelial lymphocytes, suppressor/cytotoxic T cells) W3/13 (most intraepithelial lymphocytes, all T cells) and MRC OX7 (anti-Thy-1.1) cells. However, MRC OX1 (anti-leukocyte-common antigen) and W3/25 (helper/inducer T cells) antibodies stained weakly compared to their performance on cryostat sections, while MRC OX4 and MRC OX6 (anti-Ia) antibodies did not react at all with Carnoy-fixed material.

It therefore appears that fixatives must be tried empirically for their ability to preserve individual antigens. Ethanol-formalin (see above) has been used to fix rat intestine for subsequent preparation of frozen sections (Mayrhofer, unpublished results). While granule staining and granule enzyme activities of intraepithelial lymphocytes are preserved by this technique, it has not been tested for preservation of antigens. It appears to be a fairly mild fixative and may be useful, whereas MFAA has been unsuccessful for all antigens tried to date.

Finally, a word of caution should be added because in the hands of some workers the granules of mast cells have bound immunoglobulin non-specifically (35,36). To some extent, this appears to be determined by the method of fixation (36). On the other hand, staining of rat intestinal mucosal mast cells by anti-IgE was shown to be specific by the use of appropriate controls (26). It is important to be aware of non-specific binding of immunoglobulin to granules of mast cells and to include specificity controls when examining them by immuno-histochemical methods.

3. PROSPECTS FOR FUTURE DEVELOPMENTS

Further progress in the research on mast cell heterogeneity is critically dependent on the development of specific, reproducible and easily applicable methods for the definition of mast cell subsets. Histochemical methods will probably be essential for this purpose but must be further developed. The results presented at this meeting have demonstrated the existence of distinct proteoglycans and/or proteinases, which may be used as histochemical markers. Future development in this field should therefore include improved methods for the identification of individual GAG such as heparin and chondroitin sulphate in situ. Work in progress suggests that this may be at least partly achieved by a combination of suitable staining methods, with specific enzymatic and chemical degradation (3). Hopefully, further development in the biochemical characterization of mast cell proteoglycans will be followed by efforts to produce specific antibodies, which may be used as immunohistochemical markers.

REFERENCES

1. **Enerback, L.** 1985. The mast cells. In: Applications of

Histochemistry to Pathologic Diagnosis. Edited by S.S. Spicer, A.J. Garvin, and G.R. Hennigar. New York, Marcel Dekker.

2. **Enerback, L.** 1966. Mast cells in rat gastrointestinal mucosa. 1. Effects of fixation. Acta Pathol. Microbiol. Scand. 66:289.

3. **Enerback, L.** Unpublished observations.

4. **Mayrhofer, G.** 1980. Fixation and staining of granules in mucosal mast cells and intraepithelial lymphocytes in the rat jejunum, with special reference to the relationship between the acid glycosaminoglycans in the two cell types. Histochem. J. 12:513.

5. **Newlands, G.F., J.F. Huntley, and H.R.P. Miller.** 1984. Concomitant detection of mucosal mast cells and eosinophils in the intestines of normal and Nippostrongylus-immune rats. A re-evaluation of histochemical and immunocytochemical techniques. Histochemistry 81:585.

6. **Wingren, U., and L. Enerback.** 1983. Mucosal mast cells of the rat intestine: a re-evaluation of fixation and staining properties, with special reference to protein blocking and solubility of the granular glycosaminoglycan. Histochem. J. 15:571.

7. **Michels, N.A.** 1938. The mast cells. In: Downey's Handbook of Haematology 1:232. Republished in: Mast Cells and Basophils. Edited by J. Padawer. Ann. N.Y. Acad. Sci. 103, Appendix.

8. **Enerback, L.** 1966. Mast cells in rat gastrointestinal mucosa. 2. Dye binding and metachromatic properties. Acta Pathol. Microbiol. Scand. 66:303.

9. **Yurt, R.W., R.W. Leid, K.F. Austen, and J.E. Silbert.** 1977. Native heparin from rat peritoneal mast cells. J. Biol. Chem. 252:518.

10. **Enerback, L., S.O. Kolset, M. Kuschke, A. Hjerpe, and U. Lindahl.** 1985. Glycosaminoglycans in rat mucosal mast cells. Biochem. J. 227:661.

11. **Strobel, S., H.R.P. Miller, and A. Ferguson.** 1981. Human intestinal mucosal mast cells: evaluation of fixation and staining techniques. J. Clin. Pathol. 34:851.

12. **Ruitenberg, E.J., L. Gustowska, A. Elgersma, and H.M. Ruitenberg.** 1982. Effects of fixation on the light micro-scopical visualization of mast cells in the mucosa and connective tissue of the human duodenum. Int. Arch Allergy Appl. Immunol. 67:233.

13. **Befus, D., R. Goodacre, N. Dyck, and J. Bienenstock.** 1985. Mast cell heterogeneity in man. 1. Histologic studies of the intestine. Int. Arch. Allergy Appl. Immunol. 76:232.

14. **Olafsson, J.H., G. Roupe, and L. Enerback.** 1985. Dermal mast cells in mastocytosis. Fixation, distribution and quantitation. Acta Dermato-Venerol. In press.

15. **Scott, J.E., and J. Dorling.** 1965. Differential staining of acid glycosaminoglycans (mucopolysaccharides) by Alcian blue

in salt solutions. Histochemie 5:221.

16. **Miller, H.R.P., and Walshaw.** 1972. Immune reactions in mucous membranes. IV. Histochemistry of intestinal mast cells during helminth expulsion in the rat. Am. J. Pathol. 69:195.

17. **Spicer, S.S.** 1960. A correlative study of the histochemical properties of rodent acid mucopolysacchardies. J. Histochem. Cytochem. 8:18.

18. **Combs, J.W., D. Lagunoff, and E.P. Benditt.** 1965. Different-iation and proliferation of embryonic mast cells of the rat. J. Cell. Biol. 25:577.

19. **Tas, J.** 1977. The alcian blue and combined alcian blue-safranin O staining of glycosaminoglycans studied in a model system and in mast cells. Histochem. J. 9:205.

20. **Befus, A.D., F.L. Pearce, J. Gauldie, P. Horsewood, and J. Bienenstock.** 1982. Mucosal mast cells. I. Isolation and functional characteristics of rat intestinal mast cells. J. Immunol. 128:2475.

21. **Carroll, S.M., Mayrhofer, G., H.J.S. Dawkins, and D.I. Grove.** 1984. Kinetics of intestinal lamina propria mast cells, gobule leucocytes, intraepithelial lymphocytes, goblet cells and eosinophils in murine strongyloidiasis. Int. Arch. Allergy Appl. Immunol. 74:311.

22. **Dobson, C.** 1966. Immunofluorescent staining of globule leucocytes in the colon of the sheep. Nature (Lond.) 211:87.

23. **Enerback, L., and I. Svensson.** 1980. Isolation of rat peritoneal mast cells by centrifugation on density gradients of Percoll. J. Immunol. Methods 39:135.

24. **Feltkamp-Vroom, T.M., P.J. Stallman, R.C. Aalberse, and E.E. Reerink-Brongers.** 1975. Immunofluorescence studies on renal tissue, tonsils, adenoids, nasal polyps, and skin of atopic and nonatopic patients, with special reference to IgE. Clin. Immunol. Immunopathol. 4:392.

25. **Halliwell, R.E.W.** 1973. The localization of IgE in canine skin: an immunofluorescent study. J. Immunol. 110:422.

26. **Hunt, S.V., D.W. Mason, and A.F. Williams.** 1977. In rat bone marrow Thy-1 antigen is present on cells with membrane immunoglobulin and on precursors of peripheral B lympho-cytes. Eur. J. Immunol. 7:817.

27. **Huntley, J.F., G. Newlands, and H.R.P. Miller.** 1984. The isolation and characterization of globule leucocytes: their derivation from mucosal mast cells in parasitized sheep. Parasite Immunol. 6:371.

28. **Huntley, J.F., B. McGorum, G.F.J. Newlands, and H.R.P. Miller.** 1984. Granulated intraepithelial lymhocytes: their relationship to mucosal mast cells and globule leucocytes in the rat. Immunology 53:525.

29. **Mayrhofer, G., H. Bazin, and J.L. Gowans.** 1976. Nature of cells binding anti-IgE in rats immunized with Nippostrongy-lus brasiliensis: IgE synthesis in regional nodes and

concentration in mucosal mast cells. Eur. J. Immunol.
30. **Mayrhofer, G., and R.J. Whately.** 1983. Granular intra-epithelial lymphocytes of the rat small intestine. I. Isolation, presence in T lymphocyte–deficient rats and bone marrow origin. Int. Arch. Allergy Appl. Immunol. 71:317.
31. **Mayrhofer, G., and R. Pitts.** In: Mast Cell Differentiation and Heterogeneity. Edited by A.D. Befus, J. Bienenstock and J. Denburg. This volume.
32. **Mayrhofer, G.** (unpublished results).
33. **Mayrhofer, G., and R. Pitts.** (unpublished results).
34. **Nijhuis–Heddes, J.M.A., J. Lindeman, A.J. Otto, M.W. Snieders, P.A. Kievit–Tyson, and J.H. Dijkamn.** 1982. Distribution of immunoglobulin–containing cells in the bronchial mucosa of patients with chronic respiratory disease. Eur. J. Resp. Dis. 63:249.
35. **Paterson, N.A.M., S.I. Wasserman, J.W. Said, and K.F. Austen.** 1976. Release of chemical mediators from partially purified human lung mast cells. J. Immunol. 117:1356.
36. **Rudzik, O., and J. Bienenstock.** 1974. Isolation and characterization of gut mucosal lymphocytes. Lab. Invest. 30:260.
37. **Ruitenberg, E.J., A. Elgersma, and C.H.J. Lamers.** 1979. Kinetics and characteristics of intestinal mast cells and globule leucocytes. In: The Mast Cell, its role in health and disease. Edited by J. Pepys and A.M. Edwards. Tunbridge, Pitman Medical Publishing, p.732.
38. **Simson, J.A.V., D.S. Hintz, A.M. Munster, and S.S. Spicer.** 1977. Immunocytochemical evidence for antibody binding to mast cell granules. Exp. Molec. Pathol. 26:85.
39. **Woodbury, R.G., G.M. Gruzenski, and D. Lagunoff.** 1978. Immunofluorescent localization of a serine protease in rat small intestine. Proc. Nat. Acad. Sci. USA 75:2785.
40. **Young, S., and D.M. Cowan.** 1965. The apparent immunofluorescence of tissue mast cells. Brit. J. Exp. Pathol. 46:649.

Subject Index

A

A23187
differential effects, 218,359–360
human intestinal mast cells, 211
maturation effects, 260
rat mast cells, 209
Abelson leukemia virus, 47–51
Acetylcholine; *see also* Cholinergic regulation
histamine release, 318
parietal cell activation, 317
Acid hydrolases
function, 195–196
heparin-containing mast cells, 186
Adenosine
differential effects, 359–360
human histamine-containing cells, 337
mouse bone marrow mast cells, 325–326
Adenosine receptors, 327
Adenosine triphosphate, 325–326
AH 9679, differential effects, 359–360
Alcian blue
human respiratory mast cells, 282
methodology, 408,410,412–413
monkey mast cells, 233
Alcian blue-safranin, 410–411
Aldehyde, 409
Allergies
basophil contribution, 379–388
mucosal mast cells, 18–21
Alpha-adrenergic agonists, 337
Alveolar mast cells, 282
Aminophylline, 218
Amputation neuromata, 393
Anaphylactic reaction
electron microscopy, 373,376
rat mast cell protease II release, 246–248
and ultrastructure, 171
Angiogenesis, 358
Antigen identification, 411–413
Antigen provocation
histamine release, 279
human respiratory tract mast cells, 277–283
Arachidonic acid
chondroitin sulfate mast cells, 192
cultured basophils, release, 88,91–92
heparin-containing mast cells, 186–188
human basophils versus mast cells, 338

human histamine-containing cells, 333–334
Arylsulfatase A
in heparin-containing mast cells, 186
rat intestinal mast cells, 241
Asbestosis, 392
Ascaris suum, 263–268
Astemizole, 280
Asthma
bronchoalveolar mast cells, 266
late reaction, mast cells, 280–281
Astroglia, 33
Atopic patients
basophil/mast cell progenitors, 73–74,381–382
histamine content, granulocytes, 75
major basic protein, 75
ATPase, 324–325
Autocrine mechanisms, 49–50
Avidin stains, 350–351
Azatadine, 339

B

B cells, 170
Basophil promoting activity, 159–165
Basophilic leukemia cells, *see* Rat basophilic leukemia cells
Basophilic myelocytes, 97–102
Basophils; *see also* Human basophils
and allergies, 379–388
in blood, 2,15
cytochemistry, 349
differentiation, 63–68,71–83,394–395
distribution, 350–351
eosinophil relationship, 77–80
growth promotion, 77
heterogeneity of, and maturation, 257–262
identification, and morphology, 95–114
in vitro culture, 85–93
lineages, and phenotype, 169
maturation, and morphology, 95–114
neuromodulation, 223–227
persistent cell factor generation, 31
precursors, 348–349
properties, 348